Green Energy and Technology

Climate change, environmental impact and the limited natural resources urge scientific research and novel technical solutions. The monograph series Green Energy and Technology serves as a publishing platform for scientific and technological approaches to "green"—i.e. environmentally friendly and sustainable—technologies. While a focus lies on energy and power supply, it also covers "green" solutions in industrial engineering and engineering design. Green Energy and Technology addresses researchers, advanced students, technical consultants as well as decision makers in industries and politics. Hence, the level of presentation spans from instructional to highly technical.

Indexed in Scopus.

Indexed in Ei Compendex.

Ana Cristina Gonçalves · Isabel Malico
Editors

Forest Bioenergy

From Wood Production to Energy Use

Springer

Editors
Ana Cristina Gonçalves
MED—Mediterranean Institute
for Agriculture, Environment
and Development and CHANGE—Global
Change and Sustainability Institute
Instituto de Investigação e Formação
Avançada
Departamento de Engenharia Rural
Escola de Ciências e Tecnologia
University of Évora
Évora, Portugal

Isabel Malico
IDMEC
Escola de Ciências e Tecnologia,
Departamento de Engenharia Mecatrónica
Universidade de Évora
Évora, Portugal

ISSN 1865-3529 ISSN 1865-3537 (electronic)
Green Energy and Technology
ISBN 978-3-031-48223-6 ISBN 978-3-031-48224-3 (eBook)
https://doi.org/10.1007/978-3-031-48224-3

This Springer imprint is published by the registered company Springer Nature Switzerland AG
The registered company address is: Gewerbestrasse 11, 6330 Cham, Switzerland

Paper in this product is recyclable.

Preface

This book provides a comprehensive analysis of forest biomass, from feedstock production to its transformation into energy and its diverse applications. It intends to present the state of the art of forest biomass production, assessment, characterisation and conversion into heat and power.

After establishing the current context and status of forest bioenergy, the book presents the various sources and worldwide distribution of forest biomass for energy. Subsequently, it explores the characterisation of the forest stands and the availability of biomass for energy per stand structure, including stands managed for timber, non-wood products and energy plantations. Then, it addresses the biomass evaluation and monitoring by considering data sources, modelling methods and existing models. Following the topics centred on forest biomass production and estimation, the book provides a comprehensive overview of established and emerging conversion technologies for the production of bioheat, biopower and fuels. It then covers the essential properties of forest biomass that play a relevant role in its transformation into energy and fuels. The subsequent three chapters cover the most common applications of forest biomass for energy and the associated technologies. These chapters specifically focus on the use of biomass for residential heat, industrial heat, district heat, power generation and combined heat and power.

This book is intended for a broad audience, from undergraduate and graduate students to academics, researchers and practitioners who aim to deepen their knowledge on the topic of forest bioenergy. It presents a multidisciplinary approach to the topic, and integrates a broad range of topics in a single book, covering many important aspects of forest biomass energy.

Évora, Portugal

Ana Cristina Gonçalves
Isabel Malico

Contents

Introduction to Forest Bioenergy

Isabel Malico and Ana Cristina Gonçalves

Abstract This chapter introduces and outlines the book "Forest Bioenergy: From Wood Production to Energy Use", dedicated to biomass, currently the most commonly used renewable energy source, which contributes to 10% of the worldwide energy supply. The majority of bioenergy comes from woody biomass, which is mainly converted into heat (mostly in households, followed by industries). Its conversion to power is also relevant, while the production of transport biofuels is a promising pathway. Modern bioenergy presents numerous advantages: it has a renewable, versatile, local and distributed nature; it helps increase energy security and meet the rising global energy demands; it easily substitutes for fossil fuels; and it presents potential environmental and economic benefits. Carbon sequestration and storage are among the several environmental services provided by forests. The amount of biomass they produce, and consequently, their bioenergy potential, is highly variable. Forest plantations provide the highest bioenergy yields per unit area, while in forest systems managed for other purposes, factors such as stand structure affect residual biomass generation. Assessing and monitoring biomass, along with determining bioenergy potentials, are essential tasks, often based on mathematical models that vary in complexity and span different spatial and temporal scales, frequently with associated cartography.

Keywords Wood · Wood residues · Energy plantations · Biofuels · Bioheat · Biopower

I. Malico (✉)
IDMEC, Escola de Ciências e Tecnologia, Departamento de Engenharia Mecatrónica, Universidade de Évora, Rua Romão Ramalho 59, 7000-671 Évora, Portugal
e-mail: imbm@uevora.pt

A. C. Gonçalves
MED—Mediterranean Institute for Agriculture, Environment and Development and CHANGE—Global Change and Sustainability Institute, Instituto de Investigação e Formação Avançada, Departamento de Engenharia Rural, Escola de Ciências e Tecnologia, Universidade de Évora, Apartado 94, 7002-544 Évora, Portugal
e-mail: acag@uevora.pt

A. C. Gonçalves and I. Malico (eds.), *Forest Bioenergy*, Green Energy and Technology,
https://doi.org/10.1007/978-3-031-48224-3_1

1 Introduction

Forest bioenergy is a term used to describe the energy obtained directly or indirectly from biodegradable, renewable raw materials from woody shrub and tree species, excluding agricultural ones. The sources of forest bioenergy are diverse and include products, residues and waste, such as fuelwood, wood pellets, residues from forest-based industries and post-consumer wood (*cf.* chapter "Sources and Distribution of Forest Biomass for Energy").

Forests have been providing mankind with fuels for heating and cooking for thousands of years, and they are still the main energy source for millions of people who do not have access to clean fuels and technologies to satisfy their basic needs. As such, they provide an essential service to these populations but, in many circumstances, at a high cost. Severe negative impacts arise when biomass is used in a traditional way through the combustion of solid biomass in inefficient and polluting equipment. Indeed, emissions from inefficient biomass burning cause adverse health problems [1–5] and impact the climate [6–9]. Additionally, the traditional use of biomass leads to a large demand for wood fuels, puts pressure on forests, contributes to forest degradation and has negative effects on gender equity [10–13] (*cf.* chapter "Biomass for Domestic Heat").

In contrast to the traditional use of biomass, modern biomass uses for energy are characterised by more efficient and cleaner technologies. On the condition that forest resources are obtained in a sustainable way and efficient, clean conversion technologies are employed, the use of forest biomass for bioenergy and biofuel production is a valuable and advantageous option to meet energy needs. Many high-income countries, which had, to some extent, forgotten bioenergy during a large part of the last century, have renewed their interest in this form of energy, influenced in part by the environmental advantages of biomass over fossil fuels and its socioeconomic benefits.

Biomass can help meet the increasing energy needs of the growing world population. From 1971 to 2020, the world total energy supply increased by around 150%, from 230.5 EJ to 584.5 EJ (Fig. 1) [14]. In this period, the largest average annual growth rate came from nuclear energy (7.2%), followed by renewable energy sources (RES) (2.2%) and fossil fuels (1.8%). However, in the last decades, increased awareness of the environmental problems caused by fossil fuels and the public perception of the risks of nuclear energy have led to a change in the growth rates of the energy supplied by the different energy sources, especially nuclear. From 2001 to 2020, the largest average annual growth rate in energy supply was associated with RES (2.5%), followed by fossil fuels (1.7%) and nuclear (0.2%).

Despite the relevance of renewable energy sources in general and biomass in particular, fossil fuels still dominate the global energy mix today. In 2020, they were the energy sources most used in the world (80%, Fig. 2). Crude oil and oil products accounted for 29.5% of the world total energy supply, closely followed by coal (27%), natural gas (24%) and renewable energy sources (15%), with biomass being the largest contributor among all the RES.

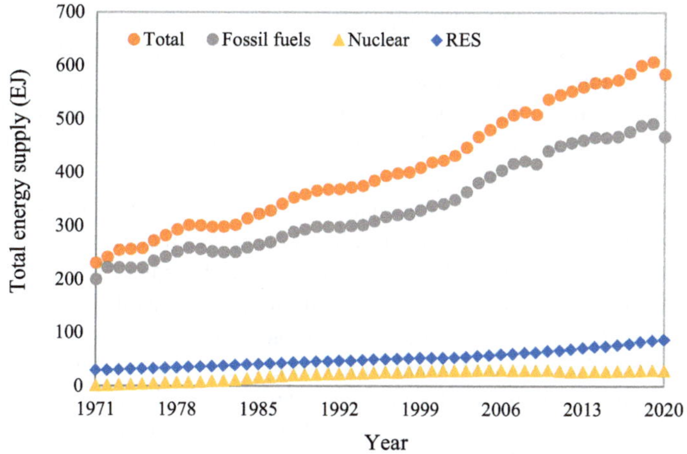

Fig. 1 World total energy supply from 1971 to 2020 by energy source. (*Data source* [14, 15])

Fig. 2 Share of energy sources in the world total energy supply in 2020. (*Data source* [14, 16])

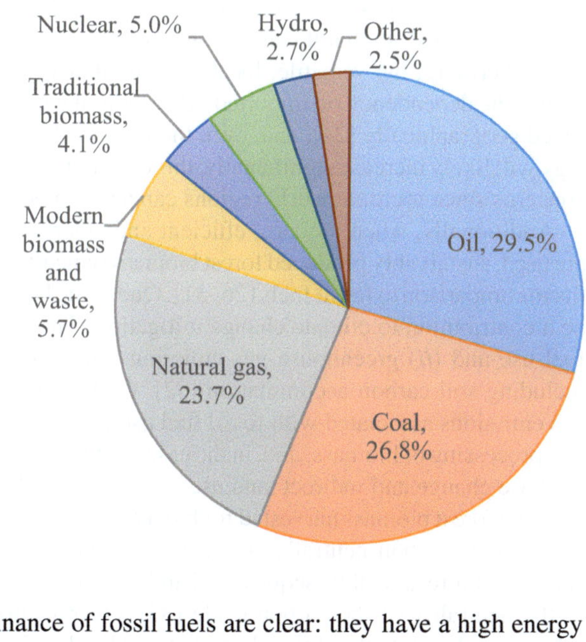

The reasons for the dominance of fossil fuels are clear: they have a high energy density and are convenient to use. Coal, the most carbon-intensive fossil fuel, is still the main source of power production worldwide (*cf.* chapter "Biomass for Power Production and Cogeneration"). It is the fossil fuel with the largest and most evenly distributed reserves around the world and with a relatively low cost [17], even though, today, in most markets, renewable energy options are the most cost-effective new sources [18]. Coal is also an important energy source for energy-intensive industries, such as the iron and steel or cement industries (*cf.* chapter "Biomass for Industrial and District Heating"). On the other hand, oil products in liquid form are more energy-dense than coal and very appropriate for use as transport fuel, being, by far, the most used energy sources in this sector (>90% in 2020, [14]). Natural gas, the least carbon-emitting fossil fuel, is very versatile and plays a relevant role in power plants, buildings and industrial facilities (*cf.* chapters "Biomass for Domestic Heat" and "Biomass for Industrial and District Heating" and "Biomass for Power Production and Cogeneration").

In the last two centuries, the large-scale consumption of fossil fuels has allowed for a rapid growth in industrial and agricultural production, enormous technological advances and improvements in the living conditions of humans. However, it also has numerous severe impacts on the environment and public health [9, 19] and affects energy security [20, 21]. As a consequence, a rapid phaseout of fossil fuels is needed, and both energy efficiency and sustainable energy sources should be promoted.

Biomass can substitute for fossil fuels relatively easily since it can be used and stored in similar ways to fossil fuels [22], therefore providing energy when needed (in power generation, biomass is an important complement to intermittent renewable energy sources like solar and wind, providing firm low-carbon electricity [23]). However, unlike fossil fuels, it is a renewable energy source when sustainably obtained, i.e., when the increment of biomass by plant regrowth is equal to or larger than the removal, so that it can be continuously available in large quantities.

Furthermore, the versatile, local and distributed nature of biomass may help to reduce the dependency on imported oil or natural gas, which are much more concentrated geographically [24], and, even though the global international trade of bioenergy will likely increase significantly, this does not necessarily lead to energy security concerns since multiple world regions can act as bioenergy suppliers [25].

Additionally, when modern, efficient and clean energy conversion technologies are used, sustainably produced forest biomass typically presents environmental benefits in comparison to fossil fuels [26–31]. One of the benefits of forest bioenergy might be a contribution to climate change mitigation. This derives from (*i*) avoided fossil fuel use and (*ii*) greenhouse gas emission mitigation during biomass production, including soil carbon accumulation [32]. On the other hand, there are greenhouse gas emissions associated with fossil fuel use in the production, harvesting, transport and processing of biomass, and, in the case of the establishment of energy plantations, land use change and indirect land use change effects have also to be considered.

If the forest biomass harvested for bioenergy is produced sustainably, it is arguably considered "carbon neutral", i.e., it is assumed that the carbon exported from the stands and forests will be sequestered and stored during tree regrowth, thus resulting in the neutrality of the carbon cycle [33]. Overall, the use of biomass for energy releases carbon into the atmosphere that will be absorbed by tree growth [34], which, depending on the species and tree growth rates, will need shorter (young and/or fast-growing species) or longer (old and/or slow-growing species) time [35, 36]. This is not, however, the only definition of carbon neutrality. The authors of a review study on carbon neutrality found eight different concepts for the term but no standardised concept or definition of carbon neutrality [34]. Bioenergy is often justified and promoted on the basis of its inherent carbon neutral status, but carbon neutrality is not an inherent property of biomass; rather, it is a relative characteristic of a bioenergy product [34], depending, for example, on the fossil fuel being displaced, the energy conversion efficiency of the bioenergy pathway followed, the growth rate of forests, the frequency and intensity of biomass harvests, the initial forest carbon stock and the forest management practices used [37].

The carbon storage capacity of forests may be enhanced by biomass removal in certain circumstances. This is the case, for example, when biomass is exported from the forest to reduce the fuel load and prevent wildfires or when thinnings are performed (see chapter "Stand Structure and Biomass" for further information). Other ecological benefits of biomass removals include the control of invasive flora and fauna species to restore natural habitats [38].

Another advantage of bioenergy might be its economic attractiveness. The generation of bioheat by combustion is often cost-competitive with fossil fuels [39, 40]. Also, the generation of bioelectricity with low-cost residual biomass might be cheaper than fossil fuel options [41]. However, the cost-effectiveness of bioenergy depends on the specific application, and high costs are a barrier, for example, for the development of many bioenergy options (e.g., advanced biofuels for transportation) or fuel switching (e.g., in high-temperature industrial applications or electricity generation) [41]. On the other hand, rural development, job creation or the promotion of social sustainability are often cited as advantages of bioenergy [39, 42–47].

Despite its many advantages, the use of biomass for energy is not without controversy. Besides the already mentioned problems associated with the traditional uses of biomass, the overall sustainability of bioenergy is often questioned [32, 48–51]. Biomass in general, and forest biomass in particular, are limited resources that require land and water for their growth. Moreover, the energy efficiency of photosynthesis is very low (the most efficient trees do not exceed solar storage efficiencies of 1% [52]). Therefore, in this regard, biomass is not the most effective way to store solar energy. However, storing energy is not the only function of forests, and if sustainably managed, forests provide other important ecosystem services. For example, photosynthesis generates oxygen, which is essential to flora and fauna, and forest areas promote soil and water protection [53] (see chapter "Stand Structure and Biomass" for further information).

To date, bioenergy is substantially sourced from residues and waste, but because their potential is limited, the supply of additional large quantities of biomass is dependent on energy plantations [49]. A large expansion of, especially agricultural, energy plantations in some countries could increase human pressure on the terrestrial biosphere, threaten the ability of global ecosystems to provide essential ecosystem services, and might be associated with substantial ecological costs, such as soil degradation, biodiversity loss or nitrogen release [54]. Furthermore, large-scale energy plantations, especially the agricultural ones, may potentially compete with food production, leading to significant socioeconomic effects [49]. However, if forest species plantations are grown specifically for energy on marginal lands [55] or on current agricultural land that can be diverted from food and feed production without further impairing food security [56], energy plantations can be a source of beneficial renewable energy. For example, Langeveld et al. [57] concluded that including large-scale short-rotation coppices in intensive, arable crop cultivations in homogeneous monocultural landscapes may have both positive and negative effects depending on the exact implementation. As referred to by Robertson et al. [32], prior land use has a significant effect on the benefits of energy plantations. The issue of bioenergy sustainability is, therefore, very complex, with several factors impacting the overall sustainability of bioenergy, among which are the location and scale of production, the type of feedstock and the conversion technology [58, 59].

2 Current Production and Consumption of Bioenergy

In 2020, the world energy supply of biomass and waste was 57.5 EJ [14], which represented around 10% of the total energy supply (Fig. 2). Most of this contribution was modern bioenergy, but a substantial part was still traditional bioenergy used for cooking and heating with basic, inefficient and pollutant technologies [16]. On the other hand, the non-renewable fraction of municipal and industrial waste represented a relatively small fraction of the share of energy supply attributed to biomass and waste.

Even if traditional bioenergy use is not considered renewable because of its severe negative impacts, modern biomass provided approximately half of the global renewable primary energy supply in 2020, making it the largest contributor among RES, followed by hydro. When accounting for traditional biomass use, biomass represented almost two-thirds of the world renewable energy supply.

In the last 30 years, the share of bioenergy and waste in the world total energy supply has remained relatively stable (according to data from the International Energy Agency, IEA [14], the average share from 1990 to 2020 was 9.7%). Nevertheless, the energy supplied by these two energy sources in this period increased slightly less (56%) than the total energy supply (60%).

The vast majority of global bioenergy and waste consumption is attributed to solid biomass. However, over the last three decades, there has been a diversification of bioenergy sources, with other forms gaining relevance (Figs. 3 and 4). This trend is particularly notable in Organisation for Economic Cooperation and Development (OECD) countries, as reported by the IEA [14].

The diversity of bioenergy sources (in solid, liquid or gaseous form) and available energy conversion technologies makes bioenergy very versatile. It is used to provide heat and power and as a transport fuel in a diversity of sectors of activity. Almost half of the biofuels and waste is used in the residential sector worldwide, but the industrial and energy sectors also consume a relevant share of biofuels and waste [14]. On the other hand, even though the transport sector represents a relatively small part of the world supply of biofuels and waste (7% in 2020 [14]), its relevance is also high since, presently, liquid biofuels dominate the renewable energy supply in this sector. Currently, these transportation fuels are essentially first-generation biofuels produced from food crops or vegetable oils [60].

Worldwide, the majority of solid biofuels, primarily consisting of forest biomass, are used in the residential sector (Fig. 5). However, the share of this sector has been

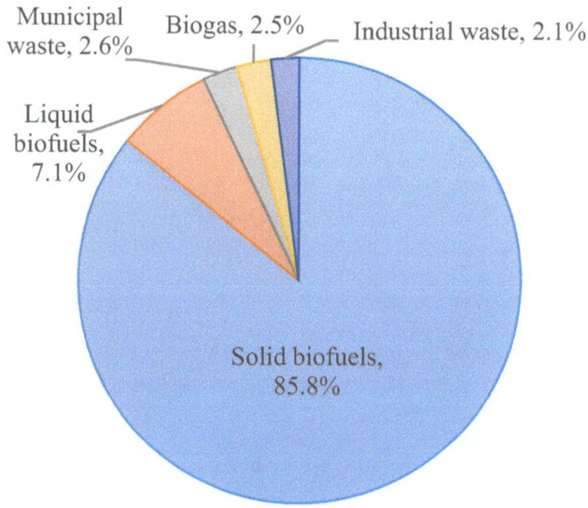

Fig. 3 Share of the various sources in the biomass and waste energy supply in 2020 in the world. (*Data source* [16])

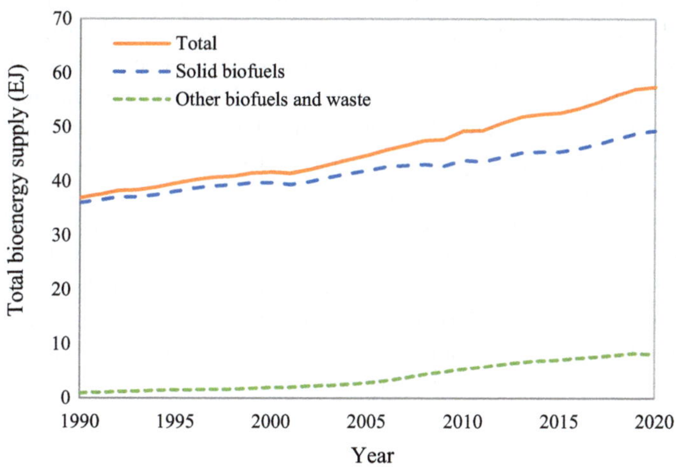

Fig. 4 World total bioenergy and waste supply from 1990 to 2020 by energy source. (*Data source* [14])

declining over the last few decades. In 1990, residential use represented 70% of solid biofuel consumption, but by 2020, it had decreased to 53%. In both years, the industrial sector was the second largest consumer of solid biofuels, followed by "other transformation", which includes charcoal production. The conversion of solid biomass into electricity in dedicated power plants has gained significance in the last 30 years. In 1990, it accounted for only 1% of solid biofuel consumption, but by 2020, it had increased to 8%. These power plants consumed around 70% more solid

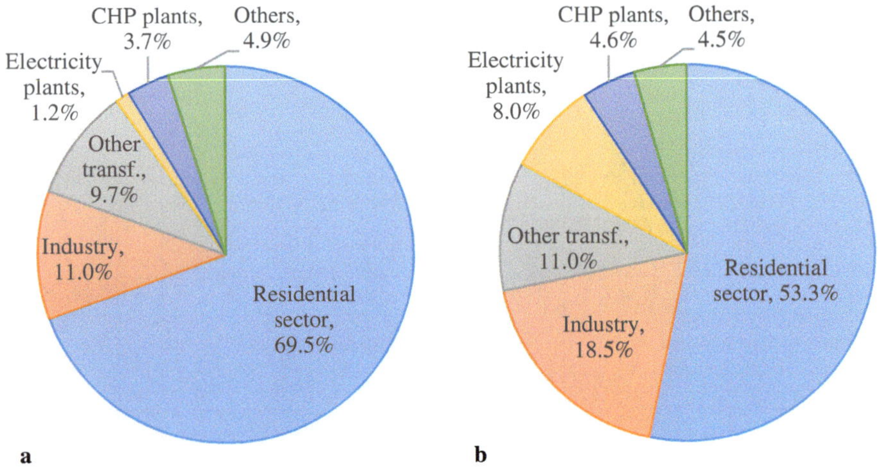

Fig. 5 Share of the various sectors in the consumption of solid biofuels in the world in **a** 1990 and **b** 2020. (*Data source* [14])

biomass in 2020 worldwide than the more efficient combined heat and power (CHP) plants. In the "others" category, the largest consumers of solid biofuels in 2020, in descending order, were commercial and public services, energy industry own use, agriculture and heat plants. In conclusion, to date, solid biofuels are mainly converted into heat and, to a lesser extent, electricity, while the production of transport biofuels from solid biomass is still not commercial and faces several challenges (*cf.* chapter "Forest Biomass as an Energy Resource").

In the group of OECD countries (Fig. 6), which accounted for 17% of the world solid biofuel use, the sector that consumed most of the solid biofuels in 2020 was industry (34%), closely followed by the residential sector (32%). As far as electricity production is concerned, CHP plants represent 15% of solid biofuel consumption, a higher share than the less efficient electricity-only plants (12%). Heat and charcoal production are included in the category "others".

3 Forest Biomass as an Energy Source

Of all the possible solid biofuel sources (e.g., firewood, forest residues, wood-based industry residues, post-consumer wood, agricultural residues, agro-industrial solid residues), forest biomass is, by far, the most used nowadays and, therefore, has a very important contribution to the world renewable energy share [61].

Forest trees, stands and forests are important to biomass (and carbon) sequestration and stocking [62] due to the longevity of trees [63], their large dimensions [64] and their worldwide distribution [65]. Stands and forests have provided for thousands of years a large set of products and services such as wood of several dimensions

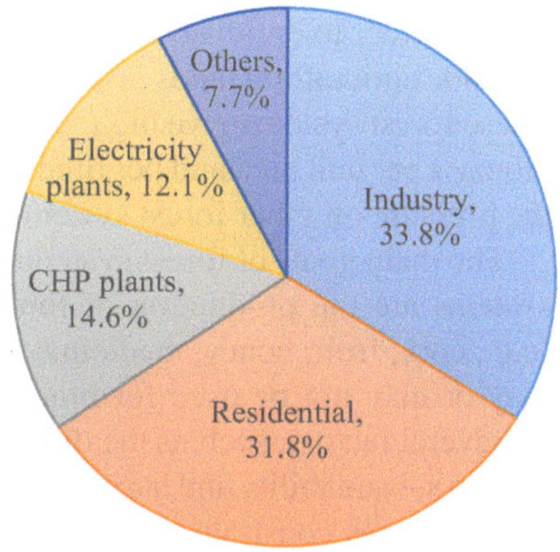

Fig. 6 Share of the various sectors in the consumption of solid biofuels in OECD countries in 2020. (*Data source* [14])

(including woody products for energy), protection of soil and water, conservation of flora, fauna and habitats, many non-woody products (e.g., honey, mushrooms, medicinal plants), aesthetics, recreation and biomass and carbon storage [66, 67].

Biomass sequestration and storage present high variability (*cf.* chapter "Sources and Distribution of Forest Biomass for Energy", Fig. 9). This results from the variability in species, stand structure and site (*cf.* chapter "Stand Structure and Biomass"). In general, biomass storage increases with age due to the increase in the dimension of trees [68], whereas sequestration decreases per unit of light intercepted due to the increase in respiration [35]. Thus, biomass and carbon sequestration are more efficient when trees are in the early stages of development, as growth rates are higher than in mature trees [36]. Yet, the biomass and carbon stocks are higher at late development stages due to tree dimensions [64]. The dynamics of biomass sequestration in stands and forests are more complex than those of individual trees due to their dependence on species, stand structure, site, silvicultural system and silvicultural practices (*cf.* chapter "Stand Structure and Biomass"). This variability is reflected in the biomass partitioning. In general, in forest stands, biomass can be broadly divided into tree and soil (organic matter). Tree biomass can be further divided into live (above and below ground) and dead (above and below ground) biomass [69].

Forest biomass for energy purposes results from harvesting. Yet, differences in biomass yields, amounts of biomass harvested and exported and the share of biomass that can be used for timber and energy vary per species, stand structure, silvicultural system, silvicultural practices, site, harvest equipment and market (e.g., [70, 71]). In terms of biomass for energy, forest systems can be broadly divided into two groups: energy plantations (*cf*. chapter "Energy Plantations"), where all above ground biomass is harvested for energy [72] and forest systems oriented for timber production or agroforestry systems, where forest residues from cuts, thinnings and/ or prunings are used for energy [71, 73].

Energy plantations are intensively managed forest systems with the main goal of producing biomass for energy [72]. These forest systems use species or clones with high growth rates, thus high rates of biomass (and carbon) sequestration [74]; are frequently established in sites (soil and climate) near the optimum of the species traits in order to optimise their growth [72]; and are established in marginal lands or set aside agricultural lands in order to avoid competition with agricultural lands [75]. These forest systems give high yields, and are thus able to provide larger amounts of biomass per unit area and per time unit than the other forest systems [76] and release the pressure on other forest systems for bioenergy supply [77].

The main goals of forest systems oriented for timber production and agroforestry systems are the production of timber and other woody and non-woody products (e.g., cork, fruit, honey, medicinal plants) and services [78, 79], and forest residues may or may not be used for bioenergy [71, 73]. Forest residue quantities depend on several factors, such as the dimensions and quality of the woody products, forest system sustainability and harvesting. The dimensions of the woody products and their market prices constrain their use. In general, woody products of large dimensions and good quality are used for timber [80], which has a higher market price than forest residues [71].

The export of forest residues is related to the sustainability of the forest stands and their products. Forest residue exports always imply the export of biomass, carbon and nutrients and have impacts on hydrology and diversity [81]. Moreover, the sustainability of forest systems is related to the biomass storage/export ratio [82]. Forest residues are generated during harvest (cut, thinning and/or pruning). Their removal, apart from the sustainability of the forest systems, is related to stand structure, function, topography, site, harvesting equipment, logistics and costs [83]. In general, in stand structures (*cf.* chapter "Stand Structure and Biomass") that are more uniform and when the removal of wood is made mainly in one cut (pure even-aged stands under clearcut systems), the export of forest residues is facilitated due to their large amounts, higher recovery rates and lower costs than in stands with higher structural diversity (pure or mixed uneven-aged stands under selection or shelterwood systems and/or under protection or conservation status) [84–86]. Furthermore, site and topography may increase forest residue quantity, mainly due to damage during tree harvesting [83].

Restrictions to the collection of forest residues and subsequent energy use may exist, imposed, for example, by regulations in protected areas or difficulties in collecting biomass in areas with difficult accessibility (e.g., steep slopes) [87–90] or because of the dispersion of the residues in the stands. Additionally, losses in the collection, transport and use stages of the feedstock need to be considered [91–95]. Another issue that may reduce the availability of residual biomass for energy is the existence of other uses for biomass and, therefore, a competition between the same biomass resources [71, 80, 96–98]. The consideration of these restrictions and the conversion of feedstock mass to energy lead to the determination of the amount of available biomass energy, that is, the biomass energy content that is potentially available for energy production [71, 97, 99]. Several authors [100–105] also consider sustainability criteria for stands, forests and productions that constrict the available amount of biomass. Batidzirai et al. [106] reviewed the key factors and drivers affecting the determination of biomass availability for energy and analysed a selected set of country-based bioenergy potential studies. They conclude that generally not all the basic elements expected in an ideal bioenergy assessment are included in the analyses, that the methods used are not always harmonised, which leads to different energy potential results, and that studies have different levels of methodological transparency. It is recommended that the analyses include all key factors that are critical determinants of bioenergy potentials, employ high-resolution georeferenced data sets and account for potential feedback effects [106].

Biomass evaluation and monitoring have to be done in order to quantify woody products and forest residues. This evaluation is frequently done with mathematical models that vary in complexity and spatial and temporal scales (*cf.* chapters "Modelling Biomass" and "Overview of the Biomass Models"), usually with associated cartography.

The availability of forest biomass for energy has been estimated with mathematical models at both regional or local scales [87, 107–116] and national scales [96, 117–122]. Yet, the above ground biomass specificity, particularly in the tropics, makes accurate generalisations at regional or landscape levels difficult [123, 124]. Moreover, biomass is not static in space or time. Several disturbances, either management-related or natural (*cf.* chapter "Stand Structure and Biomass"), can cause the reduction of biomass. In a review, natural disturbances (drought, fire, wind and bark beetle) were evaluated for their effect on biomass dynamics [125]. Although forest systems are quite resilient to disturbances, climate change can drive the systems to their turning point, especially if the regime of disturbances is outside its historical range of variation. The maintenance of the resilience of the stands and forests then requires proactive and reactive adaptive measures in the management of the forest systems [125]. Another study refers to the losses of biomass due to windstorms [126], while another evaluated the effect of wind and bark beetle disturbances on carbon sequestration based on a landscape model [127]. According to the simulations, the forest areas will be a carbon sink until the end of the twenty-first century [127]. However, climate change might result in a change in the disturbance regime, which might lead to a reduction in the ability of the forests to sequester carbon, and thus turning them from a carbon sink to a carbon source [127].

The assessment of energy potentials is challenging. Two main reasons were identified to justify the variability in the results of several studies on energy potentials for the same geographical region [116]: (*i*) the various energy potential concepts subjacent to the analyses and (*ii*) the spatial variability of the data used for their estimation. Harmonisation of data can be used to overcome this variability. Scaramuzzino et al. [116] proposed a four-step harmonisation framework (Fig. 7): (1) identify the best-suited territorial unit, which should satisfy two conditions: (*i*) data must be reliable for the territorial unit or simple to calculate, and (*ii*) the territorial unit should be easily identified (e.g., NUT2, NUT3); (2) select the sources of renewable energy (e.g., forest residues, agricultural residues, livestock residues, waste) and review their potential (e.g., with data available from databases such as Eurostat, FAO and/or Copernicus Land Monitoring Service); (3) harmonise the indicators of the energy potential per unit area (e.g., $PJ \cdot km^{-2}$, $PJ \cdot inhab^{-1}$); (4) select and harmonise the non-energy territorial indicators due to the selected territorial unit (e.g., selection of topographic and climatic data and harmonisation with the median per territorial unit).

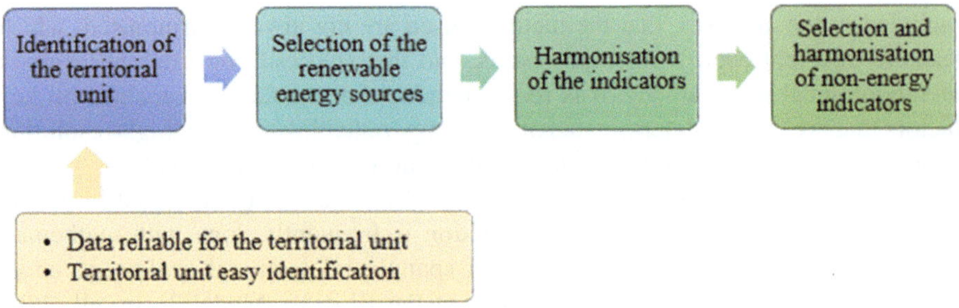

Fig. 7 Four-step harmonisation framework for energy potential assessment studies

The biomass evaluation and monitoring have been done, in particular for large areas, with geographical information systems (GIS) and with geographic decision support systems (DSS), resulting in the development of specific methodologies and techniques to identify and quantify the potential of biomass for energy purposes, namely for the installation of energy systems based on combustion [85, 114, 117, 128–135] or gasification [45, 132, 133, 136] or for the production of biofuels [135, 137–140]. The aforementioned methodologies were used to evaluate biomass in existing [84, 141–144] or potential [84, 87, 88, 112, 145, 146] forest areas, in both forests and farmland [114, 128, 131, 133] or specifically in forests for timber production and other products and services [87, 108, 117, 147]. Some studies were focused on evaluating the economic viability of implementing biomass systems [121, 128, 131, 146, 148].

When evaluating residual biomass, the residue production yields have to be known in order to relate them to the total available biomass (Fig. 8). These ratios, generally expressed in tonnes of residues per year and unit area, are dependent on species, stand structure and silvicultural operations and thus can have significant local and regional variation [87]. In some cases, they are expressed on an as-received basis, others on a dry basis [149, 150]. An alternative to explicitly reporting the residue production yields is to consider the percentage of the total mass of the tree that can be used for energy purposes [151].

Several maps of biomass have been produced [152–154]. The maps of biomass are dependent on the data input and the methods used. One important issue is the harmonisation of the input variables to reduce the uncertainties of the maps [152, 154]. The variability of the species, stand structure, site and management results in a wide variability in biomass. This may increase the errors. For example, Avitabile and Camia [152] reported, for Europe, a trend towards overestimation in forest areas with low biomass (<100 Mg·ha^{-1}) and underestimation for medium and high biomass (>100 Mg·ha^{-1}). Furthermore, the errors increased from national spatial resolutions (29–40%) to higher spatial resolutions, 58–67% [152]. Moreover, care should be taken when analysing maps as uncertainties related to site and management change forest stand dynamics, development and yield, both positively and negatively [155]. Thus, it is recommended that the maps of biomass be complemented with the quantification of the uncertainties [154].

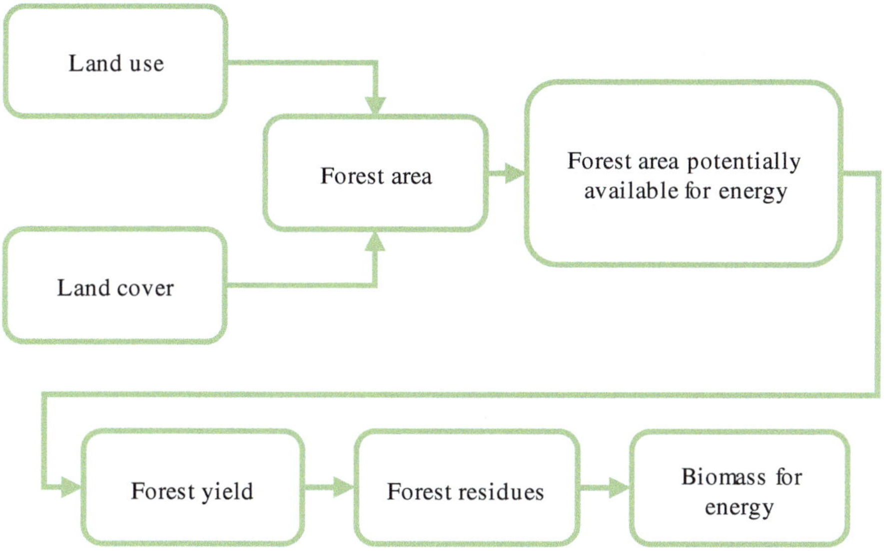

Fig. 8 Flowchart for the evaluation of the areas to produce forest biomass residues. (adapted from [87])

The determination of the areas for energy plantations (*cf*. chapter "Energy Plantations") should consider the following restrictions [146]: (*i*) guarantee food production; (*ii*) avoid losses of biodiversity; (*iii*) mitigate greenhouse gas emissions; (*iv*) minimise negative impacts on soils, water and air. This results in areas potentially available for energy plantations being those either set aside by agriculture or without suitability for agricultural crops, which are not under conservation or protection [146].

One issue related to biomass for energy is the identification of the areas available for energy plantations. These areas can be determined following a methodology in four steps [112]: (1) selection of the species to be used and assessment of their ecological and cultural characteristics; (2) determination of the suitability of the sites, which refers to the selection of a set of data, frequently in a geographical information systems environment, including soils, land morphology, climate, protected areas and administrative boundaries and a suite of assumptions and a subsequent set of operations that enable the identification of the areas where the selected species can be grown; (3) determination of the availability of land, which refers to the identification of the potential areas available, considering the existing restrictions, whether economic or social; (4) assignment of the land, which refers to the definition of a decision process that enables the determination of the areas where the energy plantations can be installed (Fig. 9). The areas identified are dependent on the initial assumptions made. If only the optimal conditions that could potentially generate higher yields are considered, the area estimation could be rather conservative [112].

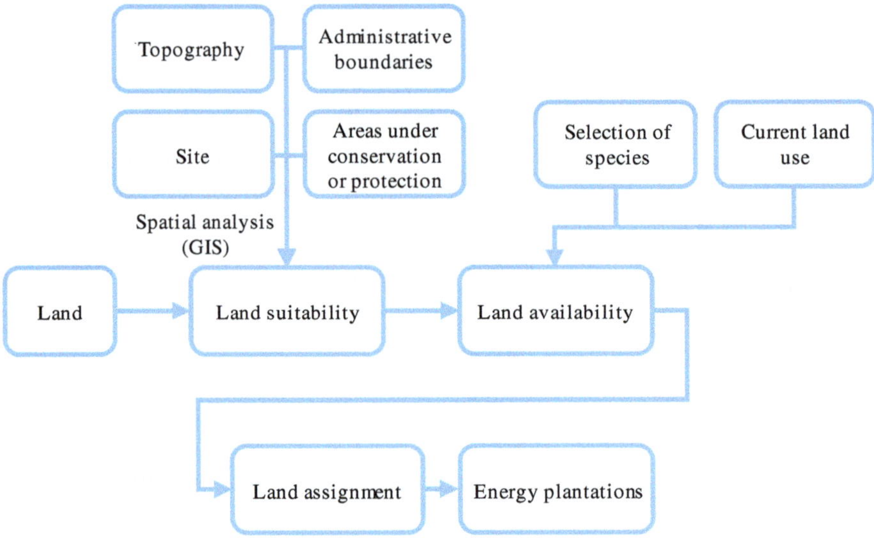

Fig. 9 Flowchart for the selection of areas to install energy plantations. (adapted from [112])

4 Outline of the Book Chapters

This book presents the state of the art of forest biomass production, assessment, characterisation and conversion into heat and power. After the overview presented in this chapter, in chapter "Sources and Distribution of Forest Biomass for Energy", forest biomass is defined and the different categories of forest biomass are characterised, starting with the forest biomass directly sourced from land use systems, passing through the residues of the wood-based industries and ending up in the residues and waste recovered from economic and social activities outside the forest sector.

Forest stands are communities of trees with competitive and facilitation interactions over a long timeframe. These interactions are dynamic in space and time and depend on stand structure, silvicultural systems and silvicultural practices, as well as species and site. In general, the biomass of a tree increases with age. Yet, the rate of biomass storage is dependent on the site (availability of growing space) and interactions between trees in a stand (competition *versus* facilitation). The dynamics of biomass at stand level are more complex to analyse as they are dependent not only on individual tree growth (and thus biomass storage) but also on the balance between live biomass and biomass exports. The intensity, frequency and quality of biomass exported influence the sustainability of the system. In chapter "Stand Structure and Biomass", definitions and concepts of silviculture that allow the analysis and discussion of biomass dynamics and stand sustainability are introduced. The stand structure, silvicultural systems, silvicultural practices, biomass partitioning and dynamics, forest system sustainability and biomass yields, harvest and exports are analysed.

Forest stands are the major sources of woody products, including biomass for energy. Energy plantations are forest stands designed to produce high quantities of biomass in short timeframes for bioenergy. These forest stands have the advantage, apart from producing large amounts of biomass for energy, of releasing the pressure on other forest stands to provide bioenergy. Chapter "Energy Plantations" characterises the energy plantations in terms of species, density, rotation, harvest cycles, site selection, management practices, harvesting and yields.

Biomass cannot be directly measured. The two basic methods for estimating biomass are the direct method, which is destructive, labour-intensive and expensive, and the indirect method, which uses mathematical functions. The use of models has the advantage of enabling the evaluation, monitoring and prediction of biomass in time and space. Yet, the models are dependent on species, site, tree biomass partitioning, stand structure and spatial and temporal scales. As a result, many models have been developed. Chapter "Modelling Biomass" reviews the data sets available for biomass modelling, the mathematical methods and techniques to fit the functions and the model uncertainties.

Modelling biomass is a challenge due to the variability of tree allometry and stand structure, which has resulted in a high number of biomass functions. At tree level, diameter at breast height and height are the most frequently used explanatory variables. However, due to the variability in tree allometry, other explanatory variables have been used, such as development stage, site or tree social status. At the area level, many explanatory variables have been used, derived from forest inventory, remote sensing and ancillary. Moreover, many mathematical models have been utilised to fit the biomass functions. There has been a constant search for models that are able to accommodate the variability of biomass. Chapter "Overview of the Biomass Models" reviews the biomass models at tree and area levels, according to the data used (destructive, forest inventory, remote sensing and ancillary) and mathematical methods and techniques (from parametric to non-parametric).

As far as the conversion of forest biomass to energy is concerned, in the last decades, the role of forest biomass for cooking and household heating has been losing importance and the energy uses of biomass have diversified. Today, different conversion technologies are commercially available or, if still in the research and development stage, considered promising. Chapter "Forest Biomass as an Energy Resource" presents an overview of the most relevant processing technologies for the conversion of forest biomass into energy and fuels, their applications and their readiness levels. Also important for the use and development of these technologies is the knowledge of the properties of biomass that are relevant for its conversion into energy and fuels. Chapter "Forest Biomass as an Energy Resource" provides an overview of the most relevant characteristics of forest biomass and reviews commonly used pre-treatment methods aimed at upgrading raw forest biomass into more suitable feedstocks for specific conversion technologies.

Chapter "Forest Biomass as an Energy Resource" is followed by three chapters that describe in more detail the most common uses of forest biomass for energy and the associated technologies. The first of these chapters, chapter "Biomass for Domestic Heat", is dedicated to residential heat production, which is where presently more solid biomass is consumed. Even though modern technologies should be promoted, the fact is that inefficient and polluting traditional technologies are still widely used, so chapter "Biomass for Domestic Heat" presents the available traditional technologies alongside the modern ones. The impacts of the traditional use of biomass are also reviewed, followed by the improvements to traditional technologies.

Chapter "Biomass for Industrial and District Heating" focuses on the sector that globally consumes more energy nowadays: the industrial sector. After a brief description of how energy is consumed by the industry, the most important technologies used to produce process heat, which is the industrial end-use that requires more energy, are described. Since the energy conversion technologies used in industrial facilities are similar to those used in district heating plants, the chapter also describes district heating systems and the role of biomass for this use.

Due to the relevance of process heat in industrial energy consumption, chapter "Biomass for Industrial and District Heating" only focuses on the production of heat. However, combined heat and power is also commonly generated in energy-intensive industrial facilities. CHP technologies are described in chapter "Biomass for Power Production and Cogeneration", which is dedicated to electricity production with forest biomass. After a general description of the power sector, chapter "Biomass for Power Production and Cogeneration" describes the technologies commonly used to produce electricity from forest biomass in dedicated biomass plants and in co-combustion plants.

The last chapter of this book, chapter "Conclusions and Future Research Needs", is dedicated to some conclusions and a description of possible future research.

Funding This work was funded by National Funds through FCT—Foundation for Science and Technology, under the Project UIDB/05183/2020, through MED, and Project UIDB/50022/2020, through IDMEC, under LAETA.

References

1. Naeher LP, Brauer M, Lipsett M et al (2007) Woodsmoke health effects: a review. Inhal Toxicol 19:67–106. https://doi.org/10.1080/08958370600985875
2. Fullerton DG, Bruce N, Gordon SB (2008) Indoor air pollution from biomass fuel smoke is a major health concern in the developing world. Trans R Soc Trop Med Hyg 102:843–851. https://doi.org/10.1016/j.trstmh.2008.05.028
3. Noubiap JJN, Essouma M, Bigna JJR (2015) Targeting household air pollution for curbing the cardiovascular disease burden: a health priority in sub-Saharan Africa. J Clin Hypertens 17:825–829. https://doi.org/10.1111/jch.12610
4. Hanna R, Duflo E, Greenstone M (2016) Up in smoke: the influence of household behavior on the long-run impact of improved cooking stoves. Am Econ J Econ Policy 8:80–114. https://doi.org/10.1257/pol.20140008
5. Bede-Ojimadu O, Orisakwe OE (2020) Exposure to wood smoke and associated health effects in sub-Saharan Africa: a systematic review. Ann Glob Health 86:32. https://doi.org/10.5334/aogh.2725

6. Smith KR, Uma R, Kishore VVN et al (2000) Greenhouse implications of household stoves: an analysis for India. Annu Rev Energy Environ 25:741–763. https://doi.org/10.1146/annurev.energy.25.1.741
7. Masera OR, Bailis R, Drigo R et al (2015) Environmental burden of traditional bioenergy use. Annu Rev Environ Resour 40:121–150. https://doi.org/10.1146/annurev-environ-102014-021318
8. Malico I, Nepomuceno Pereira S, Costa MJ (2017) Black carbon trends in southwestern Iberia in the context of the financial and economic crisis. The role of bioenergy. Environ Sci Pollut Res 24:476–488. https://doi.org/10.1007/s11356-016-7805-8
9. Lelieveld J, Klingmüller K, Pozzer A et al (2019) Effects of fossil fuel and total anthropogenic emission removal on public health and climate. Proc Natl Acad Sci 116:7192–7197. https://doi.org/10.1073/pnas.1819989116
10. Arnold M, Köhlin G, Persson R, Shepherd G (2003) Fuelwood revisited: what has changed in the last decade? Center for International Forestry Research, Bogor
11. Kammila S, Kappen JF, Rysankova D et al (2014) Clean and improved cooking in Sub-Sahran Africa, 2nd edn. The World Bank, Washington, DC
12. Bailis R, Wang Y, Drigo R et al (2017) Getting the numbers right: revisiting woodfuel sustainability in the developing world. Environ Res Lett 12:115002. https://doi.org/10.1088/1748-9326/aa83ed
13. Sedano F, Mizu-Siampale A, Duncanson L, Liang M (2022) Influence of charcoal production on forest degradation in Zambia: a remote sensing perspective. Remote Sens 14:3352. https://doi.org/10.3390/rs14143352
14. IEA (2022) Energy statistics data browser. https://www.iea.org/data-and-statistics/data-tools/energy-statistics-data-browser. Accessed 7 Apr 2023
15. IEA (2023) Interactive Sankey diagram. https://www.iea.org/sankey. Accessed 7 Apr 2023
16. IEA (2022) World energy outlook 2022. International Energy Agency, Paris
17. Balat M (2009) Coal in the global energy scene. Energy Sources Part B Econ Plan Policy 5:50–62. https://doi.org/10.1080/15567240701758927
18. IEA (2021) World energy outlook 2021. International Energy Agency, Paris
19. Dincer I (1999) Environmental impacts of energy. Energy Policy 27:845–854. https://doi.org/10.1016/S0301-4215(99)00068-3
20. Lefèvre N (2010) Measuring the energy security implications of fossil fuel resource concentration. Energy Policy 38:1635–1644. https://doi.org/10.1016/j.enpol.2009.02.003
21. Chu S, Majumdar A (2012) Opportunities and challenges for a sustainable energy future. Nature 488:294–303. https://doi.org/10.1038/nature11475
22. Demirbaş A (2006) Global renewable energy resources. Energy Sources Part Recovery Util Environ Eff 28:779–792. https://doi.org/10.1080/00908310600718742
23. Sepulveda NA, Jenkins JD, de Sisternes FJ, Lester RK (2018) The role of firm low-carbon electricity resources in deep decarbonization of power generation. Joule 2:2403–2420. https://doi.org/10.1016/j.joule.2018.08.006
24. Mandley SJ, Daioglou V, Junginger HM et al (2020) EU bioenergy development to 2050. Renew Sustain Energy Rev 127:109858. https://doi.org/10.1016/j.rser.2020.109858
25. Daioglou V, Muratori M, Lamers P et al (2020) Implications of climate change mitigation strategies on international bioenergy trade. Clim Change 163:1639–1658. https://doi.org/10.1007/s10584-020-02877-1
26. Elsayed MA, Matthews R, Mortimer ND (2003) Carbon and energy balances for a range of biofuels. Sheffield Hallam University
27. Jeswani HK, Gujba H, Azapagic A (2011) Assessing options for electricity generation from biomass on a life cycle basis: environmental and economic evaluation. Waste Biomass Valoriz 2:33–42. https://doi.org/10.1007/s12649-010-9057-z
28. Amponsah NY, Troldborg M, Kington B et al (2014) Greenhouse gas emissions from renewable energy sources: a review of lifecycle considerations. Renew Sustain Energy Rev 39:461–475. https://doi.org/10.1016/j.rser.2014.07.087

29. Timmons DS, Buchholz T, Veeneman CH (2016) Forest biomass energy: assessing atmospheric carbon impacts by discounting future carbon flows. GCB Bioenergy 8:631–643. https://doi.org/10.1111/gcbb.12276
30. Jeswani HK, Whiting A, Azapagic A (2020) Environmental and economic sustainability of biomass heat in the UK. Energy Technol 8:1901044. https://doi.org/10.1002/ente.201901044
31. Akbari M, Oyedun AO, Gemechu E, Kumar A (2021) Comparative life cycle energy and greenhouse gas footprints of dry and wet torrefaction processes of various biomass feedstocks. J Environ Chem Eng 9:105415. https://doi.org/10.1016/j.jece.2021.105415
32. Robertson GP, Hamilton SK, Barham BL, et al (2017) Cellulosic biofuel contributions to a sustainable energy future: choices and outcomes. Science 356:eaal2324. https://doi.org/10.1126/science.aal2324
33. McKendry P (2002) Energy production from biomass (part 2): conversion technologies. Bioresour Technol 83:47–54. https://doi.org/10.1016/S0960-8524(01)00119-5
34. Selivanov E, Cudlín P, Horáček P (2023) Carbon neutrality of forest biomass for bioenergy: a scoping review. IForest Biogeosci For 16:70–77. https://doi.org/10.3832/ifor4160-015
35. Albaugh TJ, Albaugh JM, Fox TR et al (2016) Tamm review: light use efficiency and carbon storage in nutrient and water experiments on major forest plantation species. For Ecol Manag 376:333–342. https://doi.org/10.1016/j.foreco.2016.05.031
36. Ameray A, Bergeron Y, Valeria O et al (2021) Forest carbon management: a review of silvicultural practices and management strategies across boreal, temperate and tropical forests. Curr For Rep 7:245–266. https://doi.org/10.1007/s40725-021-00151-w
37. Agostini A, Giuntoli J, Boulamanti A (2014) Carbon accounting of forest bioenergy: conclusions and recommendations from a critical literature review. Publications Office of the European Union, Luxembourg
38. Schroeder LM (2008) Insect pests and forest biomass for energy. In: Röser D, Asikainen A, Raulund-Rasmussen K, Stupak I (eds) Sustainable use pf forest biomass for energy. A Synthesis with Focus on the Balticand Nordic Region. Springer, Dordrecht, pp 109–128
39. Bauen A, Berndes G, Junginger M, et al (2009) Bioenergy—a sustainable and reliable energy source. A review of status and prospects. IEA Bioenergy
40. Lovegrove K, Alexander D, Bader R, et al (2019) Renewable energy options for industrial process heat. Australian Renewable Energy Agency, O'Connor
41. IRENA (2020) Recycle: bioenergy. IRENA
42. Wan M, Lähtinen K, Toppinen A, Toivio M (2012) Opportunities and challenges in the emerging bioenergy business: the case of the finnish sawmill industry. Int J For Eng 23:89–101. https://doi.org/10.1080/14942119.2012.10739965
43. Martire S, Tuomasjukka D, Lindner M et al (2015) Sustainability impact assessment for local energy supplies' development—the case of the alpine area of Lake Como, Italy. Biomass Bioenergy 83:60–76. https://doi.org/10.1016/j.biombioe.2015.08.020
44. Berndes G, Abt B, Asikainen A, et al (2016) Forest biomass, carbon neutrality and climate change mitigation. European Forest Institute
45. Cardoso J, Silva V, Eusébio D (2019) Techno-economic analysis of a biomass gasification power plant dealing with forestry residues blends for electricity production in Portugal. J Clean Prod 212:741–753. https://doi.org/10.1016/j.jclepro.2018.12.054
46. Anca-Couce A, Hochenauer C, Scharler R (2021) Bioenergy technologies, uses, market and future trends with Austria as a case study. Renew Sustain Energy Rev 135:110237. https://doi.org/10.1016/j.rser.2020.110237
47. Locoh A, Thiffault É, Barnabé S (2022) Sustainability impact assessment of forest bioenergy value chains in Quebec (Canada)—a ToSIA approach. Energies 15:6676. https://doi.org/10.3390/en15186676
48. Tilman D, Socolow R, Foley JA et al (2009) Beneficial biofuels—the food, energy, and environment trilemma. Science 325:270–271. https://doi.org/10.1126/science.1177970
49. Haberl H, Erb K-H, Krausmann F et al (2013) Bioenergy: how much can we expect for 2050? Environ Res Lett 8:031004. https://doi.org/10.1088/1748-9326/8/3/031004

50. Beringer T, Lucht W, Schaphoff S (2011) Bioenergy production potential of global biomass plantations under environmental and agricultural constraints. GCB Bioenergy 3:299–312. https://doi.org/10.1111/j.1757-1707.2010.01088.x
51. Reid WV, Ali MK, Field CB (2020) The future of bioenergy. Glob Change Biol 26:274–286. https://doi.org/10.1111/gcb.14883
52. Dogutan DK, Nocera DG (2019) Artificial photosynthesis at efficiencies greatly exceeding that of natural photosynthesis. Acc Chem Res 52:3143–3148. https://doi.org/10.1021/acs.accounts.9b00380
53. Rodrigues AR, Botequim B, Tavares C et al (2020) Addressing soil protection concerns in forest ecosystem management under climate change. For Ecosyst 7:34. https://doi.org/10.1186/s40663-020-00247-y
54. Krausmann F, Erb K-H, Gingrich S et al (2013) Global human appropriation of net primary production doubled in the 20th century. Proc Natl Acad Sci 110:10324–10329. https://doi.org/10.1073/pnas.1211349110
55. Schweier J, Molina-Herrera S, Ghirardo A et al (2017) Environmental impacts of bioenergy wood production from poplar short-rotation coppice grown at a marginal agricultural site in Germany. GCB Bioenergy 9:1207–1221. https://doi.org/10.1111/gcbb.12423
56. Kalt G, Mayer A, Theurl MC et al (2019) Natural climate solutions versus bioenergy: can carbon benefits of natural succession compete with bioenergy from short rotation coppice? GCB Bioenergy 11:1283–1297. https://doi.org/10.1111/gcbb.12626
57. Langeveld H, Quist-Wessel F, Dimitriou I et al (2012) Assessing environmental impacts of short rotation coppice (SRC) expansion: model definition and preliminary results. BioEnergy Res 5:621–635. https://doi.org/10.1007/s12155-012-9235-x
58. Röder M, Thiffault E, Martínez-Alonso C et al (2019) Understanding the timing and variation of greenhouse gas emissions of forest bioenergy systems. Biomass Bioenergy 121:99–114. https://doi.org/10.1016/j.biombioe.2018.12.019
59. Welfle A, Röder M (2022) Mapping the sustainability of bioenergy to maximise benefits, mitigate risks and drive progress toward the sustainable development goals. Renew Energy 191:493–509. https://doi.org/10.1016/j.renene.2022.03.150
60. IEA (2022) Renewables 2022. Analysis and forecast to 2027. International Energy Agency, Paris
61. WBA (2022) Global bioenergy statistics 2022. World Bioenergy Association, Stockholm
62. Santoro M, Cartus O (2019) ESA biomass climate change initiative (Biomass_cci): global datasets of forest above-ground biomass for the year 2017, vol 1. Centre for Environmental Data Analysis
63. Pan Y, Birdsey RA, Fang J et al (2011) A large and persistent carbon sink in the world's forests. Science 333:988–993. https://doi.org/10.1126/science.1201609
64. Urbano AR, Keeton WS (2017) Carbon dynamics and structural development in recovering secondary forests of the northeastern U.S. For Ecol Manag 392:21–35. https://doi.org/10.1016/j.foreco.2017.02.037
65. van Asselen S, Verburg PH (2012) A land system representation for global assessments and land-use modeling. Glob Change Biol 18:3125–3148
66. Smith DM, Larson BC, Kelty MJ, Ashton PMS (1997) The practice of silviculture. Applied forest ecology, 9th ed. John Wiley & Sons, Inc, New York
67. Röser D, Asikainen A, Stupak I, Pasanen K (2008) Forest energy sources and potentials. In: Röser D, Asikainen A, Raulund-Rasmussen K, Stupak I (eds) Sustainable use pf forest biomass for energy. A synthesis with focus on the Balticand Nordic Region. Springer, Dordrecht, pp 9–28
68. Chai RK, Andrus RA, Rodman K et al (2019) Stand dynamics and topographic setting influence changes in live tree biomass over a 34-year permanent plot record in a subalpine forest in the Colorado Front Range. Can J For Res 49:1256–1264. https://doi.org/10.1139/cjfr-2019-0023
69. Bradford JB, Fraver S, Milo AM et al (2012) Effects of multiple interacting disturbances and salvage logging on forest carbon stocks. For Ecol Manag 267:209–214. https://doi.org/10.1016/j.foreco.2011.12.010

70. Berndes G, Hoogwijk M, van den Broek R (2003) The contribution of biomass in the future global energy supply: a review of 17 studies. Biomass Bioenergy 25:1–28. https://doi.org/10.1016/S0961-9534(02)00185-X
71. Martire S, Castellani V, Sala S (2015) Carrying capacity assessment of forest resources: enhancing environmental sustainability in energy production at local scale. Resour Conserv Recycl 94:11–20. https://doi.org/10.1016/j.resconrec.2014.11.002
72. Dickmann D (2006) Silviculture and biology of short-rotation woody crops in temperate regions: then and now. Biomass Bioenergy 30:696–705. https://doi.org/10.1016/j.biombioe.2005.02.008
73. Egnell G, Paré D, Thiffault E, Lamers P (2016) Environmental sustainability aspects of forest biomass mobilisation. In: Thiffault E, Berndes G, Junginger M et al (eds) Mobilisation of forest bioenergy in the boreal and temperate biomes. Elsevier, London, pp 50–67
74. Dimitriou I, Rutz D (2015) Sustainable short rotation coppice a handbook. WIP Renewable Energies, Munich
75. McKendry P (2002) Energy production from biomass (part 1): overview of biomass. Bioresour Technol 83:37–46. https://doi.org/10.1016/S0960-8524(01)00118-3
76. Guidi W, Pitre EF, Labrecque M (2013) Short-rotation coppice of willows for the production of biomass in Eastern Canada. In: Matovic MD (ed) Biomass now–sustainable growth and use. InTech, Rijeka, Croatia, pp 421–448
77. Stojanović M, Sánchez-Salguero R, Levanič T et al (2017) Forecasting tree growth in coppiced and high forests in the Czech Republic. The legacy of management drives the coming Quercus petraea climate responses. For Ecol Manag 405:56–68. https://doi.org/10.1016/j.foreco.2017.09.021
78. Bugalho MN, Dias FS, Briñas B, Cerdeira JO (2016) Using the high conservation value forest concept and Pareto optimization to identify areas maximizing biodiversity and ecosystem services in cork oak landscapes. Agrofor Syst 90:35–44. https://doi.org/10.1007/s10457-015-9814-x
79. Plieninger T, Hartel T, Martín-López B et al (2015) Wood-pastures of Europe: geographic coverage, social–ecological values, conservation management, and policy implications. Biol Conserv 190:70–79
80. Durocher C, Thiffault E, Achim A et al (2019) Untapped volume of surplus forest growth as feedstock for bioenergy. Biomass Bioenergy 120:376–386. https://doi.org/10.1016/j.biombioe.2018.11.024
81. Shepard JP (2006) Water quality protection in bioenergy production: the US system of forestry best management practices. Biomass Bioenergy 30:378–384. https://doi.org/10.1016/j.biombioe.2005.07.018
82. Schulze ED, Bouriaud O, Irslinger R, Valentini R (2022) The role of wood harvest from sustainably managed forests in the carbon cycle. Ann For Sci 79:17. https://doi.org/10.1186/s13595-022-01127-x
83. Thiel M, Basiliko N, Caspersen J et al (2015) Operational biomass recovery of small trees: equations for six central Ontario tree species. Can J For Res 45:372–377. https://doi.org/10.1139/cjfr-2014-0429
84. Daioglou V, Stehfest E, Wicke B et al (2016) Projections of the availability and cost of residues from agriculture and forestry. GCB Bioenergy 8:456–470. https://doi.org/10.1111/gcbb.12285
85. Nonini L, Schillaci C, Fiala M (2022) Assessing logging residues availability for energy production by using forest management plans data and geographic information system (GIS). Eur J For Res 141:959–977. https://doi.org/10.1007/s10342-022-01484-2
86. Thiffault E, Béchard A, Paré D, Allen D (2015) Recovery rate of harvest residues for bioenergy in boreal and temperate forests: a review. WIREs Energy Environ 4:429–451. https://doi.org/10.1002/wene.157
87. Castellano PJ, Volk TA, Herrington LP (2009) Estimates of technically available woody biomass feedstock from natural forests and willow biomass crops for two locations in New York State. Biomass Bioenergy 33:393–406. https://doi.org/10.1016/j.biombioe.2008.08.001

88. Aust C, Schweier J, Brodbeck F et al (2014) Land availability and potential biomass production with poplar and willow short rotation coppices in Germany. GCB Bioenergy 6:521–533. https://doi.org/10.1111/gcbb.12083
89. Perlack RD, Wright LL, Turhollow AF, Graham RL (2005) Biomass as feedstock for a bioenergy and bioproducts industry: the technical feasibility of a billion-ton annual supply. Oak Ridge National Laboratory, Oak Ridge
90. Mola-Yudego B, Arevalo J, Díaz-Yáñez O et al (2017) Wood biomass potentials for energy in northern Europe: forest or plantations? Biomass Bioenergy 106:95–103. https://doi.org/10.1016/j.biombioe.2017.08.021
91. Hamelinck CN, Suurs RAA, Faaij APC (2005) International bioenergy transport costs and energy balance. Biomass Bioenergy 29:114–134. https://doi.org/10.1016/j.biombioe.2005.04.002
92. Rentizelas AA, Tolis AJ, Tatsiopoulos IP (2009) Logistics issues of biomass: the storage problem and the multi-biomass supply chain. Renew Sustain Energy Rev 13:887–894. https://doi.org/10.1016/j.rser.2008.01.003
93. Jäppinen E, Korpinen O-J, Laitila J, Ranta T (2014) Greenhouse gas emissions of forest bioenergy supply and utilization in Finland. Renew Sustain Energy Rev 29:369–382. https://doi.org/10.1016/j.rser.2013.08.101
94. Röder M, Whittaker C, Thornley P (2015) How certain are greenhouse gas reductions from bioenergy? Life cycle assessment and uncertainty analysis of wood pellet-to-electricity supply chains from forest residues. Biomass Bioenergy 79:50–63. https://doi.org/10.1016/j.biombioe.2015.03.030
95. Mitchell CP, Bridgwater AV, Stevens DJ et al (1995) Technoeconomic assessment of biomass to energy. Biomass Bioenergy 9:205–226
96. Steubing B, Zah R, Waeger P, Ludwig C (2010) Bioenergy in Switzerland: assessing the domestic sustainable biomass potential. Renew Sustain Energy Rev 14:2256–2265. https://doi.org/10.1016/j.rser.2010.03.036
97. Farine DR, O'Connell DA, John Raison R et al (2012) An assessment of biomass for bioelectricity and biofuel, and for greenhouse gas emission reduction in Australia. GCB Bioenergy 4:148–175. https://doi.org/10.1111/j.1757-1707.2011.01115.x
98. Scarlat N, Blujdea V, Dallemand J-F (2011) Assessment of the availability of agricultural and forest residues for bioenergy production in Romania. Biomass Bioenergy 35:1995–2005. https://doi.org/10.1016/j.biombioe.2011.01.057
99. Rothe A, Moroni M, Neyland M, Wilnhammer M (2015) Current and potential use of forest biomass for energy in Tasmania. Biomass Bioenergy 80:162–172. https://doi.org/10.1016/j.biombioe.2015.04.021
100. Egnell G (2017) A review of Nordic trials studying effects of biomass harvest intensity on subsequent forest production. For Ecol Manag 383:27–36. https://doi.org/10.1016/j.foreco.2016.09.019
101. McMinn JW, Crossley DA Jr (1996) Biodiversity and coarse woody debris in southern forests. Department of Agriculture, Forest Service, Southern Research Station, Asheville, North Carolina
102. U.S. Department of Energy (2011) U.S. Billion-Ton update: biomass supply for a bioenergy and bioproducts industry. U.S. Department of Energy
103. U.S. Department of Energy (2016) 2016 Billion-Ton Report: advancing domestic resources for a thriving bioeconomy, vol 1: economic availability of feedstocks. U.S. Department of Energy, Oak Ridge National Laboratory, Oak Ridge, TN
104. Raulund-Rasmussen K, Stupak I, Clarke N, et al (2008) Effects of very intensive forest biomass harvesting and long term site productivity. In: Röser D, Asikainen A, Raulund-Rasmussen K, Stupak I (eds) Sustainable use pf forest biomass for energy. A synthesis with focus on the Balticand Nordic Region. Springer, Dordrecht, pp 29–78
105. Pyttel PL, Köhn M, Bauhus J (2015) Effects of different harvesting intensities on the macro nutrient pools in aged oak coppice forests. For Ecol Manag 349:94–105. https://doi.org/10.1016/j.foreco.2015.03.037

106. Batidzirai B, Smeets EMW, Faaij APC (2012) Harmonising bioenergy resource potentials—methodological lessons from review of state of the art bioenergy potential assessments. Renew Sustain Energy Rev 16:6598–6630. https://doi.org/10.1016/j.rser.2012.09.002
107. Gonçalves AC, Sousa AMO, Mesquita PG (2017) Estimation and dynamics of above ground biomass with very high resolution satellite images in *Pinus pinaster* stands. Biomass Bioenergy 106:146–154. https://doi.org/10.1016/j.biombioe.2017.08.026
108. Viana H, Aranha J, Lopes D, Cohen WB (2012) Estimation of crown biomass of Pinus pinaster stands and shrubland above-ground biomass using forest inventory data, remotely sensed imagery and spatial prediction models. Ecol Model 226:22–35. https://doi.org/10.1016/j.ecolmodel.2011.11.027
109. Frombo F, Minciardi R, Robba M et al (2009) Planning woody biomass logistics for energy production: a strategic decision model. Biomass Bioenergy 33:372–383. https://doi.org/10.1016/j.biombioe.2008.09.008
110. Kinoshita T, Inoue K, Iwao K et al (2009) A spatial evaluation of forest biomass usage using GIS. Appl Energy 86:1–8. https://doi.org/10.1016/j.apenergy.2008.03.017
111. Vasco H, Costa M (2009) Quantification and use of forest biomass residues in Maputo province, Mozambique. Biomass Bioenergy 33:1221–1228. https://doi.org/10.1016/j.biombioe.2009.05.008
112. Fiorese G, Guariso G (2010) A GIS-based approach to evaluate biomass potential from energy crops at regional scale. Environ Model Softw 25:702–711. https://doi.org/10.1016/j.envsoft.2009.11.008
113. Zambelli P, Lora C, Spinelli R et al (2012) A GIS decision support system for regional forest management to assess biomass availability for renewable energy production. Environ Model Softw 38:203–213. https://doi.org/10.1016/j.envsoft.2012.05.016
114. Lourinho G, Brito P (2015) Assessment of biomass energy potential in a region of Portugal (Alto Alentejo). Energy 81:189–201. https://doi.org/10.1016/j.energy.2014.12.021
115. Hamelin L, Borzęcka M, Kozak M, Pudełko R (2019) A spatial approach to bioeconomy: quantifying the residual biomass potential in the EU-27. Renew Sustain Energy Rev 100:127–142. https://doi.org/10.1016/j.rser.2018.10.017
116. Scaramuzzino C, Garegnani G, Zambelli P (2019) Integrated approach for the identification of spatial patterns related to renewable energy potential in European territories. Renew Sustain Energy Rev 101:1–13. https://doi.org/10.1016/j.rser.2018.10.024
117. Viana H, Cohen WB, Lopes D, Aranha J (2010) Assessment of forest biomass for use as energy. GIS-based analysis of geographical availability and locations of wood-fired power plants in Portugal. Appl Energy 87:2551–2560
118. Ranta T (2005) Logging residues from regeneration fellings for biofuel production–a GIS-based availability analysis in Finland. Biomass Bioenergy 28:171–182. https://doi.org/10.1016/j.biombioe.2004.08.010
119. Verkerk PJ, Lindner M, Zanchi G, Zudin S (2011) Assessing impacts of intensified biomass removal on deadwood in European forests. Ecol Indic 11:27–35. https://doi.org/10.1016/j.ecolind.2009.04.004
120. Verkerk PJ, Anttila P, Eggers J et al (2011) The realisable potential supply of woody biomass from forests in the European Union. For Ecol Manag 261:2007–2015. https://doi.org/10.1016/j.foreco.2011.02.027
121. Lundmark R, Athanassiadis D, Wetterlund E (2015) Supply assessment of forest biomass—a bottom-up approach for Sweden. Biomass Bioenergy 75:213–226. https://doi.org/10.1016/j.biombioe.2015.02.022
122. Halperin J, LeMay V, Chidumayo E, et al (2016) Model-based estimation of above-ground biomass in the miombo ecoregion of Zambia. For Ecosyst 3. https://doi.org/10.1186/s40663-016-0077-4
123. Saatchi SS, Harris NL, Brown S et al (2011) Benchmark map of forest carbon stocks in tropical regions across three continents. Proc Natl Acad Sci 108:9899–9904
124. Baccini A, Goetz SJ, Walker WS et al (2012) Estimated carbon dioxide emissions from tropical deforestation improved by carbon-density maps. Nat Clim Change 2:182–185. https://doi.org/10.1038/nclimate1354

125. Thom D (2023) Natural disturbances as drivers of tipping points in forest ecosystems under climate change—implications for adaptive management. Forestry 96:305–315. https://doi.org/10.1093/forestry/cpad011
126. Vaglio Laurin G, Puletti N, Tattoni C et al (2021) Estimated biomass loss caused by the vaia windthrow in Northern Italy: evaluation of active and passive remote sensing options. Remote Sens 13:4924. https://doi.org/10.3390/rs13234924
127. Albrich K, Seidl R, Rammer W, Thom D (2023) From sink to source: changing climate and disturbance regimes could tip the 21st century carbon balance of an unmanaged mountain forest landscape. Forestry 96:399–409. https://doi.org/10.1093/forestry/cpac022
128. Fernandes U, Costa M (2010) Potential of biomass residues for energy production and utilization in a region of Portugal. Biomass Bioenergy 34:661–666. https://doi.org/10.1016/j.biombioe.2010.01.009
129. Vainio P, Tokola T, Palander T, Kangas A (2009) A GIS-based stand management system for estimating local energy wood supplies. Biomass Bioenergy 33:1278–1288. https://doi.org/10.1016/j.biombioe.2009.05.021
130. Boccardi A, Perina N, Ghidorzi S, et al (2012) Un approccio integrato per i Bilanci Energetico-Ambientali Regionali: il caso della Regione Lombardia. 14
131. Malico I, Carrajola J, Gomes CP, Lima JC (2016) Biomass residues for energy production and habitat preservation. Case study in a montado area in Southwestern Europe. J Clean Prod 112:3676–3683. https://doi.org/10.1016/j.jclepro.2015.07.131
132. Gómez A, Rodrigues M, Montañés C et al (2010) The potential for electricity generation from crop and forestry residues in Spain. Biomass Bioenergy 34:703–719. https://doi.org/10.1016/j.biombioe.2010.01.013
133. Alfonso D, Perpiñá C, Pérez-Navarro A et al (2009) Methodology for optimization of distributed biomass resources evaluation, management and final energy use. Biomass Bioenergy 33:1070–1079. https://doi.org/10.1016/j.biombioe.2009.04.002
134. Battuvshin B, Matsuoka Y, Shirasawa H et al (2020) Supply potential and annual availability of timber and forest biomass resources for energy considering inter-prefectural trade in Japan. Land Use Policy 97:104780. https://doi.org/10.1016/j.landusepol.2020.104780
135. Cintas O, Berndes G, Englund O, Johnsson F (2021) Geospatial supply-demand modeling of lignocellulosic biomass for electricity and biofuels in the European Union. Biomass Bioenergy 144:105870. https://doi.org/10.1016/j.biombioe.2020.105870
136. de Souza L, Hamedani SR, Lora EES et al (2021) Theoretical and technical assessment of agroforestry residue potential for electricity generation in Brazil towards 2050. Energy Rep 7:2574–2587. https://doi.org/10.1016/j.egyr.2021.04.026
137. Tagomori IS, Rochedo PRR, Szklo A (2019) Techno-economic and georeferenced analysis of forestry residues-based Fischer-Tropsch diesel with carbon capture in Brazil. Biomass Bioenergy 123:134–148. https://doi.org/10.1016/j.biombioe.2019.02.018
138. Ahlström JM, Zetterholm J, Pettersson K et al (2020) Economic potential for substitution of fossil fuels with liquefied biomethane in Swedish iron and steel industry—synergy and competition with other sectors. Energy Convers Manag 209:112641. https://doi.org/10.1016/j.enconman.2020.112641
139. Singlitico A, Kilgallon I, Goggins J, Monaghan RFD (2019) GIS-based techno-economic optimisation of a regional supply chain for large-scale deployment of bio-SNG in a natural gas network. Appl Energy 250:1036–1052. https://doi.org/10.1016/j.apenergy.2019.05.026
140. Natarajan K, Leduc S, Pelkonen P et al (2014) Optimal locations for second generation Fischer tropsch biodiesel production in Finland. Renew Energy 62:319–330. https://doi.org/10.1016/j.renene.2013.07.013
141. Stolarski MJ, Szczukowski S, Tworkowski J et al (2011) Short rotation willow coppice biomass as an industrial and energy feedstock. Ind Crops Prod 33:217–223. https://doi.org/10.1016/j.indcrop.2010.10.013
142. Dimitriou I, Mola-Yudego B (2017) Poplar and willow plantations on agricultural land in Sweden: area, yield, groundwater quality and soil organic carbon. For Ecol Manag 383:99–107. https://doi.org/10.1016/j.foreco.2016.08.022

143. Nordborg M, Berndes G, Dimitriou I et al (2018) Energy analysis of poplar production for bioenergy in Sweden. Biomass Bioenergy 112:110–120. https://doi.org/10.1016/j.biombioe.2018.01.021
144. Nordborg M, Berndes G, Dimitriou I et al (2018) Energy analysis of willow production for bioenergy in Sweden. Renew Sustain Energy Rev 93:473–482. https://doi.org/10.1016/j.rser.2018.05.045
145. Deng YY, Koper M, Haigh M, Dornburg V (2015) Country-level assessment of long-term global bioenergy potential. Biomass Bioenergy 74:253–267. https://doi.org/10.1016/j.biombioe.2014.12.003
146. Kluts I, Wicke B, Leemans R, Faaij A (2017) Sustainability constraints in determining European bioenergy potential: a review of existing studies and steps forward. Renew Sustain Energy Rev 69:719–734. https://doi.org/10.1016/j.rser.2016.11.036
147. Gil MV, Blanco D, Carballo MT, Calvo LF (2011) Carbon stock estimates for forests in the Castilla y León region, Spain. a GIS based method for evaluating spatial distribution of residual biomass for bio-energy. Biomass Bioenergy 35:243–252. https://doi.org/10.1016/j.biombioe.2010.08.004
148. Noon CE, Daly MJ (1996) GIS-based biomass resource assessment with BRAVO. Biomass Bioenergy 10:101–109
149. Baumgras JE (1980) Biomass yields from allegheny hardwood thinning. Forest Service, USDA
150. Dias J, Azevedo JLT (2004) Evaluation of biomass residuals in Portugal Mainland. In: Afgan H, de Carvalho N, Ramos MC (eds) New and renewable energy technologies for sustainable development. Swets & Zeitlinger, Lisse, pp 215–228
151. Ćosić B, Stanić Z, Duić N (2011) Geographic distribution of economic potential of agricultural and forest biomass residual for energy use: case study Croatia. Energy 36:2017–2028. https://doi.org/10.1016/j.energy.2010.10.009
152. Avitabile V, Camia A (2018) An assessment of forest biomass maps in Europe using harmonized national statistics and inventory plots. For Ecol Manag 409:489–498. https://doi.org/10.1016/j.foreco.2017.11.047
153. Blackard J, Finco M, Helmer E et al (2008) Mapping U.S. forest biomass using nationwide forest inventory data and moderate resolution information. Remote Sens Environ 112:1658–1677. https://doi.org/10.1016/j.rse.2007.08.021
154. Spawn SA, Sullivan CC, Lark TJ, Gibbs HK (2020) Harmonized global maps of above and belowground biomass carbon density in the year 2010. Sci Data 7:112. https://doi.org/10.1038/s41597-020-0444-4
155. Schulze K, Malek Ž, Verburg PH (2019) Towards better mapping of forest management patterns: a global allocation approach. For Ecol Manag 432:776–785. https://doi.org/10.1016/j.foreco.2018.10.001

Sources and Distribution of Forest Biomass for Energy

Isabel Malico and Ana Cristina Gonçalves

Abstract Forest biomass used for energy or biofuels can be sourced directly from land-use systems, indirectly from wood-based industries or recovered from other human activities outside the forest sector. The former, referring to primary biomass from forests, includes organic products or residues derived directly from living or recently dead trees or other forest vegetation. It constituted nearly half of the world's harvested forest biomass in 2021 and holds particular importance in the Global South, where traditional biomass remains a vital energy source for many people. Besides direct wood fuel, secondary wood residues represent another substantial source of forest bioenergy. These organic residues, such as wood chips, sawdust or black liquor, are generated by the industries processing wood, especially primary forest industries. A large amount of these residues is well-suited for further material use and energy generation. However, wood suitable for energy is not solely generated by forest-based industries. Various other activities use wood products that eventually reach the end of their usable life and are discarded, such as wood waste from construction or demolition, furniture waste or end-of-life pallets and packaging used to transport goods. This chapter presents and characterises the different woody biomass streams that can provide feedstock for energy.

Keywords Forest systems · Forest residues · Industrial by-products · Wood residues · Wood waste

---

I. Malico (✉)
IDMEC, Escola de Ciências e Tecnologia, Departamento de Engenharia Mecatrónica, Universidade de Évora, Rua Romão Ramalho 59, 7000-671 Évora, Portugal
e-mail: imbm@uevora.pt

A. C. Gonçalves
MED—Mediterranean Institute for Agriculture, Environment and Development and CHANGE—Global Change and Sustainability Institute, Instituto de Investigação e Formação Avançada, Departamento de Engenharia Rural, Escola de Ciências e Tecnologia, Universidade de Évora, Apartado 94, 7002-544 Évora, Portugal
e-mail: acag@uevora.pt

A. C. Gonçalves and I. Malico (eds.), *Forest Bioenergy*, Green Energy and Technology,
https://doi.org/10.1007/978-3-031-48224-3_2

1 Introduction

Forest biomass is the accumulated mass, above and below ground, of living and dead woody tree and shrub species [1]. It can be grouped in three different categories (Table 1): primary, secondary and tertiary [2], respectively called direct, indirect and recovered in the classification by the Food and Agriculture Organisation, FAO [3]. These three categories reflect the supply source: land-use systems (primary biomass), wood-based industries (secondary biomass) and economic and social activities outside the forest sector (tertiary biomass).

Production technologies used and environmental, economic and social sustainability vary greatly among different production systems (and within each type of production system) [6–8]. Therefore, other than the supply source, it is important to distinguish forest biomass from the perspective of the production system (*cf.* chapter "Stand Structure and Biomass"). Forest biomass can be deliberately cultivated and grown with the purpose of producing biomass for energy in the so-called energy plantations (*cf.* chapter "Energy Plantations") or it can be obtained from other sources (e.g., natural forests, forests grown for timber, agroforestry systems, trees outside forests, wood-processing industries or other industries, municipal waste).

In some specific contexts, several non-wood materials are removed from forests or generated by forest-based industries and used as fuels. Examples are pine needles and cones [9], bamboo [10] or cork powder [11]. However, worldwide, wood is the

Table 1 Forest biomass classification (based on [2–5])

Category	Type	Description
Primary/ direct	Products of energy plantations	Biomass harvested from forest species plantations grown specifically for energy
	Products and residues	Biomass directly removed from natural forests and plantations not specifically grown for energy, other wooded lands and other lands (e.g., (*i*) wood residues generated by silvicultural activities, such as thinning, pruning, harvesting and logging, (*ii*) trees affected by natural disturbances, (*iii*) traditional fuelwood). Examples are logs, tree tops, stumps, branches and leaves
Secondary/ indirect	Residues	Industrial residues derived from primary (e.g., sawmills, pulp and paper mills) and secondary (e.g., joinery, carpentry) forest industries. Examples are sawdust, woodchips, bark, wood shavings, trimmings, cork powder and black liquor
Tertiary/ recovered	Residues and waste	Wood derived from all economic and social activities outside the forest sector. Examples are wood waste from construction sites or demolition of buildings, and end-of-life pallets, wooden containers and boxes and wood consumer durables

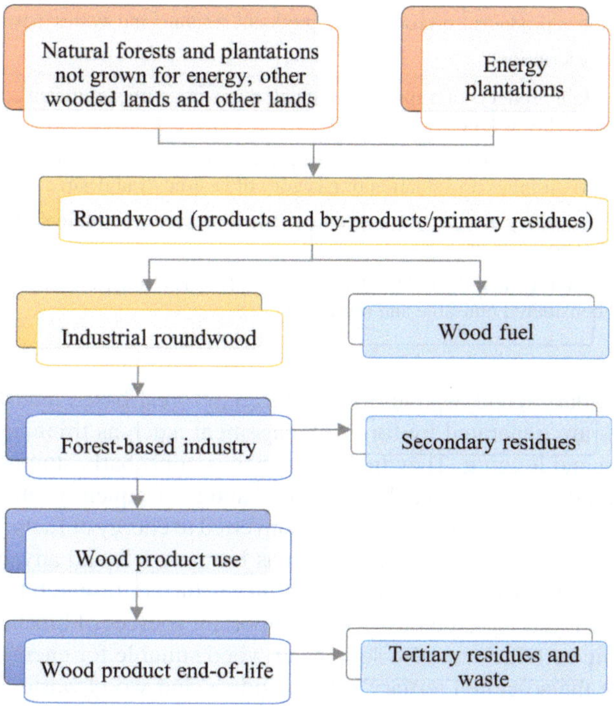

Fig. 1 Forest woody biomass streams

most commonly used forest biomass energy source [12]. Figure 1 presents the woody biomass streams from forests, woodlands and other land uses.

Wood fuel, as defined by FAO, encompasses all wood harvested and removed from forests and from trees outside forests that will be used as fuel. "It includes wood harvested from main stems, branches and other parts of trees (where these are harvested for fuel), round or split, and wood that will be used for the production of charcoal (e.g., in pit kilns or portable ovens), wood pellets and other agglomerates. It also includes wood chips to be used for fuel that are made directly (i.e., in the forest) from roundwood. It excludes wood charcoal, pellets and other agglomerates" [13]. On the other hand, industrial roundwood refers to all wood in the rough (roundwood) other than wood fuel. Therefore, roundwood, a measure of a forest harvest over a given period, is the sum of wood fuel and industrial roundwood.

Forest-based industries use industrial roundwood for the manufacture of a broad range of products. In the process, (secondary) wood residues are generated. Some are used as raw materials by other industries or converted to energy or biofuels, but a part will not be valorised. For example, sawmill residues may be used, not exclusively, for the manufacture of wood-based panels [14], for the production of pellets [15] or simply piled and burned at the sawmill [16] or landfilled [17].

Table 2 Definition of products, co-products, by-products, residues and wastes (based on [5, 27–29])

Category	Description
Product/ co-product	Main products of a specific production process with significant economic value and elastic supply
Residues	Secondary materials of a specific production process with inelastic supply. The term implies no valuation or category of desired or undesired
By-products	Secondary products of a specific production process with inelastic supply and economic value
Wastes	Materials that the holder discards, intends or is required to discard. They are considered unusable and unsalable

While secondary wood residues are produced in forest-based industries, primary wood residues are generated by forest management, such as thinning and pruning and harvesting and logging. They include branches, tops, bark, stumps, roots, small trees and generally unmerchantable stem wood and are frequently left to decompose naturally [18, 19], burned onsite [19, 20] or converted to energy or fuels [18, 21]. Their conversion into value-added products, such as biomaterials and advanced biofuels, seems to be an attractive solution that still requires further research [22, 23].

Forestry and forest-based industries are important sources of bioenergy, but other socio-economic activities are also sources of wood suitable for energy valorisation. They generate the so-called tertiary wood residues and wood waste, which consist of wood products at their end-of-life and other wood residues or waste generated by activities outside the forest sector, such as construction and demolition wood, packaging and pallets. Wood waste may be used, for example, as feedstock for the production of wood products [24, 25], landfilled [26] or burned to produce energy [25].

Before continuing, it is important to define terms that are used throughout this book and are often used in different contexts and with different meanings (Table 2). Products and co-products are the end-products that a certain process intends to obtain and whose production is elastic to changes in demand (i.e., if demand increases, production also increases). Residues and by-products are not primary products and are inelastic to demand. While residues may or may not have economic value, by-products do. Wastes, on the other hand, are materials that the holder intends to discard.

In the next sections, the three categories of forest biomass (*cf*. Table 1) will be described in detail, along with statistics that reflect their availability worldwide.

2 Primary Sources of Biomass

Land use is diverse, including agricultural systems, settlements and forest systems (Fig. 2). Forests are distributed worldwide, though with rather high variability in terms of species, stand structure and productivity [30, 31]. Forest systems can be

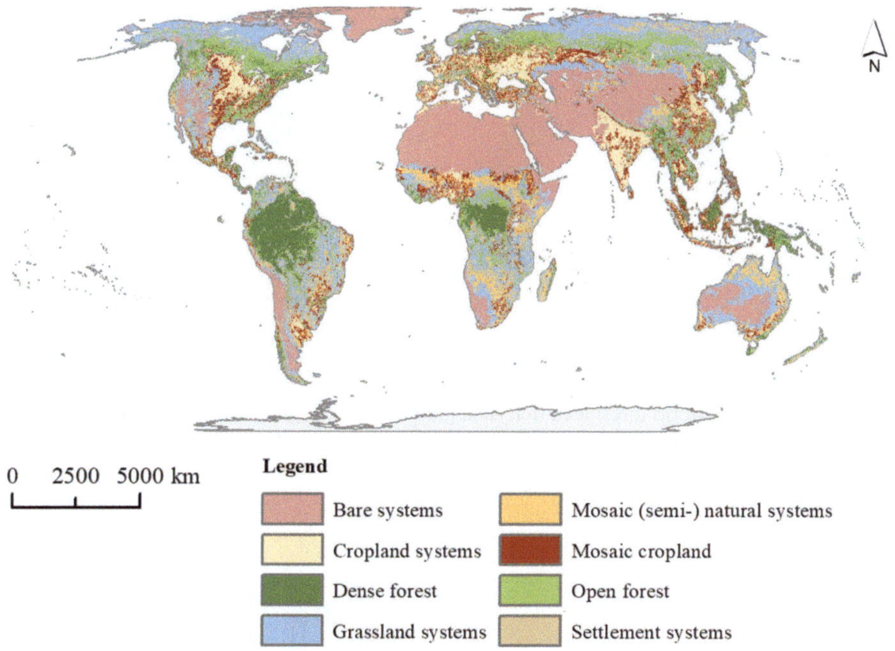

Fig. 2 Land-use systems in the world. (*Data source* [31])

grouped as primarily used for production (frequently timber), multiple uses (e.g., agroforestry systems) and other or non-use (associated with the protection and conservation of forests, habitats, flora and fauna). Alternatively, forest systems can also be divided in primary forests, forests primarily used for production and naturally regrown forests [30]. This is related to management objectives and silvicultural systems and the amount of biomass exported (*cf*. chapter "Stand Structure and Biomass").

The share of the world forest area is the largest in Europe (including the Russian Federation, according to FAO's country groups) followed by South America, North America, Africa, Asia and the smallest in Oceania (Fig. 3 left). Growing stock (in volume, $m^3 \cdot ha^{-1}$) is the largest in South America, followed by Europe, North America, Africa, Asia and Oceania (Fig. 3 right). There seems to be a trend towards a decrease in the forest area of the world (Fig. 4 left). Yet, this decrease is mainly observed in African and South American countries, whereas an increase is observed in most European, Asian and Oceanian countries and in the United States of America (Fig. 5). This is also reflected in the share of the world forest area (Fig. 4 right, Table 3), which increased from 1990 to 2020 in Europe (+1.6%), Asia (+1.5%), North America (+0.7%) and Oceania (+0.2%) and decreased in South America (−2.2%) and Africa (−1.8%).

Growing stock ($m^3 \cdot ha^{-1}$) shows a more irregular pattern in time, with a decrease from 1990 to 2010, followed by an increase from 2010 to 2015 (with values similar to

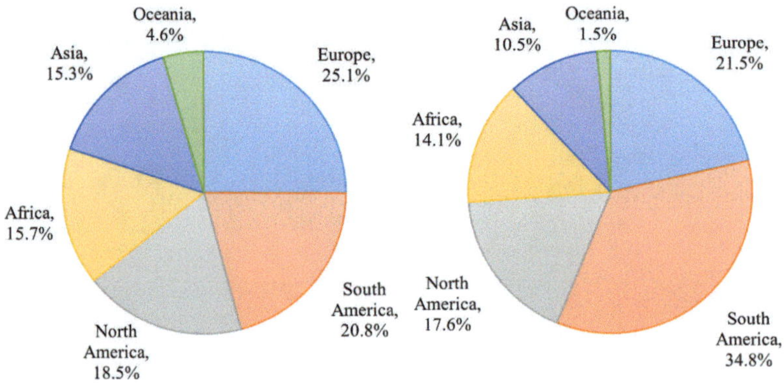

Fig. 3 Share of the world forest area (left) and growing stock (right) per FAO's country groups in 2020. (*Data source* [32])

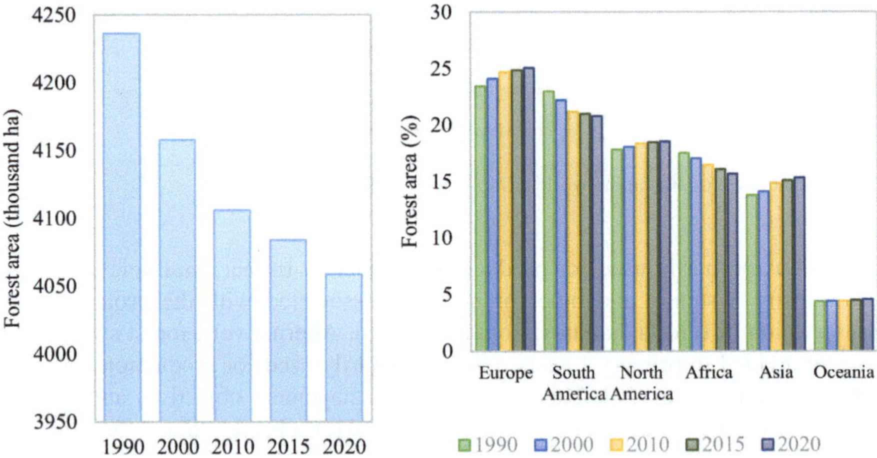

Fig. 4 Evolution of the forest area in the world (left) and share per FAO's country groups (right) from 1990 to 2020. (*Data source* [32])

those in 2000) and subsequently another decrease by 2020 (Fig. 6 left). The decrease occurs mainly in countries in South America, Africa and North America, but also in a few European and Asian countries (Fig. 7). The share of the world's growing stock increased continuously in Europe, Asia and North America, was approximately constant in Oceania, and decreased in South America and Africa (Fig. 6 right, Table 4). The largest increase from 1990 to 2020 in the share of the world's growing stock was observed in Europe (+2.4%), followed by Asia (+1.7%), North America (+1.0%) and Oceania (+0.1%), whereas the strongest decrease occurred in South America (−3.3%) followed by Africa (−2.0%) (Table 4).

Growing stock is higher than 1000 $m^3 \cdot ha^{-1}$ in most of America, Europe, Asia and central Africa, whereas it is lower than 100 $m^3 \cdot ha^{-1}$ in most of northern Africa

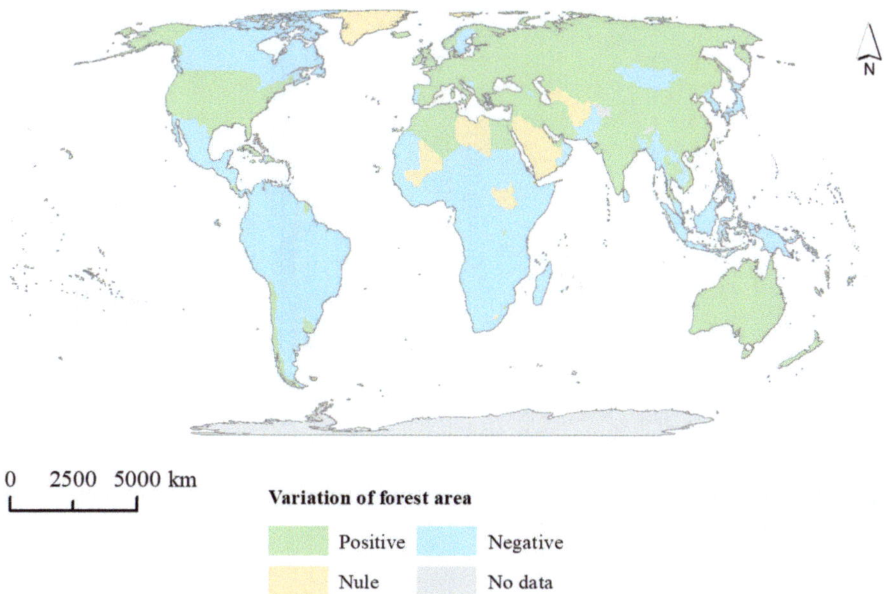

Fig. 5 Forest area variation between 1990 and 2020. (*Data source* [32])

Table 3 Share of the world forest area per FAO's country groups from 1990 to 2020. (*Data source* [32])

Forest area (%)	1990	2000	2010	2015	2020
Africa	17.5	17.1	16.5	16.1	15.7
Asia	13.8	14.1	14.9	15.1	15.3
Europe	23.5	24.1	24.7	24.9	25.1
North America	17.8	18.1	18.4	18.5	18.5
Oceania	4.4	4.4	4.4	4.5	4.6
South America	23.0	22.2	21.2	21.0	20.8

(Fig. 8). Biomass in mass ($t \cdot ha^{-1}$) is the largest in South America and central Africa, followed by Europe and Asia (Fig. 9).

Forests store large amounts of biomass, both above and below ground [33, 34]. Moreover, carbon stored in forests corresponds to more than 80% of the total aerial terrestrial carbon and 70% of the below-ground soil organic carbon ([35] and references therein). Globally, the forests sequester *circa* one third of the CO_2 emissions caused by anthropogenic actions [36].

The amount of biomass (or carbon) stored in forest ecosystems varies according to the biome, site, species, stand structure, silvicultural system and management. The three main world biomes are the boreal, the temperate and the tropical. Due to the interactions between biomes, site (soil and climate) conditions and species (arboreal,

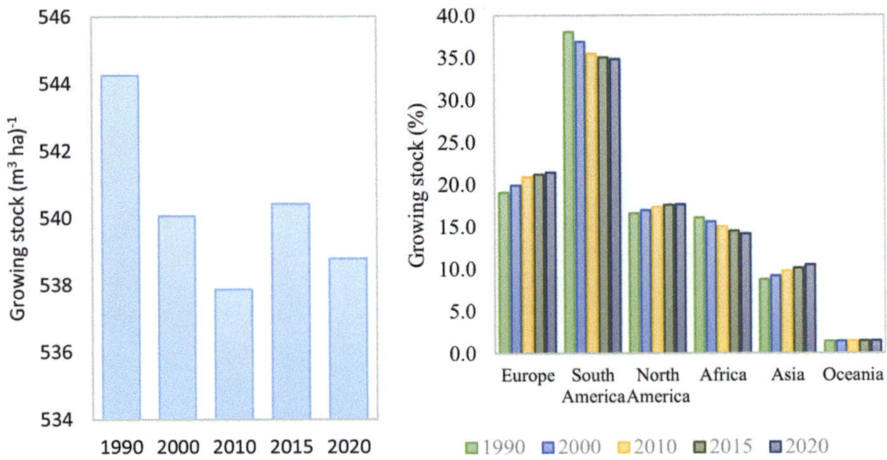

Fig. 6 Evolution of the growing stock in the world (left) and share per FAO's country groups (right) from 1990 to 2020. (*Data source* [32])

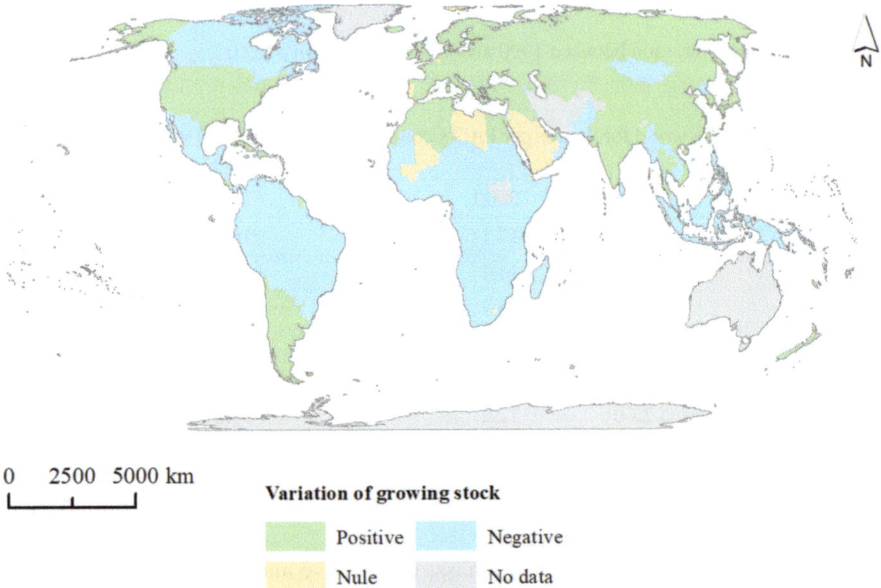

Fig. 7 Growing stock variation between 1990 and 2020. (*Data source* [32])

shrub and herbaceous), forest systems vary in area and capacity to sequester and store carbon. Management practices also influence biomass (and carbon) sequestration and storage [35]. Furthermore, net production is related to management. Forests with intensive management (in which the management practices envision the highest possible productivity in the shortest possible time, and include a set of silvicultural

Table 4 Share of the world growing stock per region from 1990 to 2020. (*Data source* [32])

Growing stock (%)	1990	2000	2010	2015	2020
Africa	16.1	15.6	15.0	14.5	14.1
Asia	8.8	9.2	9.8	10.1	10.5
Europe	19.1	19.9	20.9	21.2	21.5
North America	16.6	17.0	17.3	17.6	17.6
Oceania	1.4	1.5	1.5	1.5	1.5
South America	38.1	36.9	35.5	35.0	34.8

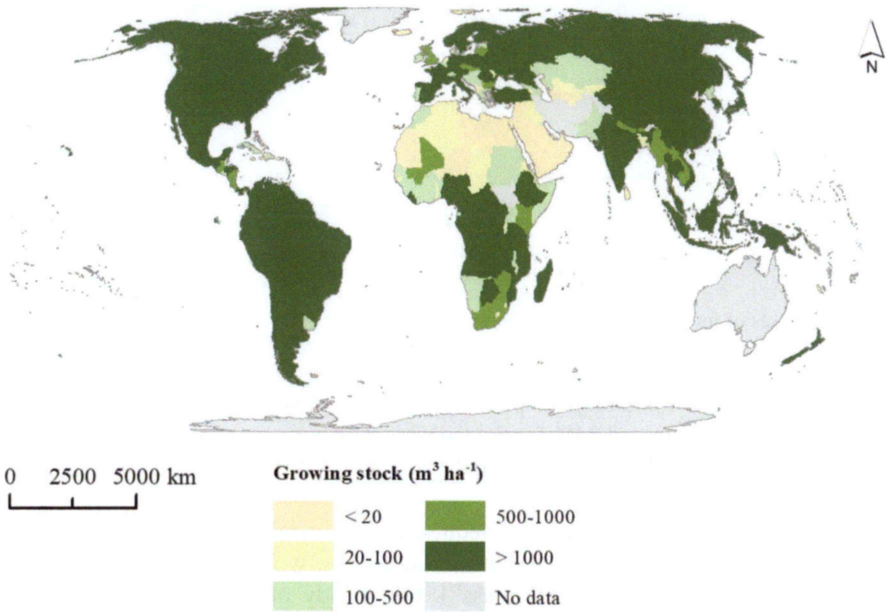

Fig. 8 Classes of growing stock. (*Data source* [32])

practices from genetic improvement to site preparation and fertilisation, thinning and pruning) tend to have higher production of biomass than forests with extensive management (in which the emphasis is to lower management intensity and costs, and include thinning and harvests of moderate intensity and long production cycles). The forests with intensive management frequently correspond to pure even-aged stands of short production cycles, while the forests with extensive management correspond to pure or mixed uneven-aged stands and long production cycles [35].

Considering the forest biomes (boreal, temperate and tropical), temperate and tropical biomes have a share of carbon stored in biomass and in the soil of 55% and 45%, respectively, whereas for boreal forest the share is 16% of carbon stored in biomass and 84% in the soil [35]. The share of production efficiency (ratio between

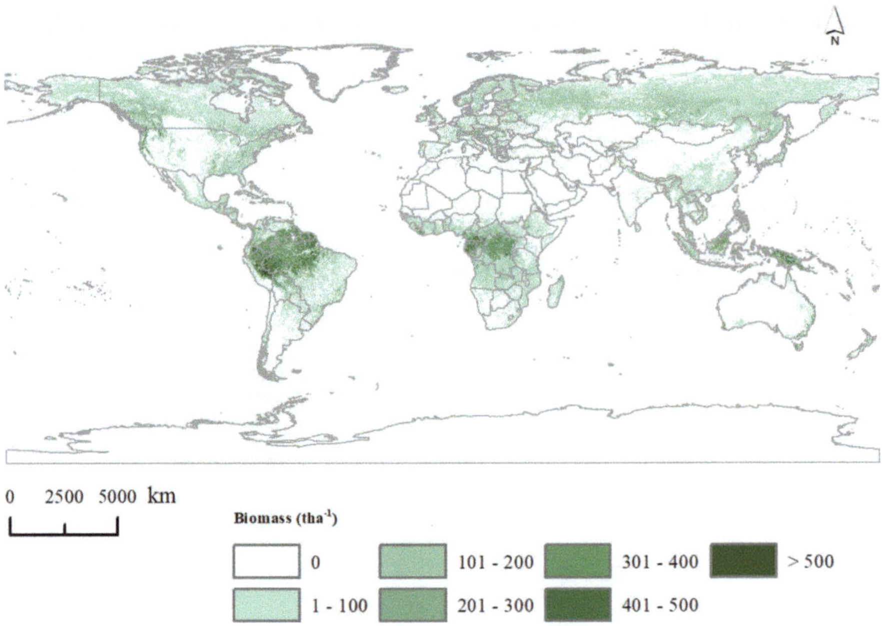

Fig. 9 Classes of forest biomass. (*Data source* [37])

net primary production and gross primary production) is 38% for the boreal biome, 46% for the temperate and 34% for the tropical. The harvests of woody products per biome are 0.3 $Gt{\cdot}Cy^{-1}$, 0.7 $Gt{\cdot}Cy^{-1}$ and 1.8 $Gt{\cdot}Cy^{-1}$, for the boreal, temperate and tropical biomes, respectively [35].

The biomass storage and carbon sequestration are related to the species, stand structure, silvicultural system, silvicultural practices, woody products and site (*cf.* chapter "Stand Structure and Biomass"). For woody products (timber, pulp and paper) with medium to long life cycles (more than 35 years), biomass (and carbon) is partially reallocated from the forests to these woody products. This results in a reduction of CO_2 emissions when compared to the use of materials based on fossil fuels [38]. Inversely, the use of biomass for energy releases CO_2 into the atmosphere in the short term, both as a consequence of the combustion of biomass [38] and to the decrease of soil organic carbon caused by organic matter decomposition [39]. However, in the medium and long term, forest growth results in the increase of above and below ground biomass (and carbon), as well as the increase of soil organic matter [38, 39], thus compensating for the removal of biomass from the forest stands in the medium and long term, as well as the carbon emissions from the utilisation of the woody products [38, 40]. Moreover, the development of stands and forests is subject to disturbances of different intensities and frequencies. In general, low-intensity and short-frequency disturbances are, most usually (but not always), related to silvicultural practices. These practices result in the export of biomass or its reallocation to deadwood, which incorporates carbon into the soil through decomposition. In a short period after the disturbance, there is the release of carbon into the atmosphere through soil organic matter decomposition, which is compensated by the growth of trees, stands and forests [39–41]. On the other hand, disturbances of high intensity and long periodicity (e.g., fires, storms, pests and diseases) tend to originate strong reductions of live biomass, converting it to dead (standing or downed) biomass, which results in an overall biomass loss, whether as live biomass or woody products with market value, or a loss of diversity [30, 42]. Thus, the management of stands and forests enables to balance the exports of woody products and the biomass storage and carbon sequestration, both in the trees and in the soil, and consequently contributing to the mitigation of greenhouse gas emissions [38–40]. Overall, the maintenance or promotion of biomass storage or carbon sequestration is related to the sustainability of forest systems. This resulted in a set of approaches that were converted, for example, in forest management strategies towards carbon stocks [41], sustainable biomass harvesting guidelines [43] and adaptation of silvicultural practices to promote biomass and carbon stocks in the forests [35].

The long-term production cycles of forest stands imply that market demand will not be met in the short time; that is, the effect of silvicultural practices, forestations or afforestations will take decades before woody products can be explored. There are silvicultural practices that promote the increase in tree growth (e.g., site preparation, control of spontaneous vegetation, use of species of fast growth, genetic improvement of species, fertilisation, irrigation, control of pest and diseases). Energy plantations (*cf*. chapter "Energy Plantations") can increase the supply of bioenergy in a shorter term than stands managed for timberwood and pulpwood [44] and may also release the pressure to provide biomass for energy in stands oriented for timberwood, pulpwood and/or under protection and conservation status [45, 46]. They might be a source of beneficial renewable energy, if established in current agricultural land that can be diverted from food and feed production without further impairing food security [47] or in marginal lands [48]. However, the potential competition of energy plantations with food for agricultural soils has ethical implications and energy plantations at a large enough scale pose environmental risks that need consideration [49–51]. This is the reason why several authors consider that energy plantations should be established in set aside agricultural lands and/or forest areas [52, 53].

Stands and forests have long harvesting cycles, in particular those whose management is oriented for sawtimber (>20 years) [44], shorter when oriented for pulpwood (8–12 years), and even shorter when oriented for energy (2–6 years) [54, 55]. According to Egnell et al. [44], most woody products for energy in boreal and temperate forests in the next decades will result from timberwood and pulpwood-oriented stands. Stands oriented for energy (or energy plantations, *cf*. chapter "Energy Plantations"), though having an important role in the supply of bioenergy, will not be able to fulfil market demand in the near future. Even though they are not the most common source of forest biomass for energy today [56–58], energy plantations are projected to become more important with the transition to a low carbon economy while meeting the increasing energy demand [49, 59]. Moreover, it is not expected that the biomass for energy will be the result of harvests in unmanaged stands [44].

The removal of biomass is related to carbon stocks through their use. If used for sawtimber, the carbon stocks are exported from the forest stands, but carbon is not emitted for a long time, corresponding to the lifespan of the objects made from it (e.g., furniture, wooden houses). The pulpwood is transformed into a wide variety of types of paper that can have a shorter (e.g., newspapers) or longer (e.g., books) lifespan. Additionally, if the biomass residues remain in the forest stands, carbon is not lost but rather reallocated mainly to the soil and tree growth [60, 61].

Figure 10 shows the amount of roundwood felled or otherwise harvested and removed in the world in 2021. The United States of America was the country with most of the roundwood production in 2021 (10.5%), followed by India (8.1%), China (7.8%), Brazil (6.2%) and the Russian Federation (5%). In terms of the country groups defined by FAO, Asia produced more roundwood in 2021, followed by Europe (the Russian Federation included) and Africa (Fig. 11). While Europe produced most of its roundwood for industrial production (78.7%), Africa and Asia produced it as fuel (90.2% and 60.5%, respectively).

Figure 12 shows the wood fuel, i.e., the part of roundwood that was harvested as fuel, produced in the world in 2021. The top producers are India (14.3%), China (7.4%), Brazil (5.9%) and Ethiopia (5.5%). Africa was where more roundwood was harvested as fuel (36.9%), closely followed by Asia (36.2%) (Fig. 13). Overall, 49% of the roundwood harvested in the world in 2021 was harvested as fuel [62], additionally, part of the industrial roundwood produced was also used as fuel (*cf.*

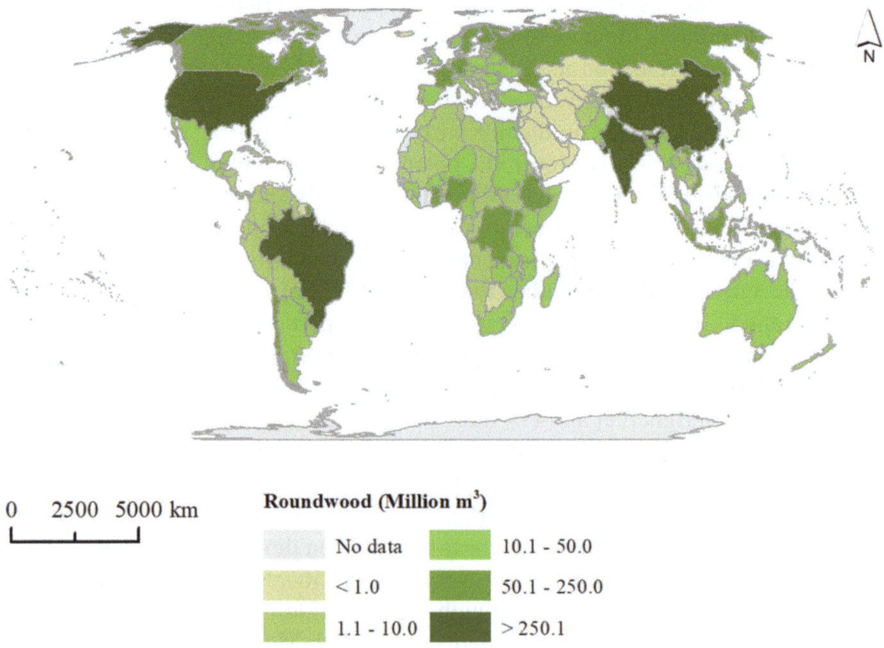

Fig. 10 Global production of roundwood in 2021. (*Data source* [62])

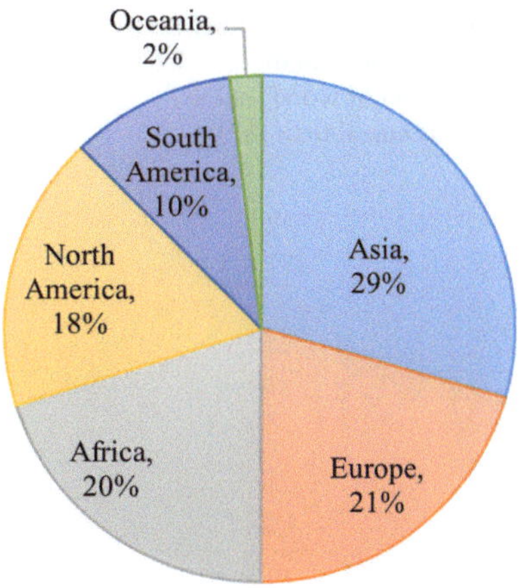

Fig. 11 Share of FAO's country groups in the production of roundwood in 2021. (*Data source* [62])

Section 3). This reflects the importance of wood fuel for human societies and the fact that, still today, traditional biomass remains an important energy source for many people around the world, particularly in rural areas of the Global South where access to modern energy sources may be limited [63].

The estimations for the production of roundwood and wood fuel (as well as industrial roundwood and recovered wood wastes presented later on) are those reported

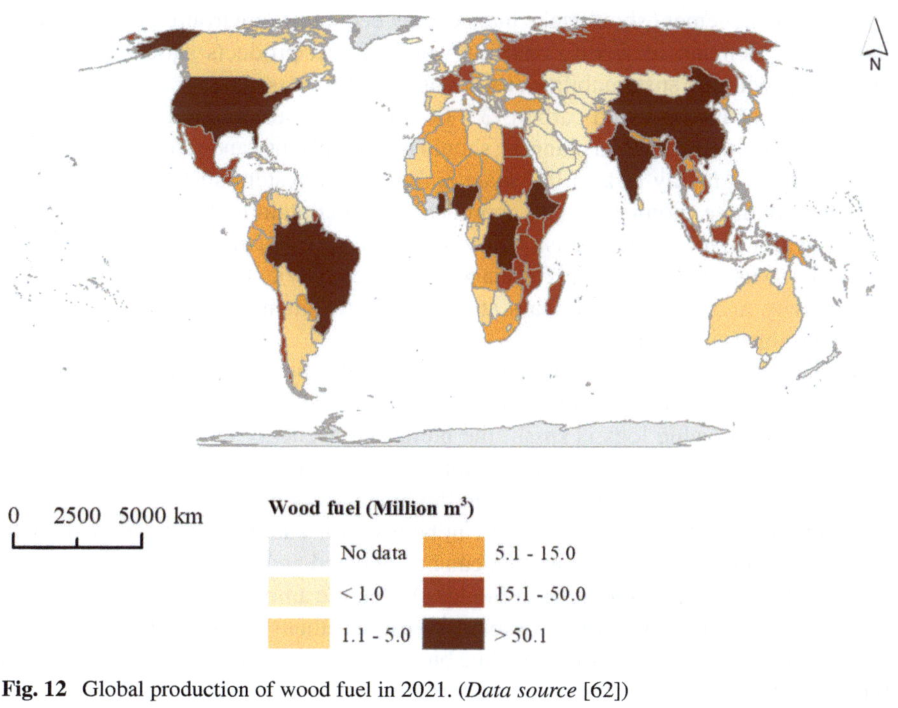

Fig. 12 Global production of wood fuel in 2021. (*Data source* [62])

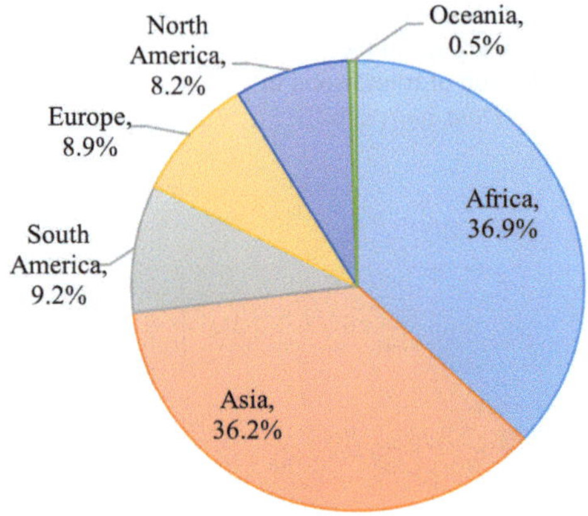

Fig. 13 Share of the FAO's country groups in the production of wood fuel in 2021. (*Data source* [62])

by FAO [62]. It is worth mentioning that FAO estimates the wood fuel production for several countries where no official statistics exist (many of these, important producers and consumers of wood fuel). Additionally, even for those countries that report wood fuel statistics, the share of roundwood used for energy may be underestimated because of existing informal ways of getting the biomass (see, e.g., [64]). Data on the informal collection of wood fuel is sparse, and a comparison between FAO estimates and other data suggests that wood fuel production in Africa and Asia will be revised upwards in the future [65].

Forest management should take into account economic, environmental and social issues. The demand of woody and non-woody forest products should be considered along with the sustainability of the stands and forests, as well as sequestration and storage of biomass and carbon [35, 66]. To attain sustainability in stands and forests, their monitoring and planning should be careful ([66] and references therein). According to Ameray et al. [35], maintaining forest system sustainability can be achieved by three (non-exclusive) approaches: (*i*) maintaining biomass and carbon stocks by low to moderate intensity harvests and long production cycles (old-growth forests); (*ii*) use extensive forest management where silvicultural practices increase productivity and maintain biomass stocks; (*iii*) use intensive forest management where productivity is high.

In stand management, economic issues should also be considered. In stands with logs with high market value, financial viability is ensured. Yet, in forest stands where the rate of high-value logs is small, the biomass for energy might be an option to incentivise the prescription and execution of silvicultural practices because of the increase in profitability that results from the biomass for energy [44].

Estimations of forest residues that can be used for energy have been made with the utilisation of rates of biomass that can be collected and used for energy (i.e., the percent of residues in total or aerial biomass). Yet, the amount of residues harvested is not always coincident with the amount of residues that is possible to collect in the stands ([66] and references therein). The biomass residues recovery rate varies from 0 to 80% (for details see chapter Stand Structure and Biomass, Sect. 5.3).

3 Secondary Sources of Biomass

Another significant source of forest biomass for energy and fuels comprises the residues of the forest-based industry, which includes several sub-sectors (the most relevant in terms of residues are the woodworking and pulp and paper industries as will be seen below). The residues produced by the industrial sector, understood as secondary products that are generated during the production of a main product, are mainly black liquor and wood residues, such as bark, slabs, sawdust and wood chips. Their importance is reflected in the fact that more than half of the wood used for energy in the countries of the United Nations Economic Commission for Europe (UNECE) are industrial by-products [67].

Woodworking industries include the production of sawnwood, wood-based panels, wooden construction materials and other wooden products. Sawmills are part of the primary forest industries (i.e., those that process wood directly harvested from forests). They receive industrial roundwood and transform it into various lumber pieces of various sizes and shapes, including planks, beams and boards. While these can be used directly, they are often further processed to create various wood-products, such as furniture or products used in construction.

Another sector that also receives industrial roundwood is the pulp and paper industry. Roundwood is first prepared, followed by mechanical or chemical processes to produce pulp, which is a versatile material used in a wide range of applications, such as the production of paper products, textiles or chemicals.

The potential for industrial roundwood processing, and therefore for the generation of secondary wood residues generated, is not only dependent on the overall roundwood supply, but also on the wood fuel demand, since roundwood is used for these two purposes [5].

Figure 14 shows the global production of industrial roundwood in 2021 and Fig. 15 the percentage of roundwood that was industrial roundwood [62]. The United States of America was the country that produced more industrial roundwood (17.4%), followed by the Russian Federation (9.2%), China (8.2%), Brazil (6.5%) and Canada (6.4%). In these countries, industrial roundwood constituted more than half the roundwood produced in 2021, except in India, where 85.8% of the roundwood was harvested as fuel.

The production of roundwood, wood fuel and industrial roundwood has increased in the last 60 years [62], as a result of the global population growth and increased demand for wood products (Fig. 16). However, the share of wood fuel in the total roundwood production decreased, which means that proportionally more roundwood was directed to the industrial sector. In 1961, 59.6% of the roundwood was harvested as wood fuel, while in 2021, industrial roundwood constituted roughly half (50.9%) of the roundwood.

The consumption of industrial roundwood in a specific country is calculated by adding the country's production to its net imports. The countries grouped as Europe in FAO statistics (the Russian Federation included) had the highest share of the world's industrial roundwood consumption in 2021 (30.7%), followed by Asia (26.5%) and

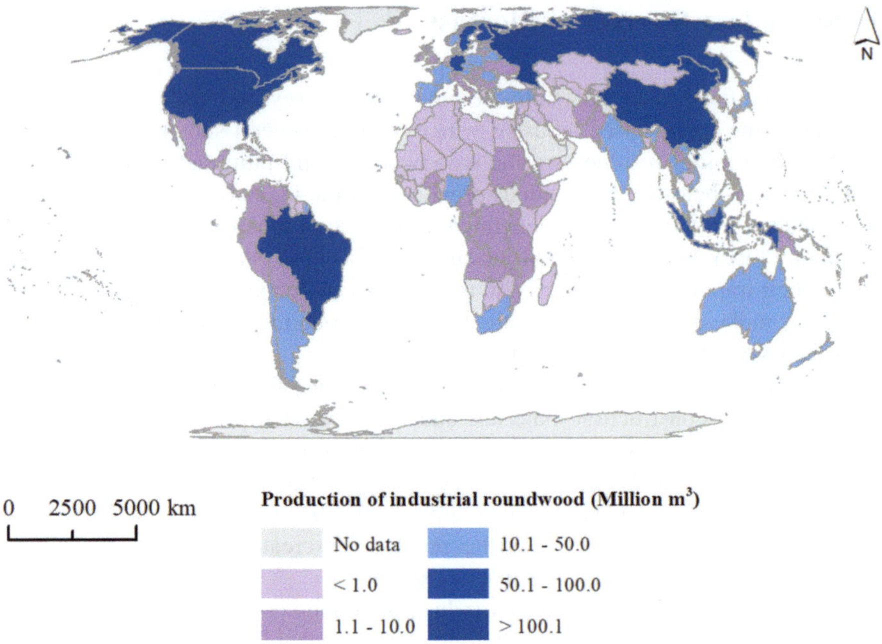

Fig. 14 Global production of industrial roundwood in 2021. (*Data source* [62])

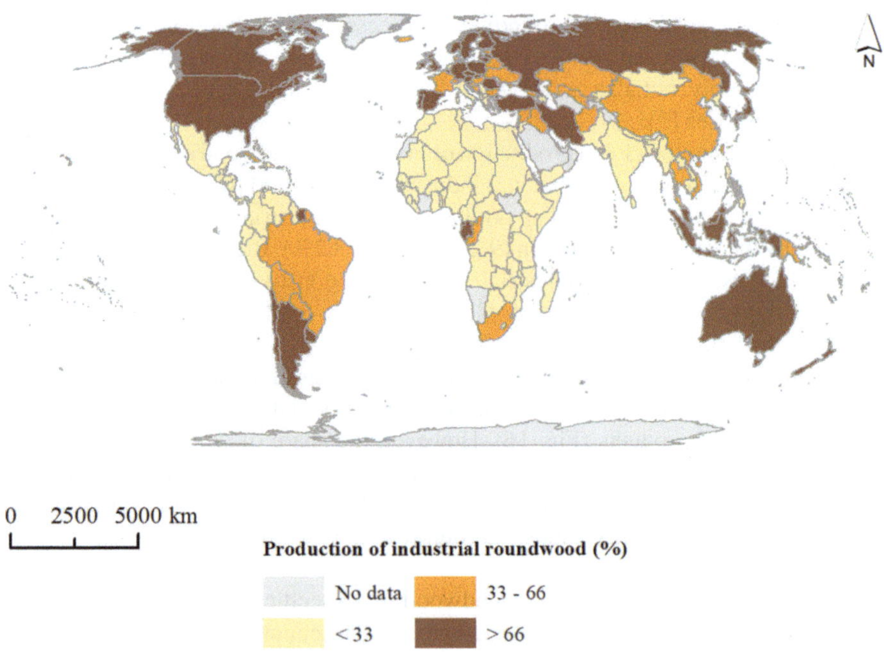

Fig. 15 Share of the production of industrial roundwood in roundwood in 2021. (*Data source* [62])

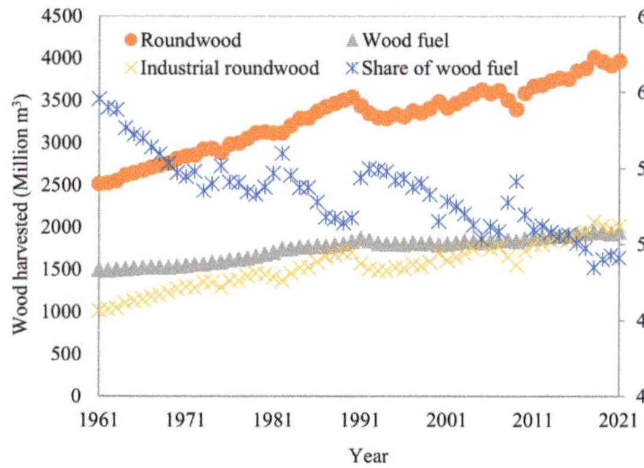

Fig. 16 Global production of roundwood, wood fuel and industrial roundwood and share of wood fuel in roundwood production from 1961 to 2021. (*Data source* [62])

North America (26%) (Figs. 17 and 18). This reflects, in part, the proportion of roundwood that was supplied to the industry (Fig. 18). Europe, northern America and Oceania used, respectively, 78.3%, 87.6% and 81.4% of the roundwood in the industrial sector, while in the other extreme, Africa and Central America (Caribbean and Mexico included) only directed, respectively, 9.3% and 12.7% of the roundwood they consumed to the industry.

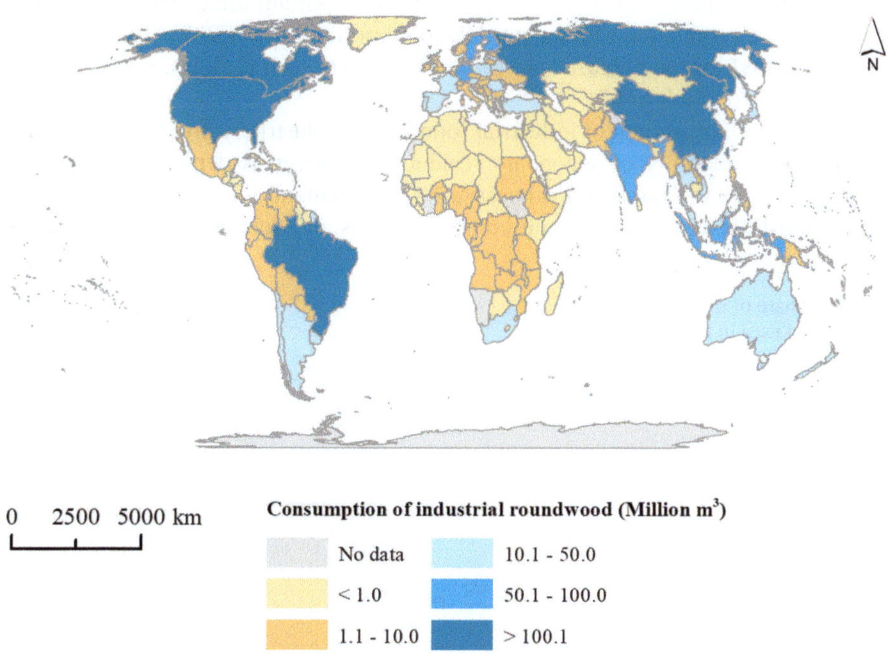

Fig. 17 Global consumption of industrial roundwood in 2021. (*Data source* [62])

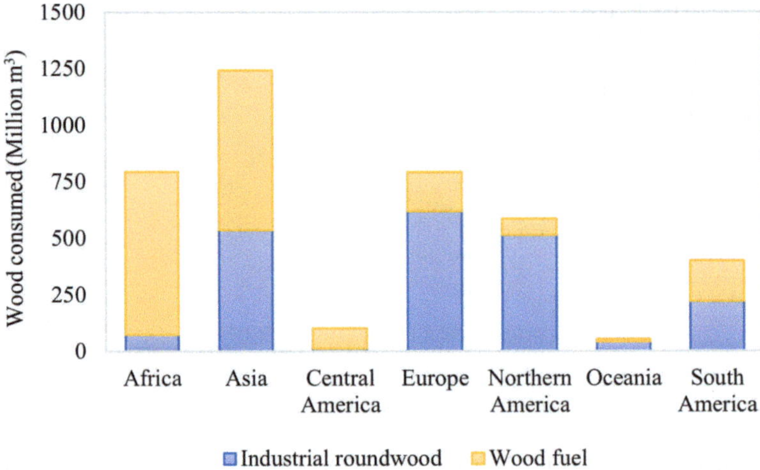

Fig. 18 Consumption of wood fuel and industrial roundwood in 2021 per FAO's country groups. (*Data source* [62])

As far as industrial use is concerned, in 2021, most roundwood (57%) was sawn lengthways for the manufacture of sawnwood or railway sleepers (ties) or used for the production of veneer (Fig. 19). The second-most important use of industrial roundwood was the production of pulp, particleboard or fibreboard (35%). The remaining (8%) was used for the manufacture of other products such as poles, posts, fencing, wood wool and tanning [62].

Given a certain amount of industrial roundwood industrially processed in a given region, the availability of secondary residues for energy is dependent on several factors, such as the industrial process itself, the demand for wood residues from other industries and the economic competitiveness and demand for bioenergy [5].

Fig. 19 Share of industrial roundwood uses in the world in 2021. (*Data source* [62])

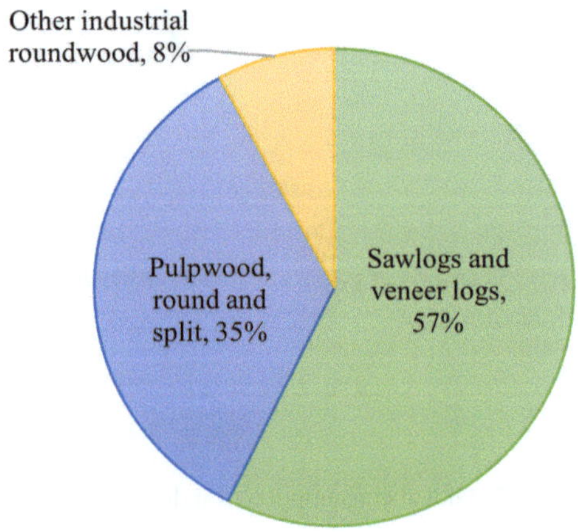

Table 5, based on data provided by [18, 28, 56, 68–75], presents some of the main sources of secondary residues and typical ranges of wood residue generation factors.

Sawmills belong to the primary forest industries and generate large amounts of woody residues in the form of bark, sawdust, slabs, edgings, trimmings and cull logs. Typically, 15 to 60% of the roundwood that enters the mills ends up as residue. The residue generation factor depends on numerous aspects, such as the type, size and quality of the logs being processed and the specific operations and equipment used. The wide range in the residue generation factor in sawmills is related to the sawnwood products (e.g., semi-processed cants, boules and flitches, planned sawnwood) [75] and the use of slab wood [56]. Sawmill residues are generally clean, uniform, concentrated and have a low moisture content (below 20%), which makes them suitable for further use [76]. As such, sawmill residues are frequently sold as raw material for the manufacture of pulp, engineered wood products and fuels (e.g., wood chips for pulp for paper, sawdust for particleboard, pellets), for landscaping applications (e.g., bark mulch), for livestock bedding and as fuel for combustion boilers (e.g., hog fuel).

The residue generation factors of plywood mills are also typically high and, as in sawmills, the residues can be valuable sources of raw materials for various industries. Typically, residues can range from 30 to 60% of the total volume of logs processed, depending on the type and quality of wood being processed, the manufacturing process used and the efficiency of the equipment and operations. Coarser residues (e.g., cores) are mainly used as raw materials for the manufacture of pulp and other fibre products, while bark and fine residues are typically used as fuel [77].

Pulp may be produced by chemical or mechanical processes, although the former dominates (90% of the total production capacity is based on the kraft process, a type of chemical pulping [78]). In mechanical pulping, bark, sludge, ash and screening residues are generated. Typically, residues account for 10 to 30% of the wood input and can be used for various purposes, such as landscaping and energy and fuel production. On the other hand, in chemical pulping, there is a comparatively wider

Table 5 Typical residue generation factors of relevant forest-based industries

Source	Residues	Residue generation factor (%)
Sawmills	Bark, sawdust, slabs, edgings, trimmings and cull logs	15–60
Plywood manufacture	Bark, sawdust, cores, green veneer, dry veneer, trimmings and sanding dust	30–60
Fiber-, particle- and strand board manufacturing	Bark and screening residues	5–20
Mechanical pulp	Bark, sludge, ash and screening residues	10–30
Chemical pulp	Bark, black liquor, sludge, screening residues and other residues	50–60

variety of residues produced, including black liquor, bark and sludge [70]. Black liquor is the most important residue in chemical pulp mills. It is usually burned in a recovery boiler to recuperate cooking chemicals contained in the black-liquor and produce steam [79, 80]. Because of this energy recovery, modern non-integrated kraft pulp mills are energy self-sufficient [79]. In certain mills, lignin is extracted from black liquor for biochemical processes and other products [70].

The residue stream generated by the production of primary wood products has been increasingly used as raw material for the production of other wood products, such as wood-based panels (particle-, fibre- and strand board) [69]. The extent to which this is done is dependent on various factors, such as the existence of industries using cascading woody material as raw feedstock, logistics, processing capacity or economic feasibility [5, 81]. The existence of a well-developed bioenergy industry with established infrastructure may affect the cascading use of forest resources [5]. For a discussion on the concepts of circular economy and cascading utilisation refer to Mair et al. [82].

The forest industries that process primary wood products generate additional residues that may be used as an energy source. For example, the production of particleboard generates screening residues and sanding dust, which, typically, account for 5 to 20% of the total volume of the feedstock processed. Because additives (e.g., binders, fillers) are used in the production of wood-based panels, these wood processing residues do not consist exclusively of primary wood fibre [28].

Further processing of wood and engineering-wood for the manufacture of finished products such as furniture, packaging or construction products results in additional residues (e.g., solid timber offcuts, dust, shavings, trims, clippings); some of which may be contaminated with adhesives and coating particles. The amount of residues generated depends on the manufacturing process. For example, Daian and Ozarska [83] assessed the wood residues generated by Australian furniture companies and concluded that the residue generation factor varied significantly, depending on the profile of the manufacture (7% to 49% of the annual supply of wooden raw material ended up as residues).

4 Tertiary Sources of Biomass

Many other industries and economic activities outside the forest-based industries generate wood residues and waste. Examples are the wood waste from the construction of buildings and wooden pallets and packaging used in the transport and storage of goods in various industries. These are classified as tertiary wood residues, along with wood products at their end-of-life.

Tertiary wood residues and waste have a wide range of origins and refer to a very heterogeneous group of materials with different levels of contaminants [84]. They can contain, for example, heavy metals originated from paints and preservatives, polycyclic aromatic hydrocarbons or volatile organic compounds [85]. The source and type of tertiary wood residues and waste determine the appropriate way to handle them. Clean wood can be used for the production of industrial and consumer products or as fuel, while hazardous waste wood requires disposal at special facilities or incineration [86]. In between, treated but non-hazardous wood varies in the level of contaminants.

To date, there is no standardised classification of wood waste streams applied internationally and their management varies among countries, which hinders the reuse of wood waste [87] and leads to different levels of wood waste energy valorisation and recycling. UNECE and FAO present a catalogue of the existing wood waste classifications in the UNECE region, where the different approaches are clearly shown [87].

The valorisation of tertiary wood residues involves (*i*) their collection at the place where they are generated, (*ii*) transportation to the place where they are valorised and (*iii*) subsequent treatment. Different countries have waste management systems with different levels of maturity, with some countries lacking collection and treatment of wood waste [88]. This is somehow reflected in the available FAO data for recovered post-consumer wood, with only a few countries officially reporting values of recovered wood (32 in 2021 [62]).

Wood waste management operations also influence the contaminant level of the waste streams. For example, Faraca et al. [89] analysed wood waste collected for recycling in Denmark and found that contaminant levels varied significantly among the materials analysed, depending on the type and source of the wood waste. The authors suggested that low-quality wood waste should be collected separately from cleaner wood waste to avoid unwanted contaminations with chemicals. Additionally, they recommended that the fractions containing fibreboard, treatments and/or composite materials from construction and demolition should be minimised in recycled material so that the lowest level of contaminants is guaranteed.

Mixed-streams of post-consumer wood pose challenges to recycling wood, and, often, the physical inspection and quality assessment required to avoid possible contaminants re-entering the wood production phase hinder the recovery of such residues [90]. Since sorting technologies are not well developed for most wood-based materials, sorting is manual, which leads to high costs, inconsistent quality and health risks for the workers [90]. Additionally, the fact that tertiary wood waste streams are usually highly dispersed and present irregular patterns also contributes to high recovery costs [5, 90]. Despite these challenges, relative to other woody biomass sources, tertiary wood residues are particularly relevant in regions with small forest production, where wood waste accounts for the largest potential of woody biomass [5].

Figure 20 presents the amount of recovered post-consumer wood that could be recycled or reused for material or energy purposes in 2021, excluding post-consumer wood that would not be reused/recycled (e.g., sent to landfills). Europe accounted for 87.7% (around 31 million t) of the total recovered post-consumer wood reported, while Asia accounted for the rest [62]. Germany was the country that recovered the most post-consumer wood, followed by France and the United Kingdom (UK). These three countries reported more than half of the post-consumer wood recovered in the world. Since in Germany, municipal and industrial solid waste must be sorted

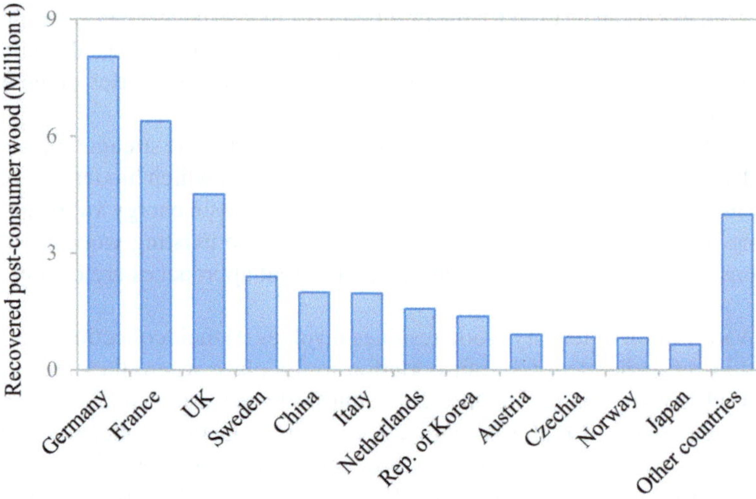

Fig. 20 Recovered post-consumer wood in the world in 2021. (*Data source* [62])

and landfilling is restricted to materials with organic matter content that does not exceed 5%, almost no wood waste is disposed of in landfills [91]. Most of it (80%) goes for energy recovery, while a fraction of the least contaminated wood waste is absorbed by the panel industry [91]. However, new waste wood fired power plants are no longer subsidised in Germany and the situation can change in the future. On the other hand, France and the United Kingdom favour material recovery (79% and 70% of the wood waste, respectively), but in France, a substantial part of the wood waste ends up in landfills [91].

5 Final Considerations

Forest biomass for energy and fuels can be classified into three groups, which reflect its origin: primary, directly sourced from energy plantations or other land-use systems (*cf*. Sect. 2); secondary, corresponding to residues of the wood-based forest industries (*cf*. Sect. 3); and tertiary, resulting from activities outside the forest sector (*cf*. Sect. 4).

The availability of primary biomass is dependent on the distribution of the forest area in the world and of the forest systems, including species, site and management (silvicultural systems and practices, harvest and logging). Overall, there seems to be a need for a balance between the biomass maintained in the forest systems (live and dead) and the biomass exported to enable the sustainability of forest systems. In general, disturbances of low intensity and short frequency (e.g., silvicultural practices) tend to maintain biomass stocks in the medium and long term [39, 41], whereas disturbances of high intensity and long periodicity (e.g., storms or fires) tend to reduce biomass storage [30, 42]. Moreover, apart from the maintenance of the sustainability of the forest systems and their productions, the maintenance of the forest area plays a key role in the amount of biomass available for woody products and energy. Forest area and growing stock increased from 1990 to 2020 in Europe, Asia, North America and Oceania, but decreased in South America and Africa (*cf*. Figs. 4 and 5). This may result, in the future, in a reduction in the availability of woody products (including biomass for energy) in the latter regions.

The availability of secondary biomass depends on various factors within the forest-based industries, including their number, size or type (e.g., woodworking, pulp and paper). The quantity of residues generated by these industries is influenced by factors such as the quality and quantity of the wood received and processed by the industry and the specific industrial process used. Globally, there was an overall increase in the production of roundwood, wood fuel and industrial roundwood from 1961 to 2021. However, the share of wood fuel in roundwood production decreased during this period. Moreover, in 2021, the share of industrial roundwood production in total roundwood production was larger in Oceania, Europe and North America than in South America, Asia and, especially, Central America and Africa, where, respectively, more than 87 and 90% of the roundwood production corresponded to wood fuel (*cf*. Fig. 15). The consumption of roundwood followed the same pattern of production. The consumption of wood fuel was larger than that of industrial roundwood in Africa, Central America and Asia (*cf*. Fig. 18). Apart from the quantity of industrial roundwood processed, the proportion of raw materials that become residues per wood-based industry type is also a factor that influences the availability of secondary forest residues for energy purposes. These generation factors are quite variable, spaning from 5 to 60% (*cf*. Table 5).

Tertiary biomass refers to the wood residues and waste generated by many non-forest industries and economic activities as well as wood products at their end-of-life. This tertiary biomass constitutes a very heterogeneous group of materials that may be clean or contaminated with different concentrations of contaminants. Its use is dependent on the content of contaminants, the available technologies to decontaminate them, logistics and economic factors. Europe reported the largest amount of wood that was recovered and can be recycled or used for material or energy purposes (*circa* 88%), with Germany, France and the United Kingdom having the largest shares (*cf*. Fig. 20).

Funding This work is funded by National Funds through FCT—Foundation for Science and Technology, under the Project UIDB/05183/2020, through MED, and Project UIDB/50022/2020, through IDMEC, under LAETA.

References

1. Hakkila P, Parikka M (2002) Fuel resources from the forest. Bioenergy from sustainable forestry. Springer, Dordrecht, pp 19–48

2. Batidzirai B, Smeets EMW, Faaij APC (2012) Harmonising bioenergy resource potentials—methodological lessons from review of state of the art bioenergy potential assessments. Renew Sustain Energy Rev 16:6598–6630. https://doi.org/10.1016/j.rser.2012.09.002
3. FAO (2004) Unified bioenergy terminology (UBET). FAO, Rome
4. Hoogwijk M, Faaij A, van den Broek R et al (2003) Exploration of the ranges of the global potential of biomass for energy. Biomass Bioenerg 25:119–133. https://doi.org/10.1016/S0961-9534(02)00191-5
5. Thiffault E, Gianvenuti A, Zuzhang X, Walter S (2023) The role of wood residues in the transition to sustainable bioenergy—analysis of good practices and recommendations for the deployment of wood residues for energy. FAO, Rome
6. Hall JP (2002) Sustainable production of forest biomass for energy. For Chron 78:391–396. https://doi.org/10.5558/tfc78391-3
7. Lattimore B, Smith CT, Titus BD et al (2009) Environmental factors in woodfuel production: opportunities, risks, and criteria and indicators for sustainable practices. Biomass Bioenerg 33:1321–1342. https://doi.org/10.1016/j.biombioe.2009.06.005
8. Van 't Veen H, Eppinga MB, Mwampamba TH, Ferreira Dos Santos MJ (2021) Long term impacts of transitions in charcoal production systems in tropical biomes. Environ Res Lett 16:034009. https://doi.org/10.1088/1748-9326/abe14d
9. Font R, Conesa JA, Moltó J, Muñoz M (2009) Kinetics of pyrolysis and combustion of pine needles and cones. J Anal Appl Pyrol 85:276–286. https://doi.org/10.1016/j.jaap.2008.11.015
10. Kumar R, Chandrashekar N (2014) Fuel properties and combustion characteristics of some promising bamboo species in India. J For Res 25:471–476. https://doi.org/10.1007/s11676-014-0478-6
11. Nepomuceno Pereira R, Malico I, Mesquita P, et al (2017) Energy use of cork residues in the Portuguese cork industry. In: Proceedings of the 12th conference on sustainable development of energy, water and environmental systems—SDEWES2017. Dubrovnik, Croatia, pp 1–20
12. Mäkelä M, Parkkinen M, Lyytimäki J, Nygrén NA (2020) Futures images of woodchips as an energy source in Finland. Futures 121:102571. https://doi.org/10.1016/j.futures.2020.102571
13. FAO (2022) Classification of forest products 2022. FAO, Rome
14. Simal Alves L, da Silva SAM, Dos Anjos AM et al (2014) Particleboard produced with sawmill waste of different wood species. AMR 884–885:689–693. https://doi.org/10.4028/www.scientific.net/AMR.884-885.689
15. Trømborg E, Ranta T, Schweinle J et al (2013) Economic sustainability for wood pellets production—a comparative study between Finland, Germany, Norway, Sweden and the US. Biomass Bioenerg 57:68–77. https://doi.org/10.1016/j.biombioe.2013.01.030
16. Sjølie HK (2012) Reducing greenhouse gas emissions from households and industry by the use of charcoal from sawmill residues in Tanzania. J Clean Prod 27:109–117. https://doi.org/10.1016/j.jclepro.2012.01.008
17. Cambero C, Hans Alexandre M, Sowlati T (2015) Life cycle greenhouse gas analysis of bioenergy generation alternatives using forest and wood residues in remote locations: a case study in British Columbia, Canada. Resour Conserv Recycl 105:59–72. https://doi.org/10.1016/j.resconrec.2015.10.014
18. Koopmans A, Koppejan J (1997) Agricultural and forest residues-generation, utilization and availability. In: Proceedings of the regional consultation on modern applications of biomass energy. FAO, Kuala Lumpur, pp 1–23
19. Pergola MT, Saulino L, Castellaneta M et al (2022) Towards sustainable management of forest residues in the southern Apennine Mediterranean mountain forests: a scenario-based approach. Ann For Sci 79:14. https://doi.org/10.1186/s13595-022-01128-w
20. Hueso-González P, Martínez-Murillo JF, Ruiz-Sinoga JD (2018) Benefits of adding forestry clearance residues for the soil and vegetation of a Mediterranean mountain forest. Sci Total Environ 615:796–804. https://doi.org/10.1016/j.scitotenv.2017.09.301
21. Díaz-Yáñez O, Mola-Yudego B, Anttila P et al (2013) Forest chips for energy in Europe: current procurement methods and potentials. Renew Sustain Energy Rev 21:562–571. https://doi.org/10.1016/j.rser.2012.12.016

22. Braghiroli FL, Passarini L (2020) Valorization of biomass residues from forest operations and wood manufacturing presents a wide range of sustainable and innovative possibilities. Curr For Rep 6:172–183. https://doi.org/10.1007/s40725-020-00112-9
23. Moriana R, Vilaplana F, Ek M (2015) Forest residues as renewable resources for bio-based polymeric materials and bioenergy: chemical composition, structure and thermal properties. Cellulose 22:3409–3423. https://doi.org/10.1007/s10570-015-0738-4
24. Azambuja RDR, Castro VGD, Trianoski R, Iwakiri S (2018) Recycling wood waste from construction and demolition to produce particleboards. Maderas, Cienc tecnol 20:381–690. https://doi.org/10.4067/S0718-221X2018005041401
25. Sommerhuber PF, Welling J, Krause A (2015) Substitution potentials of recycled HDPE and wood particles from post-consumer packaging waste in wood-plastic composites. Waste Manage 46:76–85. https://doi.org/10.1016/j.wasman.2015.09.011
26. Vis M, Mantau U, Allen B (2016) Study on the optimised cascading use of wood. Publications Office, Brussels
27. EC (2008) Directive 2008/98/EC of the European Parliament and of the Council of 19 November 2008 on waste and repealing certain directives. European Commission, Brussels, Belgium
28. Saal U, Weimar H, Mantau U (2017) Wood processing residues. In: Wagemann K, Tippkötter N (eds) Biorefineries. Springer International Publishing, Cham, pp 27–41
29. ISCC (2020) ISCC CORSIA 201-1 Waste, residues, by-products. ISCC
30. Schulze K, Malek Ž, Verburg PH (2019) Towards better mapping of forest management patterns: a global allocation approach. For Ecol Manage 432:776–785. https://doi.org/10.1016/j.foreco.2018.10.001
31. van Asselen S, Verburg PH (2012) A land system representation for global assessments and land-use modeling. Glob Change Biol 18:3125–3148
32. FAO (2020) Global forest resources assessment 2020. FAO
33. Pan Y, Hu T, Zhu X et al (2012) Mapping cropland distributions using a hard and soft classification model. IEEE Trans Geosci Remote Sens 50:4301–4312. https://doi.org/10.1109/TGRS.2012.2193403
34. Pan Y, Birdsey RA, Fang J et al (2011) A large and persistent carbon sink in the world's forests. Science 333:988–993. https://doi.org/10.1126/science.1201609
35. Ameray A, Bergeron Y, Valeria O et al (2021) Forest carbon management: a review of silvicultural practices and management strategies across boreal, temperate and tropical forests. Curr Forestry Rep 7:245–266. https://doi.org/10.1007/s40725-021-00151-w
36. Friedlingstein P, Jones MW, O'Sullivan M et al (2022) Global carbon budget 2021. Earth Syst Sci Data 14:1917–2005. https://doi.org/10.5194/essd-14-1917-2022
37. Santoro M, Cartus O (2019) ESA biomass climate change initiative (Biomass_cci): global datasets of forest above-ground biomass for the year 2017, vol 1. Centre for Environmental Data Analysis
38. Baul TK, Alam A, Strandman H, Kilpeläinen A (2017) Net climate impacts and economic profitability of forest biomass production and utilization in fossil fuel and fossil-based material substitution under alternative forest management. Biomass Bioenerg 98:291–305. https://doi.org/10.1016/j.biombioe.2017.02.007
39. Mäkipää R, Abramoff R, Adamczyk B et al (2023) How does management affect soil C sequestration and greenhouse gas fluxes in boreal and temperate forests?—A review. For Ecol Manage 529:120637. https://doi.org/10.1016/j.foreco.2022.120637
40. Gouge D, Thiffault E, Thiffault N (2021) Biomass procurement in boreal forests affected by spruce budworm: effects on regeneration, costs, and carbon balance. Can J For Res 51:1939–1952. https://doi.org/10.1139/cjfr-2021-0060
41. Ontl TA, Janowiak MK, Swanston CW et al (2020) Forest management for carbon sequestration and climate adaptation. J Forest 118:86–101. https://doi.org/10.1093/jofore/fvz062
42. Vaglio Laurin G, Puletti N, Tattoni C et al (2021) Estimated biomass loss caused by the vaia windthrow in Northern Italy: evaluation of active and passive remote sensing options. Remote Sensing 13:4924. https://doi.org/10.3390/rs13234924

43. Titus BD, Brown K, Helmisaari H-S et al (2021) Sustainable forest biomass: a review of current residue harvesting guidelines. Energy Sustain Soc 11:10. https://doi.org/10.1186/s13705-021-00281-w
44. Egnell G, Paré D, Thiffault E, Lamers P (2016) Environmental sustainability aspects of forest biomass mobilisation. In: Thiffault E, Berndes G, Junginger M et al (eds) Mobilisation of forest bioenergy in the Boreal and temperate biomes. Elsevier, London, pp 50–67
45. Bruckman VJ, Terada T, Fukuda K et al (2016) Overmature periurban Quercus-Carpinus coppice forests in Austria and Japan: a comparison of carbon stocks, stand characteristics and conversion to high forest. Eur J Forest Res 135:857–869. https://doi.org/10.1007/s10342-016-0979-2
46. Stojanović M, Sánchez-Salguero R, Levanič T et al (2017) Forecasting tree growth in coppiced and high forests in the Czech Republic. The legacy of management drives the coming Quercus petraea climate responses. For Ecol Manage 405:56–68. https://doi.org/10.1016/j.foreco.2017.09.021
47. Kalt G, Mayer A, Theurl MC et al (2019) Natural climate solutions versus bioenergy: can carbon benefits of natural succession compete with bioenergy from short rotation coppice? GCB Bioenergy 11:1283–1297. https://doi.org/10.1111/gcbb.12626
48. Schweier J, Molina-Herrera S, Ghirardo A et al (2017) Environmental impacts of bioenergy wood production from poplar short-rotation coppice grown at a marginal agricultural site in Germany. GCB Bioenergy 9:1207–1221. https://doi.org/10.1111/gcbb.12423
49. Beringer T, Lucht W, Schaphoff S (2011) Bioenergy production potential of global biomass plantations under environmental and agricultural constraints. GCB Bioenergy 3:299–312. https://doi.org/10.1111/j.1757-1707.2010.01088.x
50. Langeveld H, Quist-Wessel F, Dimitriou I et al (2012) Assessing environmental impacts of short rotation coppice (SRC) expansion: model definition and preliminary results. Bioenerg Res 5:621–635. https://doi.org/10.1007/s12155-012-9235-x
51. Rösch C, Aust C, Jörissen J (2013) Envisioning the sustainability of the production of short rotation coppice on grassland. Energ Sustain Soc 3:7. https://doi.org/10.1186/2192-0567-3-7
52. Dimitriou I, Rutz D (2015) Sustainable short rotation coppice a handbook. WIP Renewable Energies, Munich
53. Hauk S, Gandorfer M, Wittkopf S et al (2017) Ecological diversification is risk reducing and economically profitable—the case of biomass production with short rotation woody crops in south German land-use portfolios. Biomass Bioenerg 98:142–152. https://doi.org/10.1016/j.biombioe.2017.01.018
54. Arets EJMM, van der Meer PJ, Verwer CC et al (2011) Global wood production: assessment of industrial round wood supply from forest management systems in different global regions. Wageningen, Alterra
55. Bizerra Barros I, Santos Cavalcante V, Soares Moulin A et al (2021) Integrating forest residue and mineral fertilization: effects on nutrient acquisition, nutrient use efficiency and growth of eucalypt plants. For Ecol Manage 496:119461. https://doi.org/10.1016/j.foreco.2021.119461
56. Mead DJ (2005) Forests for energy and the role of planted trees. Crit Rev Plant Sci 24:407–421. https://doi.org/10.1080/07352680500316391
57. Röser D, Asikainen A, Stupak I, Pasanen K (2008) Forest energy resources and potentials. In: Röser D, Asikainen A, Raulund-Rasmussen K, Stupak I (eds) Sustainable use of forest biomass for energy. Springer, Netherlands, Dordrecht, pp 9–28
58. Stolarski MJ, Krzyżaniak M, Warmiński K et al (2019) Energy efficiency indices for lignocellulosic biomass production: short rotation coppices versus grasses and other herbaceous crops. Ind Crops Prod 135:10–20. https://doi.org/10.1016/j.indcrop.2019.04.022
59. Daioglou V, Doelman JC, Wicke B et al (2019) Integrated assessment of biomass supply and demand in climate change mitigation scenarios. Glob Environ Chang 54:88–101. https://doi.org/10.1016/j.gloenvcha.2018.11.012
60. Russell MB, Fraver S, Aakala T et al (2015) Quantifying carbon stores and decomposition in dead wood: a review. For Ecol Manage 350:107–128. https://doi.org/10.1016/j.foreco.2015.04.033

61. Shannon VL, Vanguelova EI, Morison JIL et al (2022) The contribution of deadwood to soil carbon dynamics in contrasting temperate forest ecosystems. Eur J Forest Res 141:241–252. https://doi.org/10.1007/s10342-021-01435-3
62. FAO (2023) Forest product statistics. https://www.fao.org/forestry/statistics/84922/en/. Accessed 21 Apr 2023
63. Vanegas Cantarero MM (2020) Of renewable energy, energy democracy, and sustainable development: a roadmap to accelerate the energy transition in developing countries. Energy Res Soc Sci 70:101716. https://doi.org/10.1016/j.erss.2020.101716
64. Bailis R, Drigo R, Ghilardi A, Masera O (2015) The carbon footprint of traditional woodfuels. Nat Clim Chang 5:266–272. https://doi.org/10.1038/nclimate2491
65. FAO (2022) The state of the world's forests 2022. Forest pathways for green recovery and building inclusive, resilient and sustainable economies. FAO, Rome
66. Thiel M, Basiliko N, Caspersen J et al (2015) Operational biomass recovery of small trees: equations for six central Ontario tree species. Can J For Res 45:372–377. https://doi.org/10.1139/cjfr-2014-0429
67. UNECE, FAO (2021) Forest sector outlook study 2020–2040. United Nations, Geneva
68. da Silva LE, Soares ÁAV, Goulart SL et al (2021) Challenges of the lumber production in the Amazon region: relation between sustainability of sawmills, process yield and logs quality. Environ Dev Sustain 23:4924–4948. https://doi.org/10.1007/s10668-020-00797-9
69. Ghafghazi S, Lochhead K, Mathey A-H, et al (2017) Estimating mill residue surplus in Canada: a spatial forest fiber cascade modeling approach. Forest Prod J 67:205–218. https://doi.org/10.13073/FPJ-D-16-00031
70. Hassan MdK, Villa A, Kuittinen S et al (2019) An assessment of side-stream generation from Finnish forest industry. J Mater Cycles Waste Manag 21:265–280. https://doi.org/10.1007/s10163-018-0787-5
71. Knapic S, Pereira H (2003) Rendimentos industriais de pinheiro bravo em serração. Anais do Instituto Superior de Agronomia 223–241
72. Murata K, Ikami Y, Tonosaki M (2006) Estimate of material recycling flow in the Japanese wood industry
73. Najera Luna JA, Sanchez Medrano JA, Mendez Gonzalez J (2013) Lumber yield and production time in sawmilling of pallets in Durango, Mexico. Forest Syst 22:573. https://doi.org/10.5424/fs/2013223-02693
74. Ogunwusi AA (2014) Wood waste generation in the forest industry in Nigeria and prospects for its industrial utilization. Ivil Environ Res 6:62–69
75. UNECE, FAO (2010) Forest product conversion factors for the UNECE Region United Nations. United Nations, Geneva
76. Perlack RD, Wright LL, Turhollow AF, Graham RL (2005) Biomass as feedstock for a bioenergy and bioproducts industry: the technical feasibility of a billion-ton annual supply. United States Department of Agriculture (USDA) and the United States Department of Energy (DOE), Oak Ridge National Laboratory, Oak Ridge, TN
77. McGraw B (2009) Recycling veneer-mill residues into engineered products with improved torsional rigidity. West Virginia University Libraries, MS
78. Oliveira RCP, Mateus M, Santos DMF (2018) Chronoamperometric and chronopotentiometric investigation of Kraft black liquor. Int J Hydrogen Energy 43:16817–16823. https://doi.org/10.1016/j.ijhydene.2018.01.046
79. Suhr M, Klein G, Kourti I, et al (2015) Best available techniques (BAT) reference document for the production of pulp, paper and board. Luxembourg
80. Teir S (2002) Modern boiler types and applications. Helsinki University of Technology
81. EC (2018) Guidance on cascading use of biomass with selected good practice examples on woody biomass. European Commission, Brussels
82. Mair C, Stern T (2017) Cascading utilization of wood: a matter of circular economy? Curr Forestry Rep 3:281–295. https://doi.org/10.1007/s40725-017-0067-y
83. Daian G, Ozarska B (2009) Wood waste management practices and strategies to increase sustainability standards in the Australian wooden furniture manufacturing sector. J Clean Prod 17:1594–1602. https://doi.org/10.1016/j.jclepro.2009.07.008

84. Besserer A, Troilo S, Girods P et al (2021) Cascading recycling of wood waste: a review. Polymers 13:1752. https://doi.org/10.3390/polym13111752
85. Alakangas E, Koponen K, Sokka L, Keränen J (2015) Classification of used wood to biomass fuel or solid recycled fuel and cascading use in Finland. In: Book of proceeding bioenergy. Jyväskylä, Finland, pp 79–86
86. WRA (2021) Waste wood assessment guidance for the UK waste wood industry. Wood Recycler's Association, Inkberrow
87. UNECE, FAO (2022) Catalogue of wood waste classifications in the UNECE region. FAO, Geneva
88. Junginger DM, Hennig C, Junginger EM, et al (2019) Transboundary flows of woody biomass waste streams in Europe. IEA Bioenergy
89. Faraca G, Boldrin A, Astrup T (2019) Resource quality of wood waste: the importance of physical and chemical impurities in wood waste for recycling. Waste Manage 87:135–147. https://doi.org/10.1016/j.wasman.2019.02.005
90. FAO, UNECE (2021) Circularity concepts in forest-based industries. FAO, Geneva
91. Cocchi M, Vargas M, Tokacova K (2019) Absorbing the potential of wood Waste in EU regions and industrial bio-based ecosystems—BioReg. EUBIA

Stand Structure and Biomass

Ana Cristina Gonçalves

Abstract Trees and stands store large amounts of biomass, but this storage is dynamic in time and space. It depends on the species, stand structure, silvicultural systems, and silvicultural practices. Furthermore, interactions between the trees in the stands and forests and disturbances result in biomass variability. The forest systems biomass estimation sometimes does take into account this variability. Additionally, all harvests remove biomass to a smaller or larger extent from the forest systems. Their sustainability is dependent on the amount and biomass components removed. The biomass exports are related to the management goals and the harvest type. Overall, stem biomass exports have smaller impacts than whole tree harvest on the sustainability and resilience of forest systems. However, forest residues removal can be done to maintain the forest system sustainability as long as biomass components richer in nutrients are maintained, at least partially, in the forest systems.

Keywords Structure · Composition · Yield · Removal impacts · Sustainability

1 Introduction

Trees, stands, and forests store large amounts of biomass (and carbon) due to the trees' large dimensions and long lifespans [1, 2]. In general, there is a trend toward biomass increase with ageing [3, 4]. Yet, tree and stand growth are dynamic in time and space due to factors that encompass site (edaphic and climatic conditions) and stand structure (*e.g.,* species, regime, composition, structure, density, and spatial arrangement), as well as disturbances both natural (biotic and abiotic) and artificial (*e.g.,* silvicultural practices) [5–8]. Biomass partitioning is also dynamic in time and space. It depends on the tree development stage, stand structure, and interactions

A. C. Gonçalves (✉)
MED—Mediterranean Institute for Agriculture, Environment and Development &
CHANGE—Global Change and Sustainability Institute, Instituto de Investigação e Formação Avançada, Departamento de Engenharia Rural, Escola de Ciências e Tecnologia, Universidade de Évora, Apartado 94, 7002-544 Évora, Portugal
e-mail: acag@uevora.pt

A. C. Gonçalves and I. Malico (eds.), *Forest Bioenergy*, Green Energy and Technology,
https://doi.org/10.1007/978-3-031-48224-3_3

between the individuals in a forest stand, as well as the effects of disturbances (*e.g.*, fires, storms, thinning, and pruning) [9–11].

The yields of biomass and the amount of biomass that is removed from the forest systems vary according to the stand structure and the silvicultural system. Stand structure influences the quantity of biomass (live and dead) in the forest systems, and different types of harvest result in different amounts of exported biomass [12–16]. Silvicultural systems influence the amount of biomass removed, and the timing of harvests, as well as the type of biomass exported from the forest stands [17–19]. Biomass exported can be used for several purposes, such as timber, pulp and paper, and bioenergy. The proportion of each woody product exported from forest stands depends on the wood quality, dimensions, damages during harvest, and market [20–22]. Also, the sustainability of the forest stands and productions as well as the economic and environmental constraints have to be considered [20, 22, 23].

This chapter introduces the definitions and concepts associated with biomass dynamics in space and time and discusses the factors that influence biomass and yields of woody products as well as forest system sustainability. It was divided into five sections: stand structure, silvicultural systems, and silvicultural practices (Sect. 2), biomass partitioning and dynamics (Sect. 3), forest systems sustainability (Sect. 4), and biomass yields, harvests, and exports (Sect. 5).

2 Stand Structure and Silvicultural Systems

Stand Structure

A forest stand is a community of trees that interact with each other (*e.g.*, [24, 25]), which is dynamic in space and time [26]. Traditionally, stand structure is evaluated with three criteria, namely regime, composition, and structure (Fig. 1). *Regime* can be broadly divided into two classes: high forest where regeneration is of seed origin; and coppice where regeneration is of vegetative origin. *Composition* refers to the number and proportion of forest species in the stand. It is evaluated by the number of species and with absolute density measures, such as the number of trees, basal area, volume, and crown cover, or with an index based on absolute density measures (*e.g.*, [27]). *Structure* refers to the number of cohorts (or age classes) of the stand. Two classes are considered: even-aged with one cohort or uneven-aged with two or more cohorts. It is assessed usually with diameter and height distributions (*e.g.*, [28]).

Regime, composition, and structure result in a wide range of stand structures. Moreover, other features contribute to stand structure variability (Fig. 2), namely density [29–31]; species [32–34]; species proportions [35–37]; interactions between trees [38–40]; availability of light, water, and nutrients [41–43]; niche complementarity [44, 45]; spatial distributions [46–48]; and temporal arrangements [47, 49, 50].

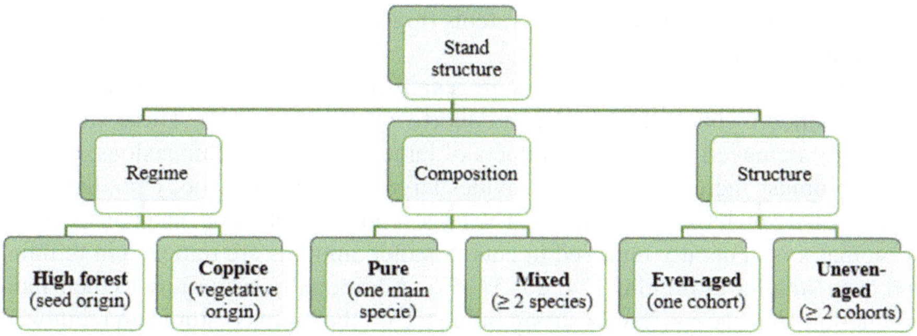

Fig. 1 Stand structure as a function of regime, composition and structure

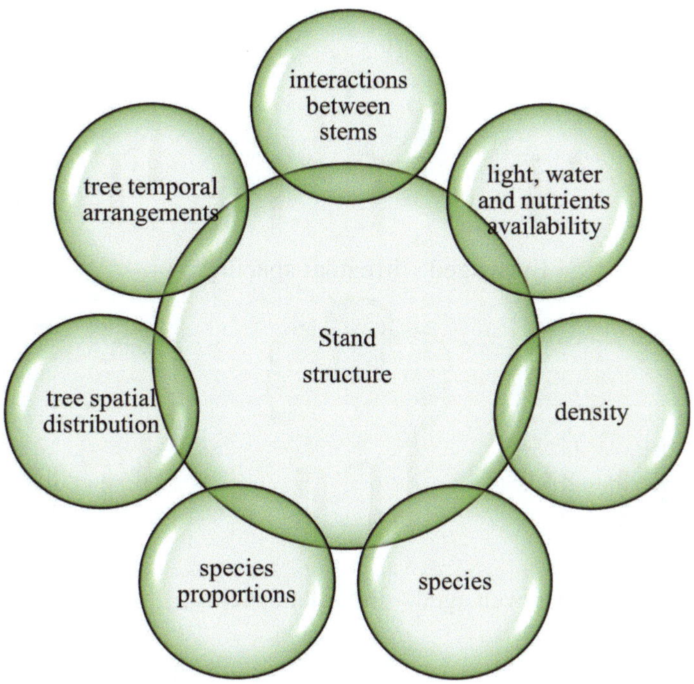

Fig. 2 Features that contribute to stand structure variability

The multitude of stand structures does not enable the identification and description of all possible alternatives (Fig. 3). Thus, it was chosen to describe a set of stand structures that enable to evaluate the potential biomass and residues, as function of its target management goals. For simplicity, six stand structures were considered, five representative of forest systems oriented for timber and/or woody products (high forest pure even-aged stands, coppice pure even-aged stands, high forest mixed even-aged stands, high forest pure uneven-aged stands, and high forest mixed uneven-aged stands) and one of the agroforestry systems (high forest pure or mixed even-aged or uneven-aged stands).

High forest pure even-aged stands are stands of seed origin, characterised by the predominance of one specie and one cohort (*e.g.*, [24, 25, 51]). These stands are frequently managed for woody products of large and medium dimensions. In these stands biomass increases in time, provides large quantities of woody products, and is more easily managed [24, 51, 52]. The largest quantity of wood, and consequently of biomass, is generally removed in cuts. Smaller amounts are removed in thinning and even smaller in pruning [24, 25, 51–53]. Additionally, according to their main production, density can have a strong influence on the biomass stored and removed,

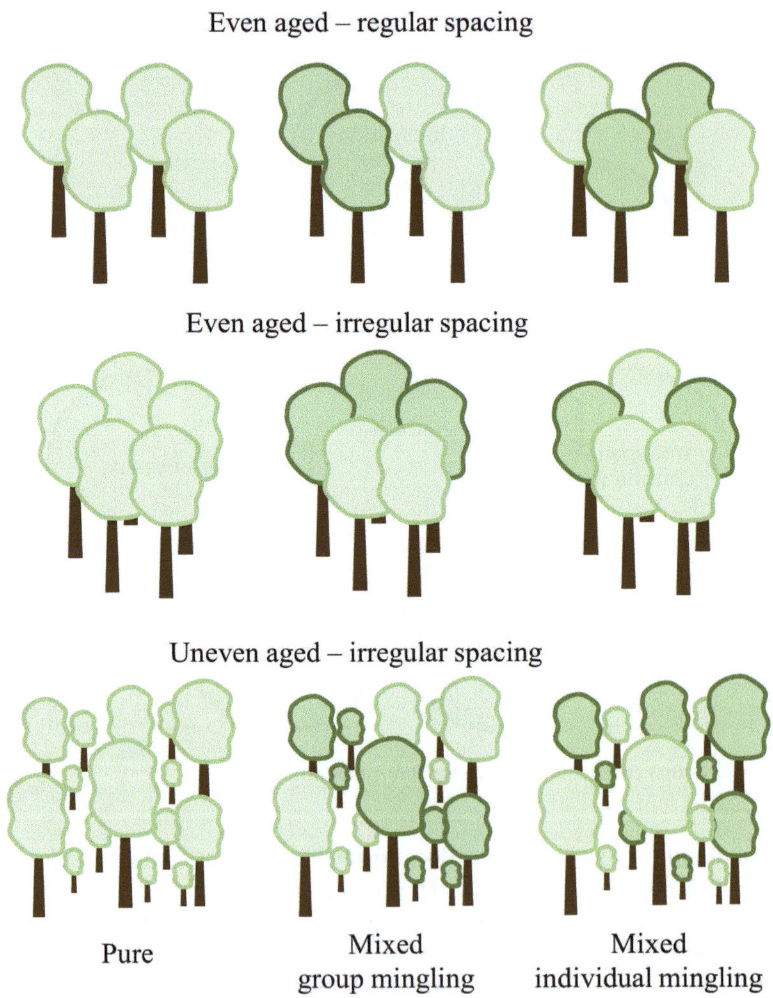

Fig. 3 Examples of stand structure

and on the proportion of residues that are generated. Stands managed for timber have frequently high densities and biomass (and volume): the higher the density the smaller the dimensions of the woody products [29, 54].

Coppice pure even-aged stands are stands of vegetative origin, characterised by having frequently only one specie and one cohort (*e.g.,* [24, 52]). These stands have frequently very high densities and are managed for woody products of small and medium dimensions (*cf*. chapter "Energy Plantations"). Density and rotation length depend on the final product; the higher the density, the shorter the rotation. Rotation can be very short (2–6 years for biomass for energy), short (6–15 years for pulp and paper), medium (15–20 years for charcoal and timber of small dimensions), or long (20–35 years for firewood and timber of small and medium dimensions). The largest quantity of wood is removed in cut [25, 51–53]. In very short and short rotations all biomass is removed [55].

In the last centuries, the extensive use of other fuels, first coal and then petroleum-based fuels, reduced the pressure on forests and woody products [53, 56, 57]. It is in this context that a change in paradigm came about and conservation issues arose. Forest management changed from being focused on their maintenance and their woody products and started to incorporate broader objectives that included other products and services of the forest systems [53, 56, 58, 59]. This change of paradigm gave birth to new approaches to forest management, where the emphasis was centred on systems with several productions (multiple use systems). These approaches guided silviculture towards high forest pure uneven-aged or mixed stands, where management is focused on emulating the natural processes while providing both wood and other products and services. To attain these stand structures, many approaches, methods, and techniques were developed [28, 56, 58, 60–68].

High forest mixed even-aged stands are stands of seed origin with two or more species and one cohort (*e.g.,* [26, 53, 69]). In these stands, productivity tends to be higher than in pure ones when species traits promote niche complementarity [8, 44, 45, 70]. The number of species and their evenness, as long as complementarity among species is observed, also have a positive effect on productivity [35, 71–73]. As in high forest pure even-aged stands, cuts generate the largest amount of biomass removal, followed by thinning and pruning [45, 74].

High forest pure uneven-aged stands are stands of seed origin with one specie and two or more cohorts. In these stands, most of the growth is concentrated in the future trees (vigorous trees of high quality for the desired productions), which should be in free growth throughout the cutting cycle (for details see [75]). Cuts and thinning frequently occur simultaneously [14, 28, 53]. The productivity of these stands, when compared with the even-aged ones, can be smaller [8, 50, 76, 77] or higher [8, 77, 78]. Another difference between even-aged and uneven-aged stands is that the periodicity of cuts and the amount of biomass (or volume) removed in the latter are, respectively, shorter and smaller than in the former. In uneven-aged stands, biomass removal per cut tends to be as similar as possible to ingrowth to maintain the target structure, that is the number of cohorts and the proportion of trees per cohort [14, 28, 53].

High forest mixed uneven-aged stands are stands of seed origin with two or more species and two or more cohorts. Similar to pure uneven-aged stands growth is concentrated on the future trees. The target number of species and cohorts, as well as their proportions, should be maintained [14, 28, 53]. Cuts and thinnings frequently occur simultaneously [14, 28, 53], and the periodicity of harvest and the quantity of biomass removed are smaller when compared with even-aged stands [14, 28, 53]. Harvests tend to remove approximately the ingrowth [12, 79].

High forest pure or mixed even-aged or uneven-aged stands managed as agroforestry systems are characterised by a low density, and canopies are open and heterogeneous. Their main productions are fruit, bark (*Quercus suber* only), timber, agriculture, grazing, ecosystem services, recreation and aesthetics [80–84]. From the many existing agroforestry systems worldwide, it was chosen to include in this review those of the Mediterranean basin whose predominant species are *Quercus suber*, *Quercus rotundifolia*, *Quercus ilex* and *Pinus pinea* [82, 85, 86]. These stands can be pure or mixed with other conifers [85] and/or other oak species [87–90]. Stands can be even-aged [82, 85, 86], or uneven-aged with 2–4 cohorts [82, 86, 91]. The most frequent removals of biomass are done in thinnings, prunings, and by the cutting of (few) dead trees [92–94]. Biomass is usually used for firewood [82, 85]. These systems' particularities are the diverse productions, partitioning of the risk, maintenance of the system's sustainability under a climate with strong annual and interannual variability, and regular annual incomes [84, 95–97]. The stands managed for bark and fruit have lower densities and biomass than those managed for timber [98].

Silvicultural Systems

The silvicultural systems are related to stand structure and the amount of biomass removed in each cut and the regeneration method. Three main silvicultural systems are described in most silviculture text books [24–26, 51, 52, 99–102], namely clearcut, shelterwood, and selection.

In the **clearcut system**, the removal of all the trees is done at the end of the production cycle (or rotation). During rotation, several tending practices are carried out to improve timber quality and quantity, such as thinnings and prunings [24, 25]. These practices, especially the thinnings, reduce standing biomass [54, 103, 104]. This biomass can either be left in the stand [16, 105] or removed [106, 107]. If biomass is left in the stand, it results in the reallocation of live biomass to dead biomass and/or soil organic matter [6]. If biomass is removed from the stand, it results in the export of biomass, *i.e.*, wood of large dimensions in cuts and smaller dimensions in thinning and pruning [45, 74].

The clearcut system advantages are the concentration of silvicultural practices and cuts in one or a few interventions (of short periodicity); the harvest is potentially more cost effective, easier as no concerns are needed with existing regeneration, and enables an easier site preparation. The disadvantages are related to the interruption of the forest microclimate, hydrological and nutrients' cycles, decomposition of organic matter, soil erosion, and the development of spontaneous vegetation [24–26, 51, 52, 99–102]. Several subtypes of clearcut systems were developed to mitigate the effects of the removal of all the trees in a stand, such as clearcut by alternate strips, by progressive strips, by patches, and with seed trees. The difference between the clearcut and its subtypes relates to the number of cuts and the regeneration method. The number of cuts in the clearcut subtypes is more than one, frequently 2 to 4, with a short time periodicity. This provides a way to reduce the negative effects of the clearcut. Yet, care should be taken so that regeneration is accomplished with only one cohort [25, 101]. The stand renewal in the clearcut system is most frequently with artificial regeneration, either by direct seeding or plantation [108]. Inversely, in the clearcut subtypes, both artificial and natural regeneration can be used. The latter is used when seed, from trees in the neighbourhood areas to regenerate, is available [109]. The species most frequently used with clearcut systems are pioneer species and the light demanding ones [25, 52]. Species tolerant to shade and sensitive to frost and/or drought are rarely used. Moreover, care should be taken to reduce the competition between the regeneration individuals and the spontaneous vegetation [25, 52].

In the **shelterwood system**, the removal of the stand is done in a sequence of cutting interventions, frequently more than four. The first intervention is the *preparatory cut*, which is carried out if it is necessary to increase crown development and thus seed/fruit production. The second cutting intervention is the *seed cut*, which has the objectives of ensuring germination and the development of the seedlings according to their traits. Its intensity and duration depend on the species traits and site quality. It is followed by the *secondary cuts*, in variable number, aiming at the removal of the main stand, as soon as regeneration is ensured. The intensity and number of the secondary cuts are dependent on species traits, site, quality and competition between the trees of the main stand and regeneration. The *final cut* removes the last trees of the main stand when regeneration does not need shelter. Silvicultural practices have the same goals and effects as those of clearcut systems. This system can be divided in two broad classes: uniform and irregular shelterwood systems. The *uniform shelterwood system* is characterised by the short periodicity of the cuts so that regeneration develops in only one cohort; and a relative uniformity in the spatial arrangement of the cuts [24, 25, 51]. The uniform shelterwood system has been further divided, according to the spatial distribution of the cuts, in uniform shelterwood by strips, by groups or patches, and by strips and patches [24, 25, 51, 101]. The main advantages of the uniform shelterwood system in relation to clearcut are that enables using natural regeneration, provides shelter for regeneration, partial maintenance of forest microclimate, and reduction of erosion risk. Its disadvantages are related to the potential damages to regeneration during harvest and take more time to implement. The *irregular shelterwood system* is characterised by a longer periodicity between cuts; the promotion of the development of more than one cohort of regeneration; and by enhancing the irregularity in the spatial arrangement of the cuts to take advantage of the advanced regeneration and/or provide the suited conditions to the species to regenerate, especially in mixed stands [28, 60]. Its main advantage is its flexibility, enabling, apart from the advantages of the uniform shelterwood system, to adapt the cuts to the site spatial variability and species traits [28]. Moreover, uniform shelterwood is more appropriated for even-aged stands, as the goal is to have only one cohort of regeneration, whereas irregular shelterwood is better suited for uneven-aged stands with a small number of cohorts as it is able to sustain or enhance the structure heterogeneity [25, 28, 60, 110].

The differentiation between the clearcut with seed tree system and the uniform shelterwood system is related to the number of trees and the number of cuts. A lower number of trees is left in the clearcut with the seed trees system, and the number of cuts is frequently two (the first cut removes the main stand and the second the seed trees). The removal of the seed trees should happen prior to the competition between seedlings/saplings and the seed trees. In this system, natural regeneration is based on seed production of 1 to 3 years and the cuts have a short periodicity. The regeneration should be as fast as possible with the development of only one regeneration cohort [111, 112]. The drawbacks of this system are related to the few seed years, low amounts of seed, adverse site conditions, and the development of spontaneous vegetation, which may limit the germination and development of seedlings [113]. In the uniform shelterwood system, the number of trees of the main stand not removed in the first cut is higher than in the clearcut with seed tree system, and the number of cuts is more than four. This enables to have seeds from several years, which have frequently a high number of seedlings as they can take advantage of site favourable conditions to germinate and develop. The disadvantages are related to the crown dimensions and their seed production [114, 115] and to the potential development of several cohorts [25].

The **selection system** (*Plenter*) is characterised by cutting cycles, where removal is done simultaneously for individuals at the target dimensions (*e.g.,* diameter at breast height) and in tending [25, 28, 53, 101]. Though there is some variability, in general, the removals tend to be similar to ingrowth to maintain the structure, *i.e.*, the number of cohorts and their proportions in time [14, 25, 28, 53, 101]. Renewal is typically by natural regeneration, and the cut's periodicity is dependent on the number of cohorts. The higher the number of cohorts, the shorter the cuts and regeneration periodicities. Two basic subtypes have been defined: *single tree* selection system, where management is based on single trees; and *group* selection system, where management is focused on clusters of trees. Its advantages are related to the continuous maintenance of hydrological and nutrients cycles and forest microclimate, reduction of erosion risk, and shelter for regeneration. Its disadvantages are related to the regeneration development, especially that of the less shade tolerant species, the damages during harvests, and the skills needed by the foresters [25, 28, 53].

Silvicultural Practices

Silvicultural practices are carried out throughout the cutting cycle to improve woody products' quality and quantity [24, 25]. These practices also influence biomass accumulation, and tree and stand biomass partitioning [116, 117]. The two silvicultural practices are thinning and pruning.

Thinning main goals are reducing competition between the individuals in a stand, selecting the most suitable individuals as a function of the target productions and yields, and redistributing the growing space amongst the individuals left in the stand after thinning [24, 25]. Thinnings are carried out in all stand structures and silvicultural systems, except in those settled at final spacing, such as coppices of very short or short cutting cycles [55]. Thinning is characterised by its method, intensity, and frequency as well as by the selection of trees to be maintained and/or removed from the stand (for details, see [75]).

Several thinning methods can be distinguished by the traits of the trees to be removed (*e.g.,* social status, stem, and crown). From the existing thinning methods, the most frequently used are thinning from below, from above, selective, of dominants, and mechanical (*e.g.,* [75] and references therein). *Thinning from below* predominantly removes individuals of the lower layers of the stand (dominated trees), and trees of the upper layers (dominant and co-dominant) are only removed if they are dead or on thinnings of heavy intensity [24, 25]. *Thinning from above* removes mainly trees of the upper canopy layers that compete with the trees with the best characteristics for potential production. The trees of the lower canopy layers are kept to promote natural pruning as well as for soil protection, habitat for flora and fauna, blowdown risk reduction, and reduction of spontaneous vegetation development [24, 25]. *Selective* (or Schädelin) *thinning* is based on the selection of future trees, from all canopy layers, that will be maintained until the end of the production cycle. The competitors of the future trees are removed so that they are, during all cutting cycle, in free growth [102, 118]. *Thinning of dominants* removes trees of the upper layers, favouring the trees of the middle and lower layers. It can have a temporary or permanent character and can be combined with thinning from below [25]. *Mechanical thinning* removes individuals by their location, either within the row or by row. It is better suited for stands where individuals are of selected material [25]. Thinning intensity is the percent of the removals in relation to stand prior to thinning. It is frequently calculated with the number of trees or basal area per hectare [25, 119].

Thinning, by cutting trees, can reallocate live biomass to dead biomass if cut trees are maintained in the stand (*e.g.,* [15]) or export partially or totally the cut trees (*e.g.,* [18]). In any case, there is always a reduction of live biomass (*e.g.,* [103]). Yet, the reduction of biomass depends on the thinning method. Thinning from above and of dominants originates a higher decrease in live standing biomass [54, 103] than thinning from below [120, 121]. This is due to the dimensions of the trees to be removed, which are larger in the thinning from above and of dominants than in the thinning from below [24, 25]. Moreover, the more intense the thinning is, the higher the decrease in live standing biomass [122, 123]. Yet, as trees that remain in the stand after thinning are released from competition, they increase their growth rates. The increase in the growth rates is dependent on the tree's development stage and position in the canopy. It is higher for young individuals in the upper canopy layers than for older ones in the lower canopy layers [116, 124–126]. Overall, the effects of thinning are a reduction of live biomass in the short term and a tendency to constant biomass in the medium and long term, due to tree growth [54, 117, 127]. Additionally, thinning, increasing the growing space per tree, may result in the development of the crown (branches and leaves), thus changing biomass partitioning [128, 129].

Pruning is a silvicultural practice that aims, in timber oriented stands, to improve the quality and quantity of timber by the removal of the branches (dead or alive) and promoting a branchless stem of at least of 4–6 m [24, 25]. For agroforestry systems, pruning to 2–4 m is recommended so that cork debarking (only for *Quercus suber*) is only made in the stem; the mechanical fruit harvest (*e.g., Pinus pinea*) is more efficient; and as a fire prevention measure [85, 130, 131]. As it is an expensive practice natural pruning should be promoted whenever possible [25, 52]. Pruning can increase or reduce biomass accumulation, depending on its intensity. In general, more intensive pruning reduces leaf area, photosynthesis, and thus growth while low intensity prunings (removal of less than 30% of the crown) have either a neutral or positive effect on growth and biomass [132].

3 Biomass Partitioning and Dynamics

In a forest stand, biomass is either in the trees or in the soil (as organic matter). The tree biomass can be in live or dead trees (Fig. 4). The dynamics of stand development result in the increase and reallocation of biomass within the stand. The cutting or death of trees results in the reallocation of live biomass to dead biomass, both above and below ground [4, 133]. As trees grow, they increase their dimensions over time. Trees grow first in height and later in diameter [24, 25]. Yet, tree growth is dynamic and dependent on growing space: the higher the growing space, the higher the tree growth [26]. The factors influencing tree biomass dynamics in space and time are age (or stage of development), density, species, site (soil and climate), the interaction between individuals and distribution in space, and stand structure.

In general, tree biomass increases with *age* [4]. With ageing biomass partitioning changes, the stem biomass increases with time, and that of the branches and leaves decreases in relation to the above ground biomass [11]. Young trees invest first in height growth to reach the upper layers of the canopy and thus light, and leaf growth to enhance photosynthesis [134]. While ageing trees' carbohydrate allocation is preferably to stem growth [9, 134]. Moreover, with ageing crown cover increases up to canopy closure, after which there is frequently a reduction of the crown volume and thus branch and leaf biomass, resulting in a reduction of the proportion of crown biomass in relation to stem biomass [11, 134, 135]. Furthermore, as trees age, the biomass produced per unit of intercepted light decreases due to respiration and hydraulic morphology [9].

Density influences the available growing space per tree (the higher the density, the lower the growing space per tree). In general, the increase in density promotes height growth and tree size variability [136], although in mixed stands, facilitation interactions promote both height and diameter growth [30]. This is linked to species characteristics complementarity and niche differentiation [70]. The proportion of species in a mixed stand influences the interactions, depending on the species traits

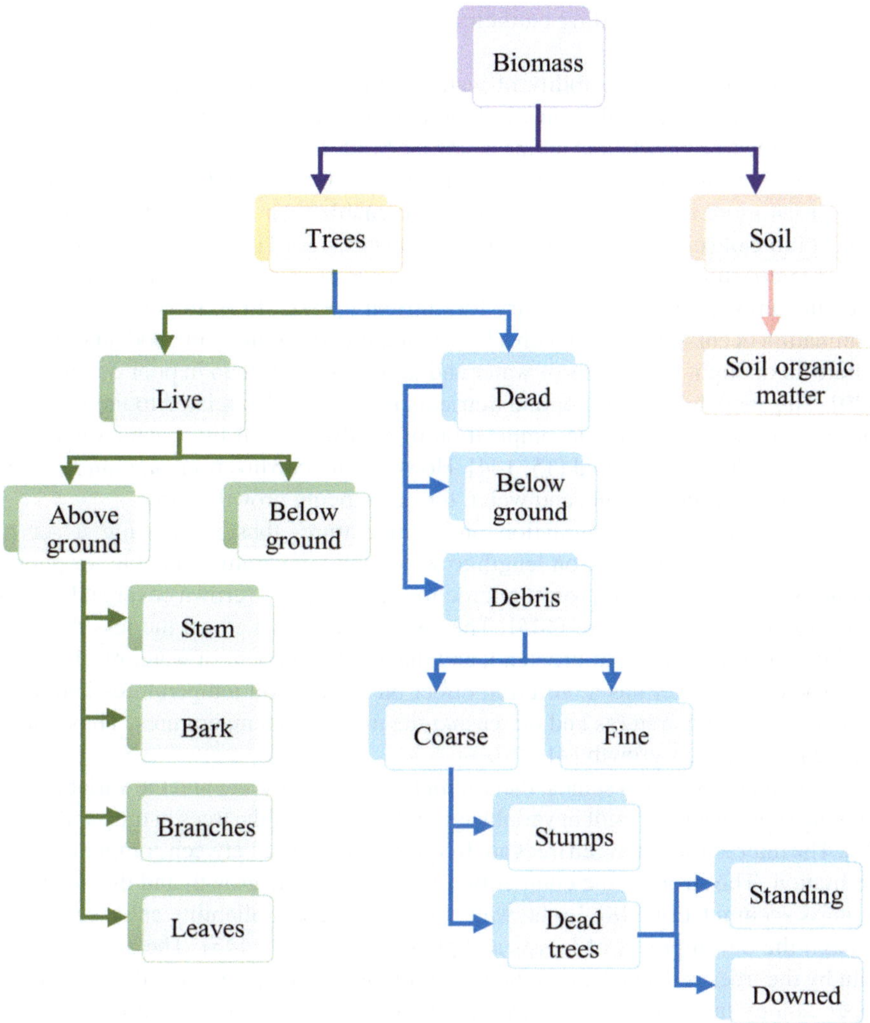

Fig. 4 Biomass distribution in forest stands

and spatial distribution. In general, competition tends to be higher within individuals of the same species than between individuals of different species [72], the former resulting in a reduction in growth and biomass storage.

Species influence biomass storage due to their traits [137]. Conifers (evergreen species) tend to have a higher investment in leaf biomass when compared to broadleaved deciduous species [138]. Species growth rate patterns also influence the dynamics of biomass over time. The species with fast growth store more biomass at the early stages of development, while for the slow growing ones, the peak is reached much later, depending on their patterns of growth and cutting cycle length [132, 139].

Soil influences the availability of water and nutrients as well as biomass partitioning. In general, the higher the availability of water and nutrients (and also light), the higher the tree growth and biomass storage [132, 137, 140], and trees invest more in the above ground biomass. Inversely, in sites with water and/or nutrient deficits, trees invest more in below ground biomass to enable their root system to explore a large volume of soil and thus water and nutrients [9, 43, 141–143]. Also, as nutrients uptake is constrained by water availability, larger root systems enable the tree to reach the amount of water necessary for nutrient uptake [143]. In general, biomass culmination occurs earlier and with a higher quantity of biomass in good quality sites (where no or slight limitations of water and nutrients exist) than in poor quality sites [139]. The availability of water and nutrients in a site is also related to the inputs of nutrients in the soil by the decomposition of deadwood and litter, which maintain the soil's productive potential [15, 144]. However, mycorrhization may improve the tree's status both in nutrients and water, thus enhancing growth [145].

Climate influences precipitation and temperature, thus influencing a set of processes from growing season length to water and temperature suitable for photosynthesis [4, 25]. Moreover, forest microclimate is related to crown cover and affects light, water, and temperature [145]. The denser the crown cover, the less the light reaching the lower layers of the stand, and the less the amount of water reaching the soil. The crown cover has a buffering effect on air and soil temperatures, reducing the maximum temperatures and increasing the minimum temperatures. Thus, forest microclimate affects growth [41, 45].

When trees develop in stands, the *interactions between trees* and the variability in growing space per tree result in variable growth patterns of the trees in time and space [26]. The interactions between trees are frequently a balance between competition and facilitation. The stronger the competition is, the lower the growth and the lower the biomass accumulation [48]. Light, water, and nutrients' availability enhance growth through the development of leaves and photosynthesis [9, 145]. The absorption of light by the trees is dependent on their position in the canopy. Generally, trees in the upper canopy layers have higher availability of light and their growth is promoted. Furthermore, the larger the crown, the greater the absorbed radiation [42]. In mixed stands, light absorption is enhanced by species phenology, foliage stratification and distribution, and the variability of crown allometry ([146] and references therein).

Water and nutrient deficits increase competition between trees for these resources and thus reduce growth. The mitigation of competition can be achieved through silvicultural practices. Thinning can mitigate the effects of the decrease in growth due to drought by allocating more growing space to each individual tree [147–149]. Similarly, the removal of spontaneous vegetation reduces competition and increases tree and stand biomass [143]. Fertilisation and irrigation, increasing nutrients and water availability, and reducing competition between trees, increase biomass (and carbon) storage (for details see [9]). Furthermore, live biomass increases with increasing rates of mortality. This may be explained by the increase in the growth of the remaining live trees which compensate for the loss of biomass due to mortality [4].

The *distribution in space and time* of the individuals in a stand is dynamic and results from tree growth, site, natural disturbances (*e.g.,* frost, drought, fires, storms, pest outbreaks), and silvicultural practices (thinning, pruning, and cuts). The dynamics of the spatial and temporal distribution of the individuals in a stand are related to stand structure (*cf*. Sect. 2), silvicultural systems (*cf*. Sect. 2), and regeneration [48]. Thinning can also increase the variability of the structure of the stand [150–152]. Moreover, the increase in water availability, along with species complementarity and silvicultural practices (small scale disturbances), promote the uneven-aged structures [153]. Inversely, storms and/or large fires (large scale disturbances) tend to promote even-aged structures [153]. This can be explained by the renewal of the stand with mainly one cohort [25]. Also, longer cutting cycles promote biomass accumulation and diversity, whereas shorter ones enhance biomass accumulation and mitigate the effect of disturbances (biotic and abiotic) [154]. The biomass exports increase with the removal of large trees (as they store large amounts of biomass). Additionally, trees may have higher or lower biomass accumulation rates, and leaving low vigour trees in the stand can reduce biomass storage [155]. Biomass growth depends on the current biomass and growing space. Yet, other features are determinant of biomass growth, such as tree internal structure, allometry, and morphology (*e.g.,* crown size, root size, stem shape). Thus, past silvicultural practices might change allometry, structure, and growth patterns. Tree crown allometry can also be modified by disturbances (*e.g.,* wind, storms) and abrasion, which are not directly linked to photosynthesis or respiration processes [139].

The biomass dynamics vary per *stand structure*. Above ground biomass, in pure even-aged stands increases with ageing until a threshold is reached, after which it flattens or has a very slow increase [117, 156]. The peak occurs earlier in dense even-aged stands as species traits are similar, and later in uneven-aged stands due to the differences in tree allometry and growth [157]. Root biomass seems to have first an increasing trend with ageing and then stabilises, with a share of *circa* 15–20% of total tree biomass [11, 158]. The above ground biomass (and carbon) sequestration and storage are higher in even-aged stands when compared to uneven-aged ones. The carbon sequestration and storage in the soil might be higher or lower in uneven-aged stands compared to even-aged ones [132]. Soil carbon increases until a threshold with tree growth and then decreases slowly due to management, overstorey tree shading, or understorey growth inhibition [156]. The high carbon soil content in uneven-aged stands is attributed to the continuous input of litter and protection of the watershed, whereas in even-aged stands after harvest, litter inputs are scarce and the soil is not or is partially covered. This may lead to the decrease in soil carbon [159].

Biomass in mixed stands can be higher or smaller than in pure stands [136, 160]. When compared to pure stands, mixed stands tend to have a smaller share of crown biomass (branches and leaves). This could be due to competition in the canopy, with some species having competitive advantages (*e.g., Fagus sylvatica*) and promoting the reduction of the crowns of other species [161–165]. The different biomass allocation is related to the species plastic morphology. This can mean that for an equal stem biomass different stands may have different crown biomass [162]. In mixed stands, the litter inputs, nutrient cycles, and litter decomposition are higher than in pure stands, thus tending to increase the soil organic matter and the storage of carbon [159] and the productivity of the stands [132, 140].

Biomass in uneven-aged stands can be higher or smaller than in even-aged stands [166, 167]. Some authors refer that the increase in structure heterogeneity (*i.e.*, the increase in the number of cohorts in uneven-aged stands) increases above ground biomass [168–170], whereas others found it decreases [170–172]. This could be due to the negative effect of disturbances, or to the different proportions of large trees that store much larger amounts of biomass [170]. As biomass accumulation is proportional to basal area, larger trees' contribution to stand biomass is greater than that of the small trees. Additionally, above ground biomass storage is higher when there is a direct proportionality between the growth and the dimension of the trees in a stand [79].

Unmanaged stands have frequently higher biomass stored than managed stands, due to the periodical removal of biomass in the harvest in the latter [107, 159]. However, mixed uneven-aged stands of broadleaved and conifer species, when compared to unmanaged stands, have similar biomass for moderate growing stocks, and slightly higher for low growing stocks. The increase in growing stock increases mortality in unmanaged stands and is higher than in mixed uneven-aged stands, regardless of the growing stock. This can be explained by management: trees are cut in mixed uneven-aged stands before they die. Furthermore, biomass (and carbon) sequestration a few years after the cuts is the highest, so harvesting cycles with short periodicities enhance biomass (and carbon) sequestration [173].

Silvicultural systems also result in differences in biomass sequestration and dynamics. In clearcut systems, at the end of the cutting cycle, all above ground biomass is removed [25]. It may alter forest stands from carbon sinks to carbon sources due to the reduction of biomass (wood export), increase of respiration (with the increase of light and temperature in the soil), and decrease of photosynthesis (reduction of leaf area) [132]. The regeneration in the clearcutting systems increases biomass gradually with ageing [54, 74]. The shelterwood and the selection systems, by not removing all the biomass in one cut, have a positive effect on biomass (and carbon) storage. The magnitude of the effect is dependent on the intensity and periodicity of the cuts as well as their spatial arrangement, which is related to post-harvest mortality, in particular to blowdown [132]. As in irregular shelterwood and selection systems, part of the trees is not harvested, it enables them to provide shelter (shade) and are also a source of litter inputs that buffer temperature and water in the soil, that is, reduces temperature ranges and tends to preserve soil moisture [174].

In a stand, apart from live trees, biomass exists also in deadwood, as below ground (dead root system) and above ground as fine or coarse dead wood, both standing and downed trees (Fig. 4). Due to their physiology and phenology, trees renovate periodically their leaves (annually in deciduous species and every 2–3 years in evergreen species) and the fine roots (usually annually). The litter produced decomposes into soil organic matter. This originates the reallocation of some of the above and below ground biomass into soil organic matter [15]. The development of the stands, due to competition, results in the death of the less competitive individuals (self-thinning), thus the reallocation of live to dead biomass [144]. Deadwood tends to increase from young to mature development stages and decrease from mature to old growth ones [15]. This above ground dead biomass can be standing or downed and decomposes at faster or slower rates, incorporating carbon and nutrients into the soil [15, 144]. Standing dead trees have lower decomposition rates than downed debris or the forest floor. Furthermore, burned trees have very slow rates of decomposition due to their content in charcoal [133]. Disturbances, whether natural or silvicultural practices influence the deadwood dynamics [133, 144]. When biomass is not exported, is promoted the reallocation of biomass from the live stand into fine and coarse woody debris, which will be incorporated into the soil through decomposition as carbon and nutrients [15, 144]. Inversely, when biomass is exported, deadwood decreases [117]. Salvage cuts reduce the dead downed biomass, which results in lower amounts of carbon in downed woody debris and slight reductions in the forest floor [133].

4 Forest Systems Sustainability

The effects of harvest on forest system sustainability are linked to harvest intensity, frequency, quantity [17, 18, 105], and the proportion of the components of biomass exported [17, 105]. Harvest affects soil, hydrology, and diversity [106, 175]. The forest system sustainability is dependent on the balance between biomass storage and export [176]. In general, the higher the biomass exports, the higher the nutrients removed [17, 177–179]. However, the allocation of nutrients in the tree components varies. The newest tissues (*i.e.*, leaves, twigs, and small branches) have a higher share of nutrients when compared to older tissues (*i.e.*, large branches and stems) [17, 180]. Moreover, the share of biomass per component varies per development stage (age) and stand structure (*cf*. Sect. 3). Biomass exports can have negative [17, 179] or positive [105, 181] effects on forest system sustainability.

The effects of harvest on soils are related to the changes in the soil's physical [174], chemical [17, 179], and biological [21] characteristics, as well as on tree growth [174] and regeneration [182] of the stands. The soil's physical properties are affected by soil compaction by the mechanical equipment [174] and, consequently, affects soil water holding capacity, and aeration [17, 21, 174]. The effects of harvesting and biomass exports on soil chemical properties are related to the changes in the inputs and outputs of carbon and nutrients on the soil and the relationship live/dead biomass [17, 174, 183, 184].

The effects of harvest are common to all stand structures, silvicultural systems, and practices [17, 174, 180]. Yet, they vary in intensity, frequency, and amount of biomass exported from the stand. In general, the higher the intensity and frequency the higher the biomass removal and the higher the export of nutrients [17, 155, 179]. Overall, the negative impacts of biomass removal increase from good to poor quality sites, affecting site and stand productivity [17, 105, 181]. Moreover, the nutrients' exports increase from stem harvest to whole tree harvest [181, 185]. This is related to the nutrients contents of each biomass component. Stem is poor in nutrients, thus their removal exports less nutrients [186] while leaves and bark removal (due to their high content in nutrients) result in larger nutrients exports [17, 21]. Yet, nutrients exports are also dependent on species, stand structure, and silvicultural systems. Exports of nutrients are higher in conifer than in broadleaved species (when the harvest is in the winter, part of the leaves of conifers are exported), which also reduces tree growth [17]. Coppice stands harvest tends to export more nutrients when compared to high forest stands [180], due to the higher density and proportion of crown biomass [17]. Inversely, high forest stands, especially those with long cutting cycles, tend to export fewer nutrients, as the largest biomass component is the stem, which is poor in nutrients [17].

Harvesting, with the reduction of crown cover, alters the forest microclimate [174]. The reduction of crown cover may result in losses of nutrients in the soil due to organic matter mineralisation, leaching, and/or reduction of soil biological activity [17, 21, 181]. Moreover, soil carbon losses are higher with the removal of the forest floor and spontaneous vegetation. Yet, deep soil layers after whole tree harvest have higher concentrations of carbon and nitrogen, suggesting leaching to deeper soil layers [185, 187], while whole tree harvest, forest floor removal, and spontaneous vegetation removal, decrease carbon and nitrogen concentrations in all depth layers of the soil [185].

Biomass removal's negative effects mitigation on the soil can maintain the soil's productive potential. One alternative is to export the stem biomass and keep the forest residues, especially those rich in nutrients, such as leaves and bark [178]. Moreover, the small and large debris left after harvest protects the forest floor [188] and increases the nutrients in the soil through decomposition [174]. Another alternative is harvesting when trees have no leaves or delaying the removal of forest residues to keep the leaves and small branches in the stand [17, 183]. Another alternative still, to maintain the system's productive potential, is fertilisation, with inorganic or organic compounds, as long as it is realistic technically and/or economically [21, 143, 178, 189]. The negative effects of residues removal are short or medium term lasting (0.2–33 years), after which tree growth promotes the recovery of the system [179, 183, 190]. This is related to the increase in crown cover and the amount of litter produced by the trees, after harvest, of the remaining stand in the case of thinning, and the new regenerated stand in cuts [191, 192]. When compared to coppice systems, high forest ones tend to originate larger quantities of litter in a longer timeframe (due to the larger crowns and longer cutting cycles). This results in the decomposition of larger amounts of litter and thus larger quantities of soil organic matter [193]. However, soil chemical properties can be affected by composition. Species with litter poor in nutrients (*e.g., Pinus sylvestris* or *Quercus robur*) may reduce nutrients availability through increasing soil acidity, whereas species with litter rich in nutrients may improve soil potential productivity (*e.g., Betula pendula* or *Carpinus betulus*). Mixed stands of species with litter rich and poor in nutrients tend to improve soil productive potential [194]. Moreover, increasing the cutting cycle length will enable the nutrients exports to be compensated by the inputs of nutrients in the soil through the decomposition of the deadwood and litter [21].

The maintenance of the forest residues in the stands is not always a viable option due to the increase in fire risk or to improve soil conditions for regeneration. In these cases, the residues should be pilled and burned. Yet, pile combustion has negative effects on the soil (compaction, temperature of soil surface, light reaching the soil); burning the piles increases soil temperature and has effects on its soil physical, chemical, and biological characteristics. The higher the fire severity, the higher the impacts on soil properties. The vegetation (trees, shrubs, and herbaceous plants) is also affected. However, while some species are tolerant to fire (the lower the intensity the higher the tolerance), others are more sensitive. In general, fires with high intensity tend to reduce the number of species and diversity. The effects are stronger in the centre of the pile than on its boundaries. The litter layer is destroyed. Large piles cause more damages than small piles, due to higher temperatures in the former. The recovery of the soil and vegetation is system dependent [195].

The forest residues harvesting has positive and negative impacts on the remaining stand and regeneration. Positive effects are due to the creation of better suited conditions for regeneration and seedling vigour. For example, the increase in temperature in cold climates, the decrease of root diseases by stump removal, and by the control of spontaneous vegetation [17, 105]. Likewise, the increase in temperature, minimisation of vapour pressure deficits, and reduction of the incidence of frosts improve the microclimatic conditions [181]. The negative impacts are related to the decrease in tree growth (3–7%) [17], the stronger the decrease of growth, the higher the intensity of forest residues removal [182, 183, 185, 187, 196, 197]. Fertilisation can mitigate the decrease in tree growth due to the increase in soil carbon and nitrogen stocks [183] and productivity [196].

Biomass is affected by hydrology. Biomass residues may have negative and positive effects on water quality [21]. The positive effects of biomass removal on water quality are the reduction of eutrophication due to the reduction of nitrogen leaching [21] and the negative effects are the runoff, nutrient leaching, and/or erosion and deposition of sediments on the water surfaces [21, 198, 199]. The risk increases with the increase in the number of machine entries in the stand, with slope, and for fragile soils when not using woody biomass to protect the roads [21].

The mitigation of the negative effects can be done by the maintenance of the biomass residues in the stand due to their mulch effect, thus keeping soil water, reducing temperature ranges, and consequently maintaining organic matter content [17, 200], not removing stumps because their decomposition is slow, incorporating nutrients in the soil and having also a protection effect in forest systems on steeper slopes and reducing the risk of deadwood and snow slides [198, 201], and/or fertilising the site [198]. The risk of runoff, nutrient leaching, and erosion increases from selection to clearcut systems due to the increase in biomass removed in one harvest from the former to the latter [200].

Forest systems' sustainability is also linked to diversity. This is related to the stand structure, silvicultural systems, and habitats for fauna and flora. In general, diversity increases from pure even-aged to mixed uneven-aged stands [153, 202]. This is related to the increase in the number of species as well as their proportions ([27] and references therein); tree dimensions' variability [28]; the arrangement of the individuals in the stands, both on the horizontal and vertical planes [203]; and resilience [7]. Additionally, longer cutting cycles tend to increase diversity [204].

The variability in stand structure results in different niches enabling the regeneration of different tree species, therefore maintaining or promoting diversity [205, 206].

During stand development, due to competition or disturbances (*e.g.,* silvicultural practices, storms, fires, pests, and diseases), some trees die, originating deadwood, either standing or downed. This deadwood provides shelter and food sources for flora and fauna [21, 207]. Moreover, old growth or unmanaged forests stands have high diversity and store large carbon stocks, both in the trees and in the soil [132, 208]. However, the removal of biomass residues can also have positive impacts on diversity, by promoting the decomposition of soil organic matter and hence increasing soil biological diversity [209].

Removal of biomass in harvest, regardless of its use, has, to a lesser or greater extent, effects on diversity. This is linked to the stand structure, silvicultural system, and intensity, frequency, and type of biomass removed. In general, diversity decreases from selection or shelterwood systems [79, 210] to clearcut systems [211]. This can be explained, at least partially, by the higher heterogeneity of the spatial [210] and temporal [198] patterns of harvest; and the maintenance of habitats, whether as individual trees or as clusters of trees, increasing the diversity of birds and small mammals [211]. Furthermore, it seems that there is a trend toward the decrease of diversity in the whole tree harvest when compared to the stem only harvest [212]. The difference between stem and whole tree harvest is the amount of residues that are maintained in the stand (larger in the stem than in whole tree harvest) that affect the amount of deadwood in the stand [198, 209]. In general, thinning effects on diversity tend to be neutral or positive due to their short lasting effects, but it depends on their method and intensity [21].

The removal of biomass in general and of residues, in particular, reduces deadwood in the stand [117, 133]. The deadwood supports a wide range of flora and fauna. On one hand, deadwood hosts a large number of insects [198], which are food sources for birds and mammals [198]. On the other hand, deadwood, standing or downed, scattered or pilled in the stand, provide shelter for several species [195, 198]. Therefore, the removal of biomass residues can reduce diversity, by reducing food sources and shelter for several species; by destroying tree regeneration; and reducing soil invertebrates and soil nutrients [195, 198].

The mitigation of the negative effects of biomass removal on diversity can be done by leaving some deadwood in the stands (the amount is dependent on the stand structure, species, and site) [198]; by retaining habitat trees (dead or alive) [207, 213]; and/or by fertilising [209].

Biomass can be affected by the increase in diversity. The maintenance of forest residues in the stand may increase pest populations that breed on dead wood, increasing the risk of attack to live trees, especially if live trees have low vigour. A measure to mitigate the pest population increase is either to remove the residues; or to make piles of residues, because the pests tend to colonise the pile's outer part and not the inner one, thus reducing the pest population [214].

5 Biomass Yields, Harvest and Exports

Biomass Yields

The quantification of the production and productivity of forest stands is traditionally calculated in volume. This is related to the main goal of many forest stands being timber products (*e.g.*, [166]). Only a few studies quantify yield in biomass (*e.g.*, [12]). Moreover, production and productivity per stand structure are uncommon in the literature, though some studies quantify yield per composition or per structure (*e.g.*, [14, 160]). Similarly, the yield in biomass in some agroforestry systems is not frequently calculated. This is related to the main goal in these systems being bark and/or fruit and not woody products [98]. Inversely, for energy plantations, there are many references to the mean annual increment in biomass (*e.g.*, [13]).

The biomass yields for the six stand structures already described (*cf*. Sect. 3) will be presented. The coppice pure even-aged stands were further divided into energy plantations and coppices of medium rotation. Energy plantations (*cf*. chapter "Energy Plantations") are coppice pure even-aged stands managed in very short rotation cycles (2–6 years), in which above ground biomass is all removed in harvest and used for energy purposes [55, 177, 215]. Coppices of medium rotation are coppice pure even-aged stands managed in short rotation cycles (6–15 years), in which all above ground biomass is removed in harvest and used for pulp and paper and energy purposes [216–218]. Thus, a total of seven stand structures are considered.

The biomass yields are presented as a range of values (Fig. 5) of the mean annual increment ($t{\cdot}ha^{-1}{\cdot}y^{-1}$). Several references were used that included productivity evaluation in biomass. However, others were used that estimated productivity in volume ($m^3{\cdot}ha^{-1}{\cdot}y^{-1}$) or carbon ($t{\cdot}ha^{-1}{\cdot}y^{-1}$). The volume mean annual increments were converted to biomass, considering that biomass was the product of volume times the wood density of the species, using the ICRAF wood density data base (ICRAF, 2021–http://db.worldagroforestry.org/). The mean annual increments in carbon were converted to biomass considering the wood density conversion factors of Wales et al. [219]. Other references had above ground biomass and age, and the mean annual increment in biomass was calculated as the ratio between above ground biomass and age [219]

Illustrative values of yield in mean annual increment per stand structure are presented in Fig. 5. In energy plantations vary between 1.0 and 66.0 $t{\cdot}ha^{-1}{\cdot}y^{-1}$ [13, 220–225], coppices pure even-aged of medium rotations among 0.6–55.9 $t{\cdot}ha^{-1}{\cdot}y^{-1}$ [216–218, 226], high forest mixed even-aged stands 0.8–14.5 $t{\cdot}ha^{-1}{\cdot}y^{-1}$ [136, 160, 227], high forest pure even-aged stands 1.0–10.0 $t{\cdot}ha^{-1}{\cdot}y^{-1}$ [12, 136, 166, 167, 228], high forest mixed uneven-aged stands 1.0–8.0 $t{\cdot}ha^{-1}{\cdot}y^{-1}$ [12, 14, 79, 173], high forest pure uneven-aged stands 0.4–3.1 $t{\cdot}ha^{-1}{\cdot}y^{-1}$ [14, 53, 166, 167, 228], and agroforestry systems between 0.5 and 2.6 $t{\cdot}ha^{-1}{\cdot}y^{-1}$ [219, 229].

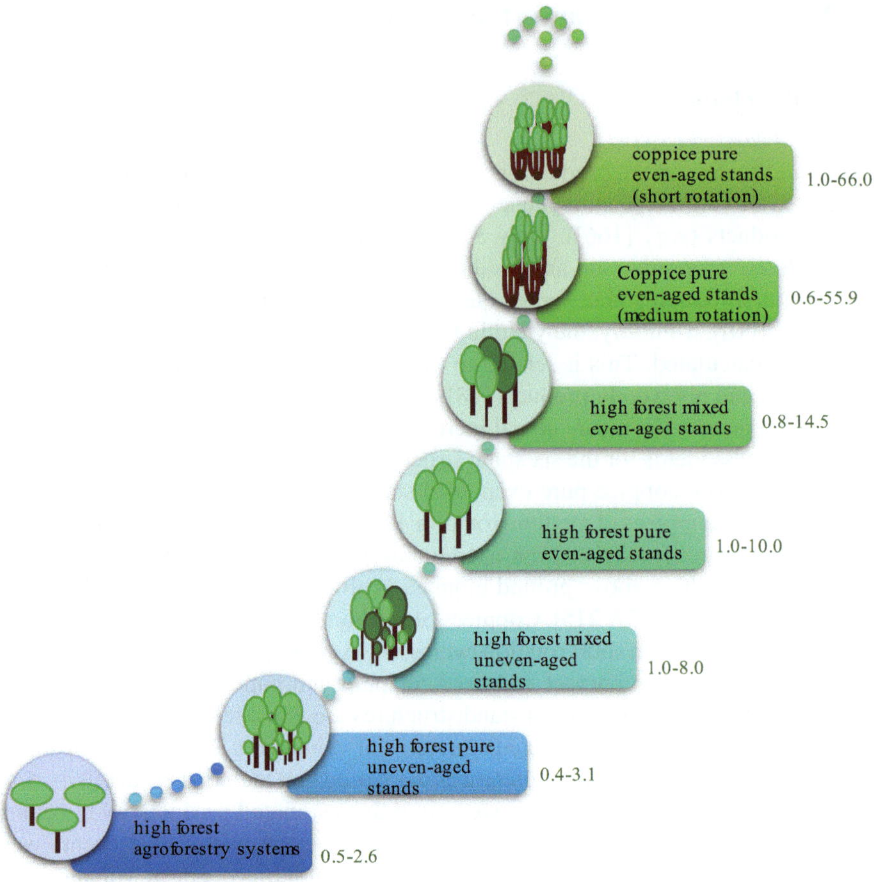

Fig. 5 Relation between stand structure and above ground biomass productivity ($t \cdot ha^{-1} \cdot y^{-1}$)

Harvest and Biomass Exports

The sustainability of the stands and forests aims to maintain constant biomass (and carbon) stocks, which are attained by the balance between storage and exports [176]. Traditionally, harvesting was centred on the timber of large, medium, and small dimensions [230]. Thus, corresponding to the export of, essentially, stem (or heartwood) and leaving forest residues in the stand. The heartwood proportion of non-active biomass is much larger than the active one. Therefore, the biomass (and carbon) stocks of the forest systems are maintained by the reallocation of biomass of the removed trees to the remaining trees and regeneration through growth and to soil organic matter through decomposition [176].

The potential of a forest stand to generate biomass for energy (Fig. 6) is directly related to the woody product's dimensions, quality, and market value [16, 231–233].

Timber has a higher market value than biomass for energy, hence, the latter generally corresponds to the part of the woody products that are not used for timber [20–22, 234]. Moreover, the market can also promote the use of whole trees for energy when there is a low demand for timberwood or pulpwood of certain species that result in a supply increase that is not absorbed by saw timber and pulp and paper industries [21]. Another source of biomass for energy is the salvage cuttings that are prescribed after disturbances such as fire, windthrow, snow storms, or pathogen outbreaks. These woody products are not suitable for timber products, and biomass for energy can be a way to reduce losses [235, 236]. One more source of biomass for energy is damaged trees during harvest. The damages in harvesting (felling and skidding) have a wide variability and are species dependent. Crown architecture and wood density are two main factors that influence the damages. Trees with wide crowns, thick and low flexibility, and/or wood of low specific gravity are more prone to damages during felling and skidding (*e.g., Populus tremuloides*) because their resistance to rupture is lower [237]. Furthermore, the increase in demand for forest woody products for energy may increase the biomass removal intensity whether by cuts and thinnings [21] or prunings [82, 238], which brings about the issue of stands' and forests' sustainability (cf. Sect. 3).

Biomass that can be used for energy (called biomass/forest residues, or sometimes by-products of harvesting) can be grouped into three broad classes (Fig. 6): (i) whole tree, corresponding to dead and disease trees (salvage cuttings), trees damaged by

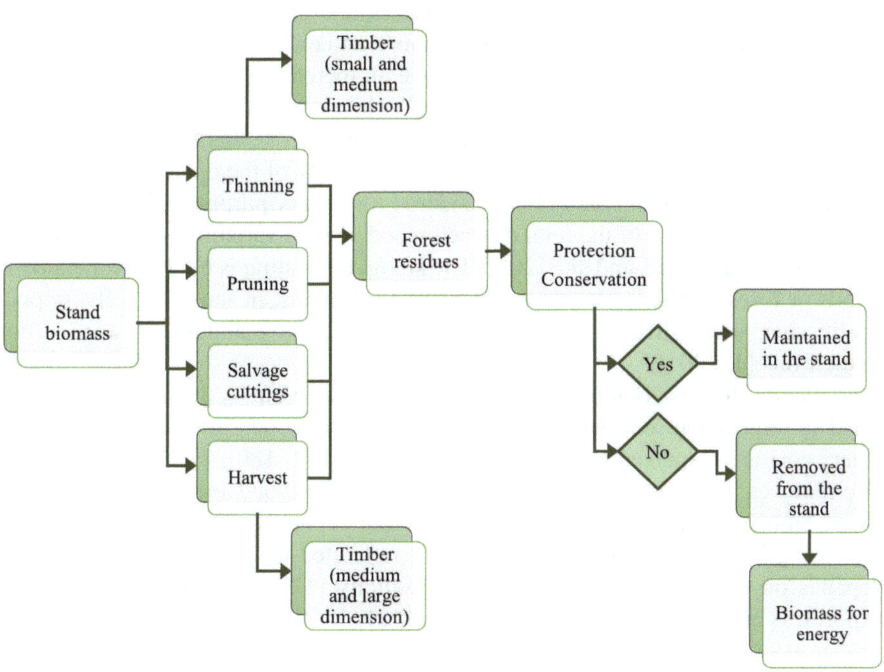

Fig. 6 Potential biomass for energy from silvicultural operations and harvests

natural disturbances, trees without quality for timber (including those removed in thinning), or trees from energy plantations [13, 16, 232, 239, 240]; (ii) trees of good quality timber damaged during harvest, which cannot be used for timber [237]; or (iii) forest residues that correspond to tops, branches, and stumps [16, 231–233]. Moreover, in terms of forest management, two options can be considered for the biomass residues (Fig. 6): (i) their maintenance in the stand, or ii) their removal. The first option is chosen when management is driven by protection and conservation objectives [105]. In this case, forest residues can be distributed across the stand or piled [16] to maintain or improve the soil's physical, chemical, and biological properties, as they have a strong mulch effect [105]. The hydrologic cycle, habitats, and diversity are also enhanced by the maintenance of forest residues in the stand [105, 241]. The second option is chosen when negative effects on the site, stand and yield sustainability, and diversity are not to fear [106, 107, 198].

Harvesting can be directed to timber or timber and forest residues. The latter has advantages as it integrates both removals of timber and forest residues, namely, reduction of costs, operational and machine productivity are higher, and there is a reduction of the number of entries in the forest stand ([19] and references therein). Studies report that most harvests are centred on timber, but forest residues are a by-product [19, 242]. A set of advantages were associated with biomass harvesting (including forest residues), namely increase of aesthetic values, reduction of the costs of site preparation when regenerating the stands, landowners' increased satisfaction, and contribution to bioenergy production. The disadvantages were related to the low forest residues' prices, instability of the market, and operational costs (increase of costs with acquisition, number of working hours, and maintenance of the equipment) [19]. Forest residues harvesting seem to have an increasing trend. The profitability of forest residues harvest is not consensual; while some authors report profitable harvests others do not ([19] and references therein). It seems that the longer the forest residues are harvested the higher the profitability of the operation [19].

Stump harvest can or cannot be considered for energy purposes. Stump biomass corresponds to 1–33% of the total downed wood, and its variability is dependent on species stand structure, and site [201]. The stump harvesting is not recommended in stands or forests under protection or conservation status, in sensitive soils, in poor quality sites, where diversity is to be maintained, and/or to maintain and store carbon stock levels [17, 201].

Typically, harvest is more concentrated in the clearcut systems, where all the stand is removed in one harvest, than in the shelterwood and selection systems, where cuts are dispersed by several interventions [132, 180]. The clearcut system also removes more biomass per harvest than the shelterwood and selection systems, where biomass removal is done in more than four cuts [132, 174]. Thinnings export increasing biomass amounts in time, which is related to tree dimensions but also depends on the method, intensity, and periodicity [127, 243]. Stem only harvest export less biomass than whole tree harvest [181, 185].

Biomass Residues Collection

Several factors influence the recovery rate of biomass residues (*i.e.*, the amount of residues that is possible to remove from the stands), namely stand structure, function (production, protection), topography, site, harvesting equipment, logistics, and costs [237, 244]. In general, pure even-aged stands (in particular plantations) tend to have higher recovery rates [200, 245] than uneven-aged stands [200]. While clearcut can have higher [200] or lower [245] recovery rates than selection or shelterwood systems. Production forest systems tend to have higher recovery rates than protection and conservation forest systems [244]. The site can affect the recovery rate, as it can influence the damages to the felled trees. For example, rocky soils or irregular topography increase the proportion of stems that are damaged in cuttings. Similarly, the cutting season also influences damages. For example, cuttings during seasons with low temperatures and/or snow tend to make trees more susceptible to damages [237].

Several references are found to the percentage of available forest residues. One study estimated the mean percentage of residues per biome and forest system [200] indicating the following values: for the boreal biome of 69% for clearcut and 78% for plantations; for the temperate biome of conifers, broadleaved species, and mixed stands of 53% for clearcut and 63% for plantations; and for savannah and tropical forests of 39% for clearcut, 18% for selection systems, and 52% for plantations. Others indicated that for *Eucalyptus* spp. plantations 25–30% of residues [239, 246]. Thinning forest residues were estimated for pure even-aged stands varying between 29 and 46% of above ground biomass, increasing with ageing. The first thinning with 29% and the fourth with 46%, whereas for mixed stands were 60% [243].

In a meta-analysis, the average values of recovery rates per biomass component were: 100% for stem; 20–80% for bark (80% for chain saw and 20% for logging machines that detach substantial quantities of bark), 60% for stump and roots, 50–60% for branches (50% for conifers species and 60% for broadleaved species), and 0–40% for leaves (0% for leafless individuals, 0% with delay of biomass residues collection for broadleaved species and 10% for conifer species, 0% when harvest is done in autumn or winter, and 40% when the aforementioned conditions are not satisfied) ([17] and references therein).

Another study evaluated the recovery rate, at the stand level, considering stand origin (natural *vs* plantation), type of species and composition (conifers, broadleaved and mixed stands), silvicultural system (clearcutting *vs* partial cutting, including shelterwood and selection systems), and harvesting equipment [245]. The average recovery rate was 52.2%, the median 54.1%, the minimum 2%, and the maximum 89.1%. The recovery rate of natural stands was 48.8% and of plantations 58.2%. Broadleaved, conifers and mixed stands had 48.1%, 57.1% and 47.0% recovery rates, respectively. The recovery rate of clearcut was 52.0% and that of the partial harvest 54.2%. The recovery rates of plantations under clearcut were 50.9% for broadleaved species and 60.0% for conifers. Recovery rates for natural stands are presented in

Table 1 Recovery rate (%) for stands of natural origin per type of species and type of harvest

Type of species/composition	Clearcut	Partial harvest	Average
Broadleaved species	46.0	50.7	47.2
Mixed stands	45.9	53.9	47.0
Conifers	51.3	68.8	52.6
Average	47.7	54.2	48.8

Table 1. The recovery rate per harvest equipment was for stem only with residues scattered in the cutting block ≈35.6%, and whole tree harvest with residues stored near the roads in the summer 48.1% and in the autumn, winter and spring 60.7%. In general, there was a trend toward higher recovery rates for conifers than broadleaved species due to the uniformity of the stands in species, size, and spacing [245]. Other empirical studies report recovery rates between 0 and 80% for several conifer and broadleaved species [22], 13–17% for *Eucalyptus* spp. stands [239, 247], 10% for *Eucalyptus* spp. plantations [239, 248], and via simulation approach, indicated recovery rates between 20.8 and 91.7% at stand level and 59.1–64.1% at landscape level [244].

6 Final Considerations

Biomass storage and dynamics in space and time are dependent on stand structure (*cf*. Sect. 2), silvicultural systems (*cf*. Sect. 2), silvicultural practices (*cf*. Sect. 2), and biomass partitioning (*cf*. Sect. 3). In general, above ground biomass storage tends to be higher in pure even-aged stands, whereas total tree biomass and carbon storage are higher in mixed and uneven-aged stands [132, 159, 160]. In clearcutting systems, biomass storage increases in time up to the end of the cutting cycle when all the trees are removed. There is an interruption in biomass storage, that restarts with the regeneration of the stand [74]. Inversely, selection and irregular shelterwood systems tend towards constant biomass storage [14]. Moreover, deadwood is an important biomass sink that, through decomposition, enables the maintenance or improvement of soil productive potential and diversity (*e.g.*, [15]).

Overall, harvest removes, to a smaller or larger extent, biomass from the forest system. Its intensity, frequency, and quality (total or per component) influence site quality, due to the impacts on soil's physical, chemical, and biological characteristics (*e.g.*, [17]), hydrology (*e.g.*, [21]) and diversity (*e.g.*, [198]). Sustainable forest management guidelines were defined for harvesting the biomass, including stem and residues, for North America, Europe and Japan. Thirty-two harvest guidelines were defined to promote the sustainability of the forest systems while providing timber and forest residues. Criteria and indicators were defined that promoted biodiversity, soil, and water protection, as well as compensatory actions to mitigate the negative effect of harvesting on the forest system's sustainability while keeping harvesting economical return [249].

Biomass yields, harvest, and exports tend to be higher in energy plantations than in agroforestry systems (*cf*. Sect. 5). Forest biomass removal for energy (forest residues) is considered when stands are not under protection or conservation status. The amount of available biomass for bioenergy is dependent on stand structure, silvicultural system [200], and its collection on the recovery rate [245].

Though many studies have been published there is still the need for further research. Many studies when estimating biomass define the composition, and less commonly structure and regime, but seldom characterise stand structure, which can derive in under or over estimates of biomass. These stresses the need to further investigate the effects of stand structure on biomass and their spatial and temporal variability, and also the availability of forest residues. The estimation of forest residues and their recovery rate also needs further studies that enable to discriminate the impact of stand structure, silvicultural system, and harvest type on the quantity of residues available for energy.

Funding This work is funded by National Funds through FCT - Foundation for Science and Technology under the Project UIDB/05183/2020.

References

1. Pan Y, Birdsey RA, Fang J et al (2011) A large and persistent carbon sink in the world's forests. Science 333:988–993. https://doi.org/10.1126/science.1201609
2. Urbano AR, Keeton WS (2017) Carbon dynamics and structural development in recovering secondary forests of the northeastern U.S. For Ecol Manage 392:21–35. https://doi.org/10.1016/j.foreco.2017.02.037
3. Cardinale BJ, Wright JP, Cadotte MW et al (2007) Impacts of plant diversity on biomass production increase through time because of species complementarity. Proc Natl Acad Sci 104:18123–18128. https://doi.org/10.1073/pnas.0709069104
4. Chai RK, Andrus RA, Rodman K et al (2019) Stand dynamics and topographic setting influence changes in live tree biomass over a 34-year permanent plot record in a subalpine forest in the Colorado Front Range. Can J For Res 49:1256–1264. https://doi.org/10.1139/cjfr-2019-0023
5. Martin M, Fenton N, Morin H (2018) Structural diversity and dynamics of boreal old-growth forests case study in Eastern Canada. For Ecol Manage 422:125–136. https://doi.org/10.1016/j.foreco.2018.04.007
6. Meigs GW, Keeton WS (2018) Intermediate-severity wind disturbance in mature temperate forests: legacy structure, carbon storage, and stand dynamics. Ecol Appl 28:798–815. https://doi.org/10.1002/eap.1691
7. Morin X, Fahse L, Jactel H et al (2018) Long-term response of forest productivity to climate change is mostly driven by change in tree species composition. Sci Rep 8:5627. https://doi.org/10.1038/s41598-018-23763-y
8. Zeller L, Pretzsch H (2019) Effect of forest structure on stand productivity in Central European forests depends on developmental stage and tree species diversity. For Ecol Manage 434:193–204. https://doi.org/10.1016/j.foreco.2018.12.024
9. Albaugh TJ, Albaugh JM, Fox TR et al (2016) Tamm review: light use efficiency and carbon storage in nutrient and water experiments on major forest plantation species. For Ecol Manage 376:333–342. https://doi.org/10.1016/j.foreco.2016.05.031

10. Mensah S, Veldtman R, Assogbadjo AE et al (2016) Tree species diversity promotes aboveground carbon storage through functional diversity and functional dominance. Ecol Evol 6:7546–7557. https://doi.org/10.1002/ece3.2525
11. Xiang W, Li L, Ouyang S et al (2021) Effects of stand age on tree biomass partitioning and allometric equations in Chinese fir (Cunninghamia lanceolata) plantations. Eur J Forest Res 140:317–332. https://doi.org/10.1007/s10342-020-01333-0
12. Bohn FJ, Huth A (2017) The importance of forest structure to biodiversity–productivity relationships. R Soc Open Sci 4:160521. https://doi.org/10.1098/rsos.160521
13. Djomo SN, Ac A, Zenone T et al (2015) Energy performances of intensive and extensive short rotation cropping systems for woody biomass production in the EU. Renew Sustain Energy Rev 41:845–854. https://doi.org/10.1016/j.rser.2014.08.058
14. Lundqvist L (2017) Tamm Review: Selection system reduces long-term volume growth in Fennoscandic uneven-aged Norway spruce forests. For Ecol Manage 391:362–375. https://doi.org/10.1016/j.foreco.2017.02.011
15. Shannon VL, Vanguelova EI, Morison JIL et al (2022) The contribution of deadwood to soil carbon dynamics in contrasting temperate forest ecosystems. Eur J Forest Res 141:241–252. https://doi.org/10.1007/s10342-021-01435-3
16. White EM (2010) Woody biomass for bioenergy and biofuels in the United States: a briefing paper. USDA Forest Service, Portland, Oregon, USA
17. Achat DL, Deleuze C, Landmann G et al (2015) Quantifying consequences of removing harvesting residues on forest soils and tree growth—a meta-analysis. For Ecol Manage 348:124–141. https://doi.org/10.1016/j.foreco.2015.03.042
18. Berger AL, Palik B, D'Amato AW et al (2013) Ecological impacts of energy-wood harvests: lessons from whole-tree harvesting and natural disturbance. J Forest 111:139–153. https://doi.org/10.5849/jof.12-020
19. Garren AM, Bolding MC, Barrett SM et al (2022) Characteristics of forest biomass harvesting operations and markets in Virginia. Biomass Bioenerg 163:106501. https://doi.org/10.1016/j.biombioe.2022.106501
20. Castellano PJ, Volk TA, Herrington LP (2009) Estimates of technically available woody biomass feedstock from natural forests and willow biomass crops for two locations in New York State. Biomass Bioenerg 33:393–406. https://doi.org/10.1016/j.biombioe.2008.08.001
21. Egnell G, Paré D, Thiffault E, Lamers P (2016) Environmental sustainability aspects of forest biomass mobilisation. In: Thiffault E, Berndes G, Junginger M et al (eds) Mobilisation of forest bioenergy in the boreal and temperate biomes. Elsevier, London, pp 50–67
22. Martire S, Castellani V, Sala S (2015) Carrying capacity assessment of forest resources: enhancing environmental sustainability in energy production at local scale. Resour Conserv Recycl 94:11–20. https://doi.org/10.1016/j.resconrec.2014.11.002
23. Mola-Yudego B, Arevalo J, Díaz-Yáñez O et al (2017) Wood biomass potentials for energy in northern Europe: forest or plantations? Biomass Bioenerg 106:95–103. https://doi.org/10.1016/j.biombioe.2017.08.021
24. Assmann E (1970) The principles of forest yield study. Pergamon Press, Oxford, UK
25. Smith DM, Larson BC, Kelty MJ, Ashton PMS (1997) The practice of silviculture. applied forest ecology, 9th edn. John Wiley & Sons, Inc, New York
26. Oliver CD, Larson BC (1996) Forest stand dynamics, Update. John Wiley & Sons Inc., New York
27. Gonçalves AC (2017) Multi-species stand classification: definition and perspectives. In: Chakravarty S, Shukla G (eds) Forest ecology and conservation. InTech, pp 3–23
28. O'Hara KL (2014) Multiaged silviculture managing for complex forest stand structures. Oxford University Press, Oxford
29. Forrester DI (2019) Linking forest growth with stand structure: tree size inequality, tree growth or resource partitioning and the asymmetry of competition. For Ecol Manage 447:139–157. https://doi.org/10.1016/j.foreco.2019.05.053
30. Pretzsch H, Zenner EK (2017) Toward managing mixed-species stands: from parametrization to prescription. Forest Ecosyst 4 https://doi.org/10.1186/s40663-017-0105-z

31. Sun H, Diao S, Liu R et al (2018) Relationship between size inequality and stand productivity is modified by self-thinning, age, site and planting density in Sassafras tzumu plantations in central China. For Ecol Manage 422:199–206. https://doi.org/10.1016/j.foreco.2018.02.003
32. Schall P, Gossner MM, Heinrichs S et al (2018) The impact of even-aged and uneven-aged forest management on regional biodiversity of multiple taxa in European beech forests. J Appl Ecol 55:267–278. https://doi.org/10.1111/1365-2664.12950
33. Toïgo M, Perot T, Courbaud B et al (2018) Difference in shade tolerance drives the mixture effect on oak productivity. J Ecol 106:1073–1082. https://doi.org/10.1111/1365-2745.12811
34. Vanhellemont M, Bijlsma R-J, De Keersmaeker L et al (2018) Species and structural diversity affect growth of oak, but not pine, in uneven-aged mature forests. Basic Appl Ecol 27:41–50. https://doi.org/10.1016/j.baae.2018.01.003
35. Fichtner A, Härdtle W, Li Y et al (2017) From competition to facilitation: how tree species respond to neighbourhood diversity. Ecol Lett 20:892–900. https://doi.org/10.1111/ele.12786
36. Grossman JJ, Cavender-Bares J, Hobbie SE et al (2017) Species richness and traits predict overyielding in stem growth in an early-successional tree diversity experiment. Ecology 98:2601–2614. https://doi.org/10.1002/ecy.1958
37. Van de Peer T, Verheyen K, Kint V et al (2017) Plasticity of tree architecture through inter-specific and intraspecific competition in a young experimental plantation. For Ecol Manage 385:1–9. https://doi.org/10.1016/j.foreco.2016.11.015
38. Binkley D, Campoe OC, Gspaltl M, Forrester DI (2013) Light absorption and use efficiency in forests: why patterns differ for trees and stands. For Ecol Manage 288:5–13. https://doi.org/10.1016/j.foreco.2011.11.002
39. Forrester DI (2014) The spatial and temporal dynamics of species interactions in mixed-species forests: from pattern to process. For Ecol Manage 312:282–292. https://doi.org/10.1016/j.foreco.2013.10.003
40. Looney CE, D'Amato AW, Fraver S et al (2016) Examining the influences of tree-to-tree competition and climate on size-growth relationships in hydric, multi-aged *Fraxinus nigra* stands. For Ecol Manage 375:238–248. https://doi.org/10.1016/j.foreco.2016.05.050
41. Ehbrecht M, Schall P, Ammer C et al (2019) Effects of structural heterogeneity on the diurnal temperature range in temperate forest ecosystems. For Ecol Manage 432:860–867. https://doi.org/10.1016/j.foreco.2018.10.008
42. Forrester DI, Ammer C, Annighöfer PJ et al (2018) Effects of crown architecture and stand structure on light absorption in mixed and monospecific *Fagus sylvatica* and *Pinus sylvestris* forests along a productivity and climate gradient through Europe. J Ecol 106:746–760. https://doi.org/10.1111/1365-2745.12803
43. Thurm EA, Biber P, Pretzsch H (2017) Stem growth is favored at expenses of root growth in mixed stands and humid conditions for Douglas-fir (*Pseudotsuga menziesii*) and European beech (*Fagus sylvatica*). Trees 31:349–365. https://doi.org/10.1007/s00468-016-1512-4
44. Forrester DI, Bauhus J (2016) A review of processes behind diversity—productivity relationships in forests. Curr Forestry Rep 2:45–61. https://doi.org/10.1007/s40725-016-0031-2
45. Thom D, Keeton WS (2019) Stand structure drives disparities in carbon storage in northern hardwood-conifer forests. For Ecol Manage 442:10–20. https://doi.org/10.1016/j.foreco.2019.03.053
46. Fraver S, D'Amato AW, Bradford JB et al (2014) Tree growth and competition in an old-growth *Picea abies* forest of boreal Sweden: influence of tree spatial patterning. J Veg Sci 25:374–385. https://doi.org/10.1111/jvs.12096
47. Larson AJ, Lutz JA, Donato DC et al (2015) Spatial aspects of tree mortality strongly differ between young and old-growth forests. Ecology 96:2855–2861. https://doi.org/10.1890/15-0628.1
48. Looney CE, D'Amato AW, Palik BJ et al (2018) Size-growth relationship, tree spatial patterns, and tree-tree competition influence tree growth and stand complexity in a 160-year red pine chronosequence. For Ecol Manage 424:85–94. https://doi.org/10.1016/j.foreco.2018.04.044

49. De Boeck HJ, Nijs I, Lemmens CMHM, Ceulemans R (2006) Underlying effects of spatial aggregation (clumping) in relationships between plant diversity and resource uptake. Oikos 113:269–278. https://doi.org/10.1111/j.2006.0030-1299.14257.x
50. Luu TC, Binkley D, Stape JL (2013) Neighborhood uniformity increases growth of individual Eucalyptus trees. For Ecol Manage 289:90–97. https://doi.org/10.1016/j.foreco.2012.09.033
51. Matthews JD (1989) Silvicultural systems. Claredon Press, Oxford
52. Boudru M (1989) Forêt et sylviculture. Presses Agronomiques de Gembloux, Gembloux, Le Traitement des Forêts
53. Schütz JPH (1997) Sylviculture 2. Presses Polytechniques et Universitaires Romandes, Lausanne La Gestion des Forêts Irrégulières et Mélangées
54. Zhang H, Zhou G, Wang Y et al (2019) Thinning and species mixing in Chinese fir monocultures improve carbon sequestration in subtropical China. Eur J Forest Res 138:433–443. https://doi.org/10.1007/s10342-019-01181-7
55. Dickmann D (2006) Silviculture and biology of short-rotation woody crops in temperate regions: then and now. Biomass Bioenerg 30:696–705. https://doi.org/10.1016/j.biombioe.2005.02.008
56. Ciancio O (2010) La teoria della selvicoltura sistemica, le "sterili disquisizioni" e il sonnambulismo dell'intellighenzia forestale. Accademia Italiana di Scienze Forestali
57. Pretzsch H (2009) Forest dynamics, growth, and yield. Springer, Berlin
58. Schütz J (1999) Close-to-nature silviculture: is this concept compatible with species diversity? Forestry 72:359–366. https://doi.org/10.1093/forestry/72.4.359
59. Pukkala T (2016) Which type of forest management provides most ecosystem services? Forest Ecosyst 3:9. https://doi.org/10.1186/s40663-016-0068-5
60. Puettmann KJ, Coates KD, Messier CC (2009) A critique of silviculture: managing for complexity. Island Press, Washington, DC
61. O'Hara KL (2001) The silviculture of transformation–a commentary. For Ecol Manage 151:81–86. https://doi.org/10.1016/S0378-1127(00)00698-8
62. Schütz J-P (2002) Silvicultural tools to develop irregular and diverse forest structures. Forestry 75:329–337. https://doi.org/10.1093/forestry/75.4.329
63. Pommerening A, Murphy ST (2004) A review of the history, definitions and methods of continuous cover forestry with special attention to afforestation and restocking. Forestry 77:27–44. https://doi.org/10.1093/forestry/77.1.27
64. Ciancio O, Nocentini S (2011) Biodiversity conservation and systemic silviculture: concepts and applications. Plant Biosyst 145:411–418. https://doi.org/10.1080/11263504.2011.558705
65. Diaci J, Kerr G, O'Hara KL (2011) Twenty-first century forestry: integrating ecologically based, uneven-aged silviculture with increased demands on forests. Forestry 84:463–465. https://doi.org/10.1093/forestry/cpr053
66. Moravcik M, Sarvaov Z, Mergani J, Kovalk M (2012) Close to nature management in high-mountain forests of norway spruce vegetation zone in Slovakia. In: Blanco JA, Lo Y-H (eds) Forest ecosystems–more than just trees. InTech, Rijeka, Croatia, pp 375–414
67. Schütz JPH (2012) Historical emergence and current application of CCF. In: Pukkala T, Gadow KV (eds) Continuous cover forestry. Springer Science+Business Media B.V., pp 1–28
68. Bauhus J, Puettmann KJ, Kühne C (2013) Close-to-nature forest management in Europe. In: Messier C, Puettmann KJ, Coates KD (eds) Managing forests as complex adaptive systems: building resilience to the challenge of global change. Routledge, Washington
69. Bravo-Oviedo A, Pretzsch H, Ammer C et al (2014) European mixed forests: definition and research perspectives. Forest Syst 23:518. https://doi.org/10.5424/fs/2014233-06256
70. Grossman JJ, Vanhellemont M, Barsoum N et al (2018) Synthesis and future research directions linking tree diversity to growth, survival, and damage in a global network of tree diversity experiments. Environ Exp Bot 152:68–89. https://doi.org/10.1016/j.envexpbot.2017.12.015
71. Bu W, Schmid B, Liu X et al (2017) Interspecific and intraspecific variation in specific root length drives aboveground biodiversity effects in young experimental forest stands. J Plant Ecol 10:158–169. https://doi.org/10.1093/jpe/rtw096

72. Collet C, Ningre F, Barbeito I et al (2014) Response of tree growth and species coexistence to density and species evenness in a young forest plantation with two competing species. Ann Bot 113:711–719. https://doi.org/10.1093/aob/mct285
73. Zhang Y, Chen HYH, Reich PB (2012) Forest productivity increases with evenness, species richness and trait variation: a global meta-analysis: diversity and productivity relationships. J Ecol 100:742–749. https://doi.org/10.1111/j.1365-2745.2011.01944.x
74. Williams NG, Powers MD (2019) Carbon storage implications of active management in mature *Pseudotsuga menziesii* forests of western Oregon. For Ecol Manage 432:761–775. https://doi.org/10.1016/j.foreco.2018.10.002
75. Gonçalves AC (2021) Thinning: an overview. In: Gonçalves AC (ed) Silviculture. IntechOpen, London, pp 41–58
76. O'Hara KL, Nagel LM (2013) The stand: revisiting a central concept in forestry. J Forest 111:335–340. https://doi.org/10.5849/jof.12-114
77. Soares AAV, Leite HG, Souza AL et al (2016) Increasing stand structural heterogeneity reduces productivity in Brazilian Eucalyptus monoclonal stands. For Ecol Manage 373:26–32. https://doi.org/10.1016/j.foreco.2016.04.035
78. Hardiman BS, Bohrer G, Gough CM et al (2011) The role of canopy structural complexity in wood net primary production of a maturing northern deciduous forest. Ecology 92:1818–1827. https://doi.org/10.1890/10-2192.1
79. Geng Y, Yue Q, Zhang C et al (2021) Dynamics and drivers of aboveground biomass accumulation during recovery from selective harvesting in an uneven-aged forest. Eur J Forest Res 140:1163–1178. https://doi.org/10.1007/s10342-021-01394-9
80. Bugalho MN, Dias FS, Briñas B, Cerdeira JO (2016) Using the high conservation value forest concept and Pareto optimization to identify areas maximizing biodiversity and ecosystem services in cork oak landscapes. Agrofor Syst 90:35–44. https://doi.org/10.1007/s10457-015-9814-x
81. Ferraz de Oliveira MI, Lamy E, Bugalho MN et al (2013) Assessing foraging strategies of herbivores in Mediterranean oak woodlands: a review of key issues and selected methodologies. Agrofor Syst 87:1421–1437. https://doi.org/10.1007/s10457-013-9648-3
82. Natividade JV (1950) Subericultura, 2nd edn. Ministério da Agricultura, Pescas e Alimentação. Direção Geral das Florestas, Lisboa
83. Plieninger T, Hartel T, Martín-López B et al (2015) Wood-pastures of Europe: geographic coverage, social–ecological values, conservation management, and policy implications. Biol Cons 190:70–79
84. Thirgood JV (1981) Man and the mediterranean forest: a history of resource depletion. Academic Press, London
85. Correia AV, Oliveira AC (1999) Principais espécies florestais com interesse para Portugal: zonas de influência mediterrânica. Direcção-Geral das Florestas, Lisboa
86. Montero G, Montoya JM (1988) Considerações sobre a densidade e produção dos povoamentos de sobro. Boletim do Instituto dos Produtos Florestais—Cortiça 600:255–259
87. Gea-Izquierdo G, Cañellas I, Montero G (2006) Acorn production in Spanish holm oak woodlands. Invest Agra Sist Recursos Forest 15:339–354. https://doi.org/10.5424/srf/2006153-00976
88. Maltez-Mouro S, García LV, Freitas H (2009) Influence of forest structure and environmental variables on recruit survival and performance of two Mediterranean tree species (Quercus faginea L. and Q. suber Lam.). Eur J Forest Res 128:27–36. https://doi.org/10.1007/s10342-008-0236-4
89. Plieninger T, Rolo V, Moreno G (2010) Large-scale patterns of Quercus ilex, Quercus suber, and Quercus pyrenaica regeneration in Central-Western Spain. Ecosystems 13:644–660. https://doi.org/10.1007/s10021-010-9345-2
90. Rego FC, Monteiro ML, Geraldes A, Mesquita J (2011) Silvicultura de Povoamentos Mistos de Quercus pyrenaica e Quercus rotundifolia: O Caso da Tapada da Nogueira. Silva Lusitana 19:135–148

91. Gonçalves AC (2022) Models of silviculture for Portuguese species: old and new guidelines. Silva Lusitana 30:17–40. https://doi.org/10.1051/silu/20223001017
92. Alejano R, Tapias R, Fernández M et al (2008) Influence of pruning and the climatic conditions on acorn production in holm oak (*Quercus ilex* L.) dehesas in SW Spain. Ann For Sci 65:209–209. https://doi.org/10.1051/forest:2007092
93. Costa R, Evaristo I, Batista D, et al (2008) Condução de Povoamentos de Pinheiro Manso e Características Nutricionais do Pinhão. Instituto Nacional dos Recursos Biológicos, Oeiras
94. Martín D, Vázquez-Piqué J, Alejano R (2015) Effect of pruning and soil treatments on stem growth of holm oak in open woodland forests. Agrofor Syst 89:599–609. https://doi.org/10.1007/s10457-015-9794-x
95. Moreno G, Aviron S, Berg S et al (2018) Agroforestry systems of high nature and cultural value in Europe: provision of commercial goods and other ecosystem services. Agrofor Syst 92:877–891. https://doi.org/10.1007/s10457-017-0126-1
96. Reisner Y, de Filippi R, Herzog F, Palma J (2007) Target regions for silvoarable agroforestry in Europe. Ecol Eng 29:401–418. https://doi.org/10.1016/j.ecoleng.2006.09.020
97. Smith JR (1929) Tree crops. A permanent agriculture. Harcourt, Brace and Company, New York
98. Gonçalves AC (2018) Effects of forest stand structure in biomass and carbon. In: Shukla G, Chakravarty S (eds) Forest biomass and carbon. InTech, Rijeka, Croatia, pp 1–21
99. Florence RG (1996) Ecology and silviculture of eucalyptus forests. Csiro Publishing, Collingwood, Victoria
100. Lanier L, Badré M, Delabraze P, et al (1986) Précis de Sylviculture. ENGREF, Nancy
101. Nyland RD (2016) Silviculture. Concepts and applications, 3rd edn. Waveland Press, Inc., Long Grove
102. Schütz JP (1990) Sylviculture 1. Presses Polytechniques et Universitaires Romandes, Lausanne, Principes d'Éducation des Forêts
103. D'Amato AW, Bradford JB, Fraver S, Palik BJ (2013) Effects of thinning on drought vulnerability and climate response in north temperate forest ecosystems. Ecol Appl 23:1735–1742. https://doi.org/10.1890/13-0677.1
104. D'Amato AW, Bradford JB, Fraver S, Palik BJ (2011) Forest management for mitigation and adaptation to climate change: insights from long-term silviculture experiments. For Ecol Manage 262:803–816. https://doi.org/10.1016/j.foreco.2011.05.014
105. Egnell G (2017) A review of Nordic trials studying effects of biomass harvest intensity on subsequent forest production. For Ecol Manage 383:27–36. https://doi.org/10.1016/j.foreco.2016.09.019
106. Belleau A, Brais S, Paré D (2006) Soil nutrient dynamics after harvesting and slash treatments in boreal aspen stands. Soil Sci Soc Am J 70:1189. https://doi.org/10.2136/sssaj2005.0186
107. Ford SE, Keeton WS (2017) Enhanced carbon storage through management for old-growth characteristics in northern hardwood-conifer forests. Ecosphere 8:e01721. https://doi.org/10.1002/ecs2.1721
108. Gonçalves AC, Fonseca TF (2023) Influence management and disturbances on the regeneration of forest stands. Front Forests Glob Change 6:1123215. https://doi.org/10.3389/ffgc.2023.1123215
109. Huss J (2004) Afforestation | Stand establishment, treatment and promotion—European experience. In: Encyclopedia of Forest Sciences. Elsevier, pp 14–27
110. Raymond P, Bedard S, Roy V et al (2009) The irregular shelterwood system: review, classification, and potential application to forests affected by partial disturbances. J Forest 107:405–413. https://doi.org/10.1093/jof/107.8.405
111. Hyppönen M, Alenius V, Valkonen S (2005) Models for the establishment and height development of naturally regenerated *Pinus sylvestris* in Finnish Lapland. Scand J For Res 20:347–357. https://doi.org/10.1080/02827580510036391
112. Valkonen S, Ruuska J, Siipilehto J (2002) Effect of retained trees on the development of young Scots pine stands in Southern Finland. For Ecol Manage 166:227–243. https://doi.org/10.1016/S0378-1127(01)00668-5

113. Miina J, Saksa T (2008) Predicting establishment of tree seedlings for evaluating methods of regeneration for *Pinus sylvestris*. Scand J For Res 23:12–27. https://doi.org/10.1080/02827580701779595
114. Béland M, Agestam E, Ekö PM et al (2000) Scarification and seedfall affects natural regeneration of scots pine under two shelterwood densities and a clear-cut in Southern Sweden. Scand J For Res 15:247–255. https://doi.org/10.1080/028275800750015064
115. Dovčiak M, Reich PB, Frelich LE (2003) Seed rain, safe sites, competing vegetation, and soil resources spatially structure white pine regeneration and recruitment. Can J For Res 33:1892–1904. https://doi.org/10.1139/x03-115
116. Crotteau JS, Rue-Johns AZ, Barnard JC (2020) Effects on understory biomass and forage 8–10 years after precommercial thinning of Sitka spruce—western hemlock stands in southeast Alaska. Can J For Res 50:215–225. https://doi.org/10.1139/cjfr-2019-0268
117. Powers MD, Kolka RK, Bradford JB et al (2012) Carbon stocks across a chronosequence of thinned and unmanaged red pine (*Pinus resinosa*) stands. Ecol Appl 22:1297–1307. https://doi.org/10.1890/11-0411.1
118. Boncina A (2000) Comparison of structure and biodiversity in the Rajhenav virgin forest remnant and managed forest in the Dinaric region of Slovenia. Glob Ecol Biogeogr 9:201–211. https://doi.org/10.1046/j.1365-2699.2000.00155.x
119. Gradel A, Ammer C, Ganbaatar B, et al (2017) On the effect of thinning on tree growth and stand structure of White Birch (*Betula platyphylla* Sukaczev) and Siberian Larch (*Larix sibirica* Ledeb.) in Mongolia. Forests 8:105. https://doi.org/10.3390/f8040105
120. Jimenez E, Vega JA, Fernandez C, Fonturbel T (2011) Is pre-commercial thinning compatible with carbon sequestration? A case study in a maritime pine stand in northwestern Spain. Forestry 84:149–157. https://doi.org/10.1093/forestry/cpr002
121. Skovsgaard JP, Stupak I, Vesterdal L (2006) Distribution of biomass and carbon in even-aged stands of Norway spruce (*Picea abies* (L.) Karst.): a case study on spacing and thinning effects in northern Denmark. Scand J For Res 21:470–488. https://doi.org/10.1080/02827580601056268
122. Finkral AJ, Evans AM (2008) The effects of a thinning treatment on carbon stocks in a northern Arizona ponderosa pine forest. For Ecol Manage 255:2743–2750. https://doi.org/10.1016/j.foreco.2008.01.041
123. Primicia I, Artázcoz R, Imbert J-B et al (2016) Influence of thinning intensity and canopy type on Scots pine stand and growth dynamics in a mixed managed forest. Forest Syst 25:e057. https://doi.org/10.5424/fs/2016252-07317
124. Juodvalkis A, Kairiukstis L, Vasiliauskas R (2005) Effects of thinning on growth of six tree species in north-temperate forests of Lithuania. Eur J Forest Res 124:187–192. https://doi.org/10.1007/s10342-005-0070-x
125. Mäkinen H, Isomäki A (2004) Thinning intensity and long-term changes in increment and stem form of Scots pine trees. For Ecol Manage 203:21–34. https://doi.org/10.1016/j.foreco.2004.07.028
126. Peltola H, Kilpeläinen A, Sauvala K, et al (2007) Effects of early thinning regime and tree status on the radial growth and wood density of Scots pine. Silva Fennica 41:489–505. https://doi.org/10.14214/sf.285
127. Trouvé R, Bontemps J-D, Collet C et al (2019) When do dendrometric rules fail? Insights from 20 years of experimental thinnings on sessile oak in the GIS Coop network. For Ecol Manage 433:276–286. https://doi.org/10.1016/j.foreco.2018.11.007
128. Liziniewicz M, Ekö PM, Klang F (2016) Effects of five tree-selection strategies when thinning spruce (*Picea abies*) stands: a case study in a field trail in southern Sweden. Scand J For Res 31:495–506. https://doi.org/10.1080/02827581.2015.1130852
129. Peri PL, Martínez Pastur G, Monelos L (2013) Natural dynamics and thinning response of young lenga (*Nothofagus pumilio*) trees in secondary forests of Southern Patagonia. Bosque (Valdivia) 34:5–6. https://doi.org/10.4067/S0717-92002013000300003
130. Correia AV, Oliveira AC (2003) Principais espécies florestais com interesse para Portugal: zonas de influência atlântica. Direcção-Geral das Florestas, Lisboa

131. Gonçalves AC, Dias AB, Afonso A et al (2016) Mechanical versus manual harvest of Pinus pinea cones. Biosys Eng 143:50–60. https://doi.org/10.1016/j.biosystemseng.2016.01.004
132. Ameray A, Bergeron Y, Valeria O et al (2021) Forest carbon management: a review of silvicultural practices and management strategies across Boreal, temperate and tropical forests. Curr Forestry Rep 7:245–266. https://doi.org/10.1007/s40725-021-00151-w
133. Bradford JB, Fraver S, Milo AM et al (2012) Effects of multiple interacting disturbances and salvage logging on forest carbon stocks. For Ecol Manage 267:209–214. https://doi.org/10.1016/j.foreco.2011.12.010
134. Mensah S, Glèlè Kakaï R, Seifert T (2016) Patterns of biomass allocation between foliage and woody structure: the effects of tree size and specific functional traits. Ann Forest Res 59:49–60. https://doi.org/10.15287/afr.2016.458
135. Hu M, Lehtonen A, Minunno F, Mäkelä A (2020) Age effect on tree structure and biomass allocation in Scots pine (Pinus sylvestris L.) and Norway spruce (Picea abies [L.] Karst.). Ann Forest Sci 77:90. https://doi.org/10.1007/s13595-020-00988-4
136. Sterba H, Del Rio M, Brunner A, Condes S (2014) Effect of species proportion definition on the evaluation of growth in pure vs. mixed stands. Forest Syst 23:547. https://doi.org/10.5424/fs/2014233-06051
137. Smith LA, Eissenstat DM, Kaye MW (2017) Variability in aboveground carbon driven by slope aspect and curvature in an eastern deciduous forest, USA. Can J For Res 47:149–158. https://doi.org/10.1139/cjfr-2016-0147
138. Poorter H, Jagodzinski AM, Ruiz-Peinado R et al (2015) How does biomass distribution change with size and differ among species? An analysis for 1200 plant species from five continents. New Phytol 208:736–749. https://doi.org/10.1111/nph.13571
139. Pretzsch H (2020) The course of tree growth. Theory and reality. Forest Ecol Manag 478:118508. https://doi.org/10.1016/j.foreco.2020.118508
140. Pretzsch H, del Río M, Ammer C et al (2015) Growth and yield of mixed versus pure stands of Scots pine (*Pinus sylvestris* L.) and European beech (*Fagus sylvatica* L.) analysed along a productivity gradient through Europe. Eur J Forest Res 134:927–947. https://doi.org/10.1007/s10342-015-0900-4
141. Lwila AS, Mund M, Ammer C, Glatthorn J (2021) Site conditions more than species identity drive fine root biomass, morphology and spatial distribution in temperate pure and mixed forests. For Ecol Manage 499:119581. https://doi.org/10.1016/j.foreco.2021.119581
142. Rog I, Jakoby G, Klein T (2021) Carbon allocation dynamics in conifers and broadleaved tree species revealed by pulse labeling and mass balance. For Ecol Manage 493:119258. https://doi.org/10.1016/j.foreco.2021.119258
143. Zhang J, Fiddler GO, Young DH et al (2021) Allometry of tree biomass and carbon partitioning in ponderosa pine plantations grown under diverse conditions. For Ecol Manage 497:119526. https://doi.org/10.1016/j.foreco.2021.119526
144. Russell MB, Fraver S, Aakala T et al (2015) Quantifying carbon stores and decomposition in dead wood: a review. For Ecol Manage 350:107–128. https://doi.org/10.1016/j.foreco.2015.04.033
145. Balandier P, Mårell A, Prévosto B, Vincenot L (2022) Tamm review: forest understorey and overstorey interactions: so much more than just light interception by trees. For Ecol Manage 526:120584. https://doi.org/10.1016/j.foreco.2022.120584
146. Gonçalves AC (2022) Influence of stand structure on forest biomass sustainability. In: Jhariya MK, Meena RS, Banerjee A, Meena SN (eds) Natural resources conservation and advances for sustainability. Elsevier, Cambridge, United States, pp 327–352
147. Bréda N, Huc R, Granier A, Dreyer E (2006) Temperate forest trees and stands under severe drought: a review of ecophysiological responses, adaptation processes and long-term consequences. Ann For Sci 63:625–644. https://doi.org/10.1051/forest:2006042
148. Sohn JA, Saha S, Bauhus J (2016) Potential of forest thinning to mitigate drought stress: a meta-analysis. For Ecol Manage 380:261–273. https://doi.org/10.1016/j.foreco.2016.07.046
149. Sohn JA, Hartig F, Kohler M et al (2016) Heavy and frequent thinning promotes drought adaptation in *Pinus sylvestris* forests. Ecol Appl 26:2190–2205. https://doi.org/10.1002/eap.1373

150. Comfort EJ, Roberts SD, Harrington CA (2010) Midcanopy growth following thinning in young-growth conifer forests on the Olympic Peninsula western Washington. For Ecol Manage 259:1606–1614. https://doi.org/10.1016/j.foreco.2010.01.038
151. O'Hara KL, Leonard LP, Keyes CR (2012) Variable-density thinning and a marking paradox: comparing prescription protocols to attain stand variability in coast redwood. West J Appl For 27:143–149. https://doi.org/10.5849/wjaf.11-042
152. O'Hara KL, Nesmith JCB, Leonard L, Porter DJ (2010) Restoration of old forest features in coast redwood forests using early-stage variable-density thinning. Restor Ecol 18:125–135. https://doi.org/10.1111/j.1526-100X.2010.00655.x
153. Ehbrecht M, Seidel D, Annighöfer P et al (2021) Global patterns and climatic controls of forest structural complexity. Nat Commun 12:519. https://doi.org/10.1038/s41467-020-20767-z
154. Repo A, Rajala T, Henttonen HM et al (2021) Age-dependence of stand biomass in managed boreal forests based on the Finnish National Forest Inventory data. For Ecol Manage 498:119507. https://doi.org/10.1016/j.foreco.2021.119507
155. Curtze AC, Muth AB, Larkin JL, Leites LP (2022) Seeing past the green: structure, composition, and biomass differences in high graded and silviculture-managed forests of similar stand density. For Ecol Manage 526:120598. https://doi.org/10.1016/j.foreco.2022.120598
156. Ahmad B, Wang Y, Hao J et al (2021) Variation of carbon density components with overstory structure of larch plantations in northwest China and its implication for optimal forest management. For Ecol Manage 496:119399. https://doi.org/10.1016/j.foreco.2021.119399
157. Foster JR, D'Amato AW, Bradford JB (2014) Looking for age-related growth decline in natural forests: unexpected biomass patterns from tree rings and simulated mortality. Oecologia 175:363–374. https://doi.org/10.1007/s00442-014-2881-2
158. Peichl M, Arain MA (2006) Above- and belowground ecosystem biomass and carbon pools in an age-sequence of temperate pine plantation forests. Agric For Meteorol 140:51–63. https://doi.org/10.1016/j.agrformet.2006.08.004
159. Ruiz-Peinado R, Bravo-Oviedo A, López-Senespleda E, et al (2017) Forest management and carbon sequestration in the Mediterranean region: a review. Forest Syst 26:eR04S. https://doi.org/10.5424/fs/2017262-11205
160. Forrester DI, Pretzsch H (2015) Tamm review: on the strength of evidence when comparing ecosystem functions of mixtures with monocultures. For Ecol Manage 356:41–53. https://doi.org/10.1016/j.foreco.2015.08.016
161. Dieler J, Pretzsch H (2013) Morphological plasticity of European beech (*Fagus sylvatica* L.) in pure and mixed-species stands. For Ecol Manage 295:97–108. https://doi.org/10.1016/j.foreco.2012.12.049
162. Dutcă I, Mather R, Ioraş F (2018) Tree biomass allometry during the early growth of Norway spruce (*Picea abies*) varies between pure stands and mixtures with European beech (*Fagus sylvatica*). Can J For Res 48:77–84. https://doi.org/10.1139/cjfr-2017-0177
163. Forrester DI, Benneter A, Bouriaud O, Bauhus J (2017) Diversity and competition influence tree allometric relationships–developing functions for mixed-species forests. J Ecol 105:761–774. https://doi.org/10.1111/1365-2745.12704
164. Pretzsch H (2014) Canopy space filling and tree crown morphology in mixed-species stands compared with monocultures. For Ecol Manage 327:251–264. https://doi.org/10.1016/j.foreco.2014.04.027
165. Pretzsch H, Block J, Dieler J et al (2010) Comparison between the productivity of pure and mixed stands of Norway spruce and European beech along an ecological gradient. Ann For Sci 67:712–712. https://doi.org/10.1051/forest/2010037
166. Hynynen J, Eerikäinen K, Mäkinen H, Valkonen S (2019) Growth response to cuttings in Norway spruce stands under even-aged and uneven-aged management. For Ecol Manage 437:314–323. https://doi.org/10.1016/j.foreco.2018.12.032
167. O'Hara KL (2005) Multiaged silviculture of Ponderosa Pine. U.S. Department of Agriculture, Forest Service

168. Ali A, Yan E-R, Chen HYH et al (2016) Stand structural diversity rather than species diversity enhancesaboveground carbon storage in secondary subtropical forests in Eastern China. Biogeosciences 13:4627–4635. https://doi.org/10.5194/bg-13-4627-2016
169. Danescu A, Albrecht AT, Bauhus J (2016) Structural diversity promotes productivity of mixed, uneven-aged forests in southwestern Germany. Oecologia 319–333. https://doi.org/10.1007/s00442-016-3623-4
170. Yuan Z, Wang S, Ali A et al (2018) Aboveground carbon storage is driven by functional trait composition and stand structural attributes rather than biodiversity in temperate mixed forests recovering from disturbances. Ann For Sci 75:67. https://doi.org/10.1007/s13595-018-0745-3
171. Binkley D, Stape JL, Bauerle WL, Ryan MG (2010) Explaining growth of individual trees: light interception and efficiency of light use by Eucalyptus at four sites in Brazil. For Ecol Manage 259:1704–1713. https://doi.org/10.1016/j.foreco.2009.05.037
172. Soares AAV, Leite HG, Cruz JP, Forrester DI (2017) Development of stand structural heterogeneity and growth dominance in thinned Eucalyptus stands in Brazil. For Ecol Manage 384:339–346. https://doi.org/10.1016/j.foreco.2016.11.010
173. Kern CC, Kenefic LS, Kuehne C et al (2021) Relative influence of stand and site factors on aboveground live-tree carbon sequestration and mortality in managed and unmanaged forests. For Ecol Manage 493:119266. https://doi.org/10.1016/j.foreco.2021.119266
174. Carpenter R, Ward EB, Wikle J et al (2021) Soil nutrient recovery after shelterwood timber harvesting in a temperate oak hardwood forest: insights using a twenty-five-year chronosequence. For Ecol Manage 499:119604. https://doi.org/10.1016/j.foreco.2021.119604
175. Shepard JP (2006) Water quality protection in bioenergy production: the US system of forestry best management practices. Biomass Bioenerg 30:378–384. https://doi.org/10.1016/j.biombioe.2005.07.018
176. Schulze ED, Bouriaud O, Irslinger R, Valentini R (2022) The role of wood harvest from sustainably managed forests in the carbon cycle. Ann For Sci 79:17. https://doi.org/10.1186/s13595-022-01127-x
177. Guidi W, Pitre EF, Labrecque M (2013) Short-rotation coppice of willows for the production of biomass in Eastern Canada. In: Matovic MD (ed) Biomass now–sustainable growth and use. InTech, Rijeka, Croatia, pp 421–448
178. Raulund-Rasmussen K, Stupak I, Clarke N, et al (2008) Effects of very intensive forest biomass harvesting and long term site productivity. In: Röser D, Asikainen A, Raulund-Rasmussen K, Stupak I (eds) Sustainable use pf forest biomass for energy. A synthesis with focus on the Balticand Nordic Region. Springer, Dordrecht, pp 29–78
179. Tattersall Smith C, Briggs RD, Stupak I, et al (2022) Effects of whole-tree and stem-only clearcutting on forest floor and soil carbon and nutrients in a balsam fir (Abies balsamea (L.) Mill.) and red spruce (Picea rubens Sarg.) dominated ecosystem. Forest Ecol Manag 519:120325. https://doi.org/10.1016/j.foreco.2022.120325
180. Pyttel PL, Köhn M, Bauhus J (2015) Effects of different harvesting intensities on the macro nutrient pools in aged oak coppice forests. For Ecol Manage 349:94–105. https://doi.org/10.1016/j.foreco.2015.03.037
181. Fleming RL, Morris DM, Hazlett PW (2021) Assessing temporal response to biomass removal: a framework for investigating evolving constraints on boreal stand development. For Ecol Manage 497:119518. https://doi.org/10.1016/j.foreco.2021.119518
182. Moore JR, Nanayakkara B, McKinley RB, Garrett LG (2021) Effects of nutrient removal by harvesting practices and fertiliser addition on end-of-rotation radiata pine wood quality. For Ecol Manage 494:119269. https://doi.org/10.1016/j.foreco.2021.119269
183. Garrett LG, Smaill SJ, Beets PN et al (2021) Impacts of forest harvest removal and fertiliser additions on end of rotation biomass, carbon and nutrient stocks of Pinus radiata. For Ecol Manage 493:119161. https://doi.org/10.1016/j.foreco.2021.119161
184. Mäkipää R, Abramoff R, Adamczyk B et al (2023) How does management affect soil C sequestration and greenhouse gas fluxes in boreal and temperate forests?—a review. For Ecol Manage 529:120637. https://doi.org/10.1016/j.foreco.2022.120637

185. James J, Page-Dumroese D, Busse M et al (2021) Effects of forest harvesting and biomass removal on soil carbon and nitrogen: two complementary meta-analyses. For Ecol Manage 485:118935. https://doi.org/10.1016/j.foreco.2021.118935
186. Tamminen P, Saarsalmi A, Smolander A et al (2012) Effects of logging residue harvest in thinnings on amounts of soil carbon and nutrients in Scots pine and Norway spruce stands. For Ecol Manage 263:31–38. https://doi.org/10.1016/j.foreco.2011.09.015
187. Lewandowski TE, Forrester JA, Mladenoff DJ et al (2019) Long term effects of intensive biomass harvesting and compaction on the forest soil ecosystem. Soil Biol Biochem 137:107572. https://doi.org/10.1016/j.soilbio.2019.107572
188. Ashton MS, Kelty MJ (2018) The practice of silviculture. Applied forest ecology, 10th edn. John Wiley & Sons Ltd, Oxford
189. Marron N (2015) Agronomic and environmental effects of land application of residues in short-rotation tree plantations: a literature review. Biomass Bioenerg 81:378–400. https://doi.org/10.1016/j.biombioe.2015.07.025
190. Jurgensen M, Tarpey R, Pickens J et al (2012) Long-term effect of silvicultural thinnings on soil carbon and nitrogen pools. Soil Sci Soc Am J 76:1418–1425. https://doi.org/10.2136/sssaj2011.0257
191. Jandl G, Acksel A, Baum C, Leinweber P (2015) Indicators for soil organic matter quality in no-till soils under perennial crops in Central Sweden. Soil Tillage Res 148:74–84. https://doi.org/10.1016/j.still.2014.12.006
192. Osei R, del Río M, Ruiz-Peinado R et al (2022) The distribution of carbon stocks between tree woody biomass and soil differs between Scots pine and broadleaved species (beech, oak) in European forests. Eur J Forest Res 141:467–480. https://doi.org/10.1007/s10342-022-01453-9
193. Hölscher D, Schade E, Leuschner C (2001) Effects of coppicing in temperate deciduous forests on ecosystem nutrient pools and soil fertility. Basic Appl Ecol 2:155–164. https://doi.org/10.1078/1439-1791-00046
194. Dhiedt E, Baeten L, De Smedt P et al (2022) Tree neighbourhood-scale variation in topsoil chemistry depends on species identity effects related to litter quality. Eur J Forest Res 141:1163–1176. https://doi.org/10.1007/s10342-022-01499-9
195. Mott CM, Hofstetter RW, Antoninka AJ (2021) Post-harvest slash burning in coniferous forests in North America: a review of ecological impacts. For Ecol Manage 493:119251. https://doi.org/10.1016/j.foreco.2021.119251
196. Garrett LG, Smith CT, Beets PN, Kimberley MO (2021) Early rotation biomass and nutrient accumulation of Pinus radiata forests after harvest residue management and fertiliser treatment on contrasting types of soil. For Ecol Manage 496:119426. https://doi.org/10.1016/j.foreco.2021.119426
197. Garrett LG, Smaill SJ, Addison SL, Clinton PW (2021) Globally relevant lessons from a long-term trial series testing universal hypothesis of the impacts of increasing biomass removal on site productivity and nutrient pools. For Ecol Manage 494:119325. https://doi.org/10.1016/j.foreco.2021.119325
198. Jonsell M (2008) The effects of forest biomass harvesting on biodiversity. In: Röser D, Asikainen A, Raulund-Rasmussen K, Stupak I (eds) Sustainable use pf forest biomass for energy. A synthesis with focus on the Balticand Nordic Region. Springer, Dordrecht, pp 129–154
199. Stupak I, Asikainen A, Röser D, Pasanen K (2008) Review of recomendations for forest energy harvesting and wood ash recycling. In: Röser D, Asikainen A, Raulund-Rasmussen K, Stupak I (eds) Sustainable use of forest biomass for energy. A synthesis with focus on the Balticand Nordic Region. Springer, Dordrecht, pp 155–196
200. Daioglou V, Stehfest E, Wicke B et al (2016) Projections of the availability and cost of residues from agriculture and forestry. GCB Bioenergy 8:456–470. https://doi.org/10.1111/gcbb.12285
201. Didion M, Abegg M (2022) Tree stumps—an important but undervalued dead wood pool. Ann For Sci 79:34. https://doi.org/10.1186/s13595-022-01155-7

202. Liang J, Crowther TW, Picard N, et al (2016) Positive biodiversity-productivity relationship predominant in global forests. Science 354:aaf8957. https://doi.org/10.1126/science.aaf8957
203. Meigs GW, Morrissey RC, Bače R et al (2017) More ways than one: mixed-severity disturbance regimes foster structural complexity via multiple developmental pathways. For Ecol Manage 406:410–426. https://doi.org/10.1016/j.foreco.2017.07.051
204. Fartmann T, Müller C, Poniatowski D (2013) Effects of coppicing on butterfly communities of woodlands. Biol Cons 159:396–404. https://doi.org/10.1016/j.biocon.2012.11.024
205. Knapp SP, Webster CR, Kern CC (2019) Can group selection with legacy retention change compositional trajectories in conventionally managed hardwoods? For Ecol Manage 448:174–186. https://doi.org/10.1016/j.foreco.2019.06.005
206. Schütz J-P (2006) Modelling the demographic sustainability of pure beech plenter forests in Eastern Germany. Ann For Sci 63:93–100. https://doi.org/10.1051/forest:2005101
207. Spînu AP, Asbeck T, Bauhus J (2022) Combined retention of large living and dead trees can improve provision of tree-related microhabitats in Central European montane forests. Eur J Forest Res 141:1105–1120. https://doi.org/10.1007/s10342-022-01493-1
208. Potapov P, Hansen MC, Laestadius L et al (2017) The last frontiers of wilderness: tracking loss of intact forest landscapes from 2000 to 2013. Sci Adv 3:e1600821. https://doi.org/10.1126/sciadv.1600821
209. Pedroli B, Elbersen B, Frederiksen P et al (2013) Is energy cropping in Europe compatible with biodiversity?—opportunities and threats to biodiversity from land-based production of biomass for bioenergy purposes. Biomass Bioenerg 55:73–86. https://doi.org/10.1016/j.biombioe.2012.09.054
210. Kirby KJ, Buckley GP, Mills J (2017) Biodiversity implications of coppice decline, transformations to high forest and coppice restoration in British woodland. Folia Geobot 52:5–13. https://doi.org/10.1007/s12224-016-9252-1
211. Grinde AR, Slesak RA, D'Amato AW, Palik BP (2020) Effects of tree retention and woody biomass removal on bird and small mammal communities. For Ecol Manage 465:118090. https://doi.org/10.1016/j.foreco.2020.118090
212. Curzon MT, D'Amato AW, Palik BJ (2016) Bioenergy harvest impacts to biodiversity and resilience vary across aspen-dominated forest ecosystems in the Lake States region, USA. Appl Veg Sci 19:667–678. https://doi.org/10.1111/avsc.12256
213. Trouvé R, Sherriff RM, Holt LM, Baker PJ (2021) Differing regeneration patterns after catastrophic fire and clearfelling: implications for future stand dynamics and forest management. For Ecol Manage 498:119555. https://doi.org/10.1016/j.foreco.2021.119555
214. Schroeder LM (2008) Insect pests and forest biomass for energy. In: Röser D, Asikainen A, Raulund-Rasmussen K, Stupak I (eds) Sustainable use pf forest biomass for energy. A Synthesis with Focus on the Balticand Nordic Region. Springer, Dordrecht, pp 109–128
215. Dimitriou I, Rutz D (2015) Sustainable short rotation coppice a handbook. WIP Renewable Energies, Munich
216. Arets EJMM, van der Meer PJ, Verwer CC et al (2011) Global wood production: assessment of industrial round wood supply from forest management systems in different global regions. Wageningen, Alterra
217. Bizerra Barros I, Santos Cavalcante V, Soares Moulin A et al (2021) Integrating forest residue and mineral fertilization: effects on nutrient acquisition, nutrient use efficiency and growth of eucalypt plants. For Ecol Manage 496:119461. https://doi.org/10.1016/j.foreco.2021.119461
218. Ugalde L, Pérez O (2001) Mean annual volume increment of selected industrial forest plantation species by forest plantation thematic papers. FAO, Rome
219. Coelho MB, Paulo JA, Palma JHN, Tomé M (2012) Contribution of cork oak plantations installed after 1990 in Portugal to the Kyoto commitments and to the landowners economy. Forest Policy Econ 17:59–68. https://doi.org/10.1016/j.forpol.2011.10.005
220. Civitarese V, Faugno S, Picchio R et al (2018) Production of selected short-rotation wood crop species and quality of obtained biomass. Eur J Forest Res 137:541–552. https://doi.org/10.1007/s10342-018-1122-3

221. Eufrade Junior HJ, de Melo RX, Sartori MMP et al (2016) Sustainable use of eucalypt biomass grown on short rotation coppice for bioenergy. Biomass Bioenerg 90:15–21. https://doi.org/10.1016/j.biombioe.2016.03.037
222. Oliveira N, Sixto H, Cañellas I et al (2015) Productivity model and reference diagram for short rotation biomass crops of poplar grown in Mediterranean environments. Biomass Bioenerg 72:309–320. https://doi.org/10.1016/j.biombioe.2014.09.019
223. Sims REH, Senelwa K, Maiava T, Bullock BT (1999) Eucalyptus species for biomass energy in New Zealand—I: growth screening trials at first harvest. Biomass Bioenerg 16:199–205. https://doi.org/10.1016/S0961-9534(98)00078-6
224. Sims REH, Senelwa K, Maiava T, Bullock BT (1999) Eucalyptus species for biomass energy in New Zealand—II: coppice performance. Biomass Bioenerg 17:333–343. https://doi.org/10.1016/S0961-9534(99)00043-4
225. Sochacki SJ, Harper RJ, Smettem KRJ et al (2013) Evaluating a sustainability index for nutrients in a short rotation energy cropping system. GCB Bioenergy 5:315–326. https://doi.org/10.1111/j.1757-1707.2012.01202.x
226. Tomé M, Almeida MH, Barreiro S et al (2021) Opportunities and challenges of Eucalyptus plantations in Europe: the Iberian Peninsula experience. Eur J Forest Res. https://doi.org/10.1007/s10342-021-01358-z
227. Wales SB, Kreider MR, Atkins J et al (2020) Stand age, disturbance history and the temporal stability of forest production. For Ecol Manage 460:117865. https://doi.org/10.1016/j.foreco.2020.117865
228. O'Hara KL, Nagel LM (2006) A functional comparison of productivity in even-aged and multiaged stands: a synthesis for Pinus ponderosa. Forest Sci 52:290–303. https://doi.org/10.1093/forestscience/52.3.290
229. del Río M, Barbeito I, Bravo-Oviedo A et al (2017) Mediterranean pine forests: management effects on carbon stocks. In: Bravo F, LeMay V, Jandl R (eds) Managing forest ecosystems: the challenge of climate change. Springer International Publishing, Cham, pp 301–327
230. Durocher C, Thiffault E, Achim A et al (2019) Untapped volume of surplus forest growth as feedstock for bioenergy. Biomass Bioenerg 120:376–386. https://doi.org/10.1016/j.biombioe.2018.11.024
231. Berndes G, Hoogwijk M, van den Broek R (2003) The contribution of biomass in the future global energy supply: a review of 17 studies. Biomass Bioenerg 25:1–28. https://doi.org/10.1016/S0961-9534(02)00185-X
232. Röser D (2008) Sustainable use of forest biomass for energy: a synthesis with focus on the Baltic and Nordic region. Springer, Dordrecht, The Netherlands
233. Smith WB, Miles PD, Perry CH, Pugh SA (2009) Forest resources fo the United States, 2007. USDA Forest Service
234. Lundmark R, Athanassiadis D, Wetterlund E (2015) Supply assessment of forest biomass—a bottom-up approach for Sweden. Biomass Bioenerg 75:213–226. https://doi.org/10.1016/j.biombioe.2015.02.022
235. Schelhaas M-J, Nabuurs G-J, Schuck A (2003) Natural disturbances in the European forests in the 19th and 20th centuries. Glob Change Biol 9:1620–1633. https://doi.org/10.1046/j.1529-8817.2003.00684.x
236. Shorohova E, Kuuluvainen T, Kangur A, Jõgiste K (2009) Natural stand structures, disturbance regimes and successional dynamics in the Eurasian boreal forests: a review with special reference to Russian studies. Ann For Sci 66:201–201. https://doi.org/10.1051/forest/2008083
237. Thiel M, Basiliko N, Caspersen J et al (2015) Operational biomass recovery of small trees: equations for six central Ontario tree species. Can J For Res 45:372–377. https://doi.org/10.1139/cjfr-2014-0429
238. Malico I, Gonçalves AC, Sousa AMO (2021) Evergreen oak biomass residues for firewood. In: Gonçalves AC, Sousa A, Malico I (eds) Forest biomass–from trees to energy. INtechOpen, London, pp 87–103

239. Rothe A, Moroni M, Neyland M, Wilnhammer M (2015) Current and potential use of forest biomass for energy in Tasmania. Biomass Bioenerg 80:162–172. https://doi.org/10.1016/j.biombioe.2015.04.021
240. Vávrová K, Knápek J, Weger J (2017) Short-term boosting of biomass energy sources—determination of biomass potential for prevention of regional crisis situations. Renew Sustain Energy Rev 67:426–436. https://doi.org/10.1016/j.rser.2016.09.015
241. McMinn JW, Crossley DA Jr (1996) Biodiversity and coarse woody debris in Southern Forests. Department of Agriculture, Forest Service, Southern Research Station, Asheville, North Carolina
242. Barrett SM, Bolding MC, Aust WM, Munsell JF (2014) Characteristics of logging businesses that harvest biomass for energy production. Forest Prod J 64:265–272. https://doi.org/10.13073/FPJ-D-14-00033
243. Leinonen A (2004) Harvesting Technology of Forest residues for fuel in the USA and Finland. VTT Tiedotteita, Helsinki
244. Nonini L, Schillaci C, Fiala M (2022) Assessing logging residues availability for energy production by using forest management plans data and geographic information system (GIS). Eur J Forest Res 141:959–977. https://doi.org/10.1007/s10342-022-01484-2
245. Thiffault E, Béchard A, Paré D, Allen D (2015) Recovery rate of harvest residues for bioenergy in boreal and temperate forests: a review. WIREs Energy Environ 4:429–451. https://doi.org/10.1002/wene.157
246. Pérez S, Renedo CJ, Ortiz A et al (2006) Energy evaluation of the Eucalyptus globulus and the Eucalyptus nitens in the north of Spain (Cantabria). Thermochim Acta 451:57–64. https://doi.org/10.1016/j.tca.2006.08.009
247. Farine DR, O'Connell DA, John Raison R et al (2012) An assessment of biomass for bioelectricity and biofuel, and for greenhouse gas emission reduction in Australia. GCB Bioenergy 4:148–175. https://doi.org/10.1111/j.1757-1707.2011.01115.x
248. Ghaffariyan MR (2013) Assessment of harvest residues from different harvesting operation sites in Australia. CRC for Forestry 1–3
249. Titus BD, Brown K, Helmisaari H-S et al (2021) Sustainable forest biomass: a review of current residue harvesting guidelines. Energy Sustain Soc 11:10. https://doi.org/10.1186/s13705-021-00281-w

Energy Plantations

Ana Cristina Gonçalves

Abstract Energy plantations have been gaining importance in the supply of biomass for energy purposes, due to their high yield in short timeframes. These forest systems also enable to reduce the pressure in other forest systems to provide biomass for energy, in particular those under protection and conservation status. This chapter reviews the state of the art of energy plantations and their yields. It addresses the selection of species, density, rotation, harvest cycles, site selection, management practices, harvesting, biomass yields, and their estimation. Overall, there is a wide set of species and management options that can be used in energy plantations. Similarly, there is a large variability in yields, that vary between and within species, due to site, density, rotation, harvest cycles, and management. Though there are many studies, further research is needed on yield optimisation, rotation length, harvest cycles, management practices, and harvesting.

Keywords Species · Clones · Regime · Site · Management · Yield

1 Introduction

Wood is considered one of the most important raw materials as it satisfies several human needs, among which is energy [1]. In the last decades, energy plantations have been gaining importance as a source of energy because of the energy crises, the concerns about the reduction of greenhouse gas emissions, the dependency on fossil fuels, the increase of carbon sequestration, and to release the pressure on other forests systems [1–8]. These forest systems date back to ancient times, but management practices have been improved to increase their yield [1, 9–11]. Their

A. C. Gonçalves (✉)
MED – Mediterranean Institute for Agriculture, Environment and Development &
CHANGE – Global Change and Sustainability Institute, Instituto de Investigação e Formação Avançada, Departamento de Engenharia Rural, Escola de Ciências e Tecnologia, Universidade de Évora, Apartado 94, 7002-544 Évora, Portugal
e-mail: acag@uevora.pt

A. C. Gonçalves and I. Malico (eds.), *Forest Bioenergy*, Green Energy and Technology,
https://doi.org/10.1007/978-3-031-48224-3_4

importance was recognised by the International Union of Forest Research Organizations (IUFRO; http://www.iufro.org) through the creation of the research group 1.03.00–Short-rotation forestry.

Dickmann [1] identifies several terms for forest energy plantations: short-rotation woody crops (SRWC), short-rotation forestry (SRF), short-rotation coppice (SRC), short-rotation intensive culture, intensive culture of forest crops, intensive plantation culture, biomass plantation culture, bioenergy plantation culture, biofuels feedstock production system, energy forestry, short-rotation fiber production system, mini-rotation forestry, silage sycamore, wood grass. The most frequently used terms are short-rotation woody crops (SRWC), short-rotation forestry (SRF), and short-rotation coppice (SRC). As no standard term has been defined in this chapter the term energy plantations will be used.

The goals of this review are to provide insights into energy plantations from the selection of species or clones to harvest and yields. This chapter is divided in two sections. One that analyses the energy plantations, including the selection of species, initial density, rotation, harvest cycles, site selection, management practices, and harvesting (Sect. 2), and another that evaluates biomass yields (Sect. 3).

2 Forest Energy Plantations

Forest energy plantations are forest systems whose main goal, frequently the only one, is producing biomass for energy, and have specific spatial and temporal features [6, 12]. The plantations are composed of very fast or fast growing tree species, many times improved hybrids. These stands have frequently very high densities (from 1000 to more than 300,000 stems·ha^{-1}), in coppice systems most of the times, with very short or short rotations (1–12 years), cutting cycles of 10 to 30 years, managed in clearcutting systems, where all aerial biomass is removed in each harvest, and are intensively managed. Their establishment and management (Fig. 1) include site selection, control of spontaneous vegetation, selection of planting techniques, fertilisation, control of pathogens, and irrigation [1, 13–24]. Planting design and management are frequently adapted to a fully mechanised system [25–27].

Biomass from forest energy plantations, when compared to other renewable energy sources, has several advantages: biomass is relatively easy to transport and store [28]; it has different uses, such as heating, electricity or biofuels [3, 28–30]; it is available worldwide [3, 13, 31, 32]; in a specific location its quantity can be increased through anticipated harvest in times of shortage or high prices of other fuels [28] or reduced through delayed harvest when its market price is low or other fuels have low prices [33]; it allows decentralisation of the energy systems [3, 28]; and it is suitable in regions with biomass availability and low population density [28, 34].

Forest energy plantations are considered economically viable when compared to other forest and agricultural productions at the management unit level. These forest systems have low risk and high economic viability. Its harvest flexibility (anticipated

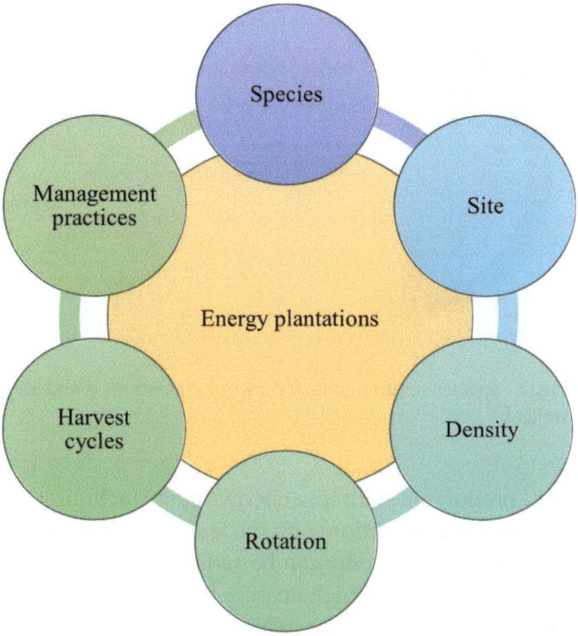

Fig. 1 Factors influencing the establishment and management of energy plantations

or delayed) promotes the reduction of risks, especially if included in agricultural crops portfolios; and it also provides ecosystem services without adding costs, especially in areas of intensive agriculture [2, 4, 5, 7, 35–37]. However, energy plantations can pose a risk when established in areas suited for agriculture, and therefore, it is recognised that they should be settled in set aside agricultural lands or marginal lands [3, 6, 12, 13, 31–33, 38], enabling simultaneously rural development and environmental benefits [1, 2, 12, 39, 40]. These plantations can also be settled for phytoremediation purposes, *i.e.*, using trees in energy plantation systems to remediate contaminated sites while using the biomass for energy [26, 41–45]. Forest energy plantations are well represented, for example, in Canada [46], China [47], United States of America [48, 49], and Europe, northern and central, and to a lesser extent in the southern [13, 31, 32].

Selection of the Species

The selection of tree species for energy plantations encompasses a set of requirements that should be fulfilled [50–53]: biomass should have high specific energy and quality as fuel, high biomass production in dry weight, good resprouting ability, fast juvenile growth, narrow crowns or large size leaves in the upper crown, adaptability to a wide range of sites, and resistance to biotic and abiotic disturbances (Fig. 2a). Ideally, the species should have [3]: maximum possible production in dry matter per stand area

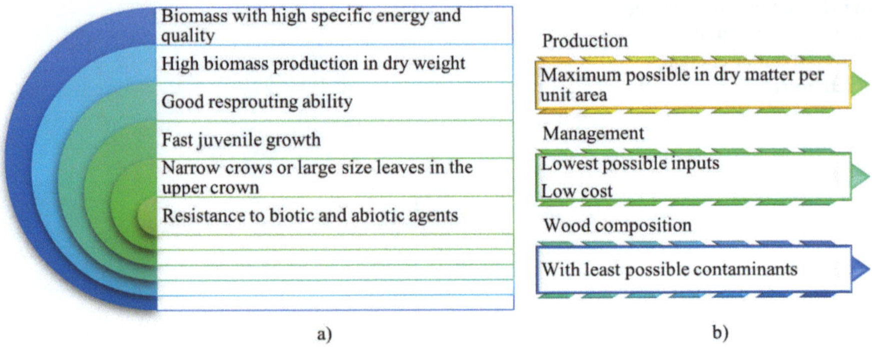

Fig. 2 Species requirements for energy plantations **a** and ideal features for management and high yields **b**

unit, production with low energy input (including nutrient requirements), low cost, and wood composition with the least possible contaminants (Fig. 2b).

These requirements can be satisfied by a large set of species, characterised by a fast initial growth which enables them to outcompete other species for the available growing space. From the many species that can be used for energy plantations some of the referred in literature are presented in Table 1.

Due to their characteristics, *i.e.*, fast or very fast growth, wide genetic base, easy propagation, short improvement cycles, easy vegetative reproduction, and ability to resprout, the aforementioned species are adapted to several climatic and soil conditions. They have also the ability to improve soil quality and to have high productions [51–53, 101]. In European Union countries three *genera* are considered to have the largest potential for energy plantations, namely *Populus* spp., *Salix* spp., and *Eucalyptus* spp. [51–53, 101].

In energy plantations, hybrids are frequently used to improve several tree species traits such as survival rate, biomass productivity, resprouting ability, adaptation to a variety of environmental conditions, and resistance to pathogens. The hybrids can be developed through genetic improvement [26, 98] and/or biotechnology [102]. For example, clones of *Populus* spp. and *Salix* spp. can differ in what regards survival rate, growth, and woody properties due to site quality and/or planting density [103] or not [104]. Relevant are also the relations genotype-environment (e.g., [86, 105]).

Density, Rotation, and Harvest Cycles

In energy plantations *density* and *rotation* length are strictly linked as the main goal is to achieve the highest possible biomass production in the shortest possible time (e.g., [26, 56, 89, 106]). Sixto et al. [6] refer to three principles that are associated with the design and management of forest energy plantations: (a) *Law of final constant yield* that states that biomass yield increases with the increase of density up to an

Table 1 Forest species used in energy plantations

Genus/Specie	References
Acacia spp.	[54, 55]
Acer pseudoplatanus	[56–59]
Ailanthus spp.	[55]
Alnus spp	[60–67],
Bambusa spp.	[68, 69]
Betula spp.	[58, 59]
Casuarina spp.	[55, 70]
Eucalyptus spp.	[1, 37, 50, 71–75]
Fraxinus spp.	[76]
Gmelina arborea	[77–79]
Leucaena spp.	[1, 55, 77]
Liquidambar styraciflua	[1, 80, 81]
Paulownia spp.	[82, 83]
Pinus taeda	[1, 84]
Platanus occidentalis	[57, 80, 81, 84–87]
Platanus spp.	[48]
Populus spp	[1, 38, 47, 50, 56, 58, 59, 63, 80, 86, 88–93]
Prosopis spp.	[94, 95]
Robinia pseudoacacia	[56, 96, 97]
Salix spp.	[1, 38, 50, 58, 59, 63, 92, 98]
Swetenia mahogany	[99]
Tectona spp.	[99, 100]
Ulmus pumila	[48]
Yushane spp.	[68]

upper threshold, above which it becomes independent of density. It can be used to determine the maximum number of stems per area unit; (b) *The development of social classes in a stand*, with dominant and dominated individuals competing among them. Harvest should be done before competition affects the growth of the individuals and the vitality of stumps; (c) *Self-thinning law* states that without mortality total biomass per area unit increases exponentially until canopy closure, after which stems tend to reduce growth. After canopy closure, some trees become dominated and eventually die unless there is a density reduction. Thus, canopy closure should be avoided.

The three aforementioned principles are the basis for trials to determine both density and rotation length in energy plantations. A wide range of densities has been studied, from 1000 stems·ha^{-1} to 310 000 stems·ha^{-1} [12, 21, 48, 58, 59, 92, 98, 101, 107–109]. Similarly, a large range of rotations has been studied, from 1 to 20 years [12, 21, 38, 48, 58, 59, 92, 98, 101, 107–109].

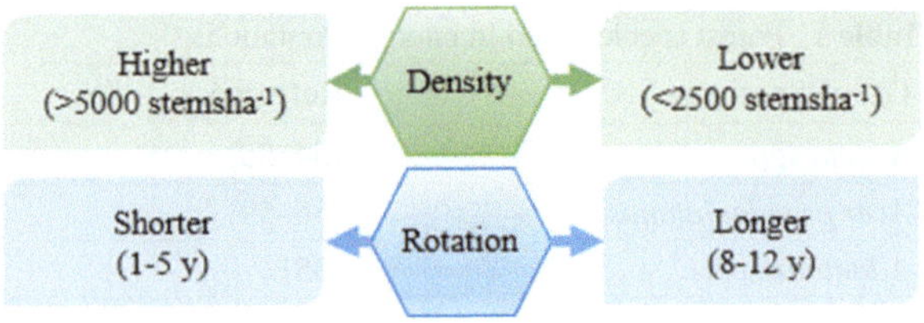

Fig. 3 Density versus rotation length

According to Dickmann [1], there seems to be a dichotomy regarding density and rotation length that is also linked with the woody products and yields to be obtained. It should be taken into consideration the production per area *versus* per individual tree. Higher densities result in higher biomass per area unit but lower biomass per individual stem [103, 110]. Thus, energy plantations can be divided into (Fig. 3): (i) higher densities and shorter rotations, and (ii) lower densities and longer rotations. The former has densities ranging from 5 000 to 200 000 stems/ha, and rotations from 1 to 5 years. Their main goal is biomass for energy where the maximum conversion of solar energy is attained and the flexibility of the biomass as raw material is not important. This strategy is also used when phytoremediation or application of vegetation as a filter of soil contaminants is needed. The latter have densities ranging from 1000 to 2500 stems·ha^{-1}, rotations from 8 to 12 years, and enables more flexibility in terms of woody products, small dimension timber, pulp and paper, and biomass for energy. Yet, a wide variety of combinations of densities and rotations can be found in the literature. Examples of stands of higher densities and shorter rotations are suggested by some authors [21, 26, 39, 86, 98, 101, 107, 108, 111, 112], while stands of lower densities and longer rotations are suggested by other authors [37, 39, 58, 59, 74, 75, 109–111, 113–115].

Though there is a wide range of literature references focused on determining the optimal density, and rotation, the results are not always coincident. This is, at least partially, explained by the constraints related to the tree species, clone, site, and climate. It is well known in silviculture that the maximum volume (or biomass) is reached when the mean annual increment equals the current annual increment [10, 116–118]. Several authors have studied the rotation that maximised biomass production as a function of density (*e.g.*, [92, 101, 106, 119]). For densities up to 10 000 stems·ha^{-1} higher yields are attained at longer rotations (*e.g.*, 4 years *versus* 2 years) [101] while densities higher than 10 000 stems·ha^{-1} the higher yields are attained at shorter rotations [119]. The wider the spacing the higher the growing space for each individual, and the higher the dimensions of the individual stems. The reduction of biomass production seems to be related to biomass allocation due to full growing space occupancy and competition among individuals [62]. The reduction of the yield with the increase of rotation length for high densities seems to be related to self-thinning. Its effects result from the increase of competition between individuals and an overall reduction of growth and, consequently, of yield. Thus, the mitigation of the self-thinning effect on yield can be attained with shorter rotations [119]. It seems that for a density equal or higher to 10 000 stems·ha^{-1} rotations of 2-years length are better suited for maximising yield while for lower density longer rotations can be used [106].

Two other aspects should be considered: one is technical, and the other is the maximization of biomass per stem or per area unit. The rotation length can be influenced by *technical aspects.* On one hand, mechanical harvest equipment has a maximum threshold cutting diameter, which can reduce rotation length [26]. On the other hand, mechanical harvest is also described as problematic for high densities, in which case the option is to reduce density and increase rotation length [1]. The other aspect relates to the *maximisation of production per stem* or *per area unit, i.e.*, fewer stems with larger dimensions or otherwise. In the former, products have a higher proportion of wood, and smaller of bark, leaves, and branches. This implies smaller densities, longer rotations, and products that can be used for energy, pulp and paper, or other small dimension timber products. Thus, the model of silviculture is more flexible in terms of products. As growth is concentrated in fewer trees, whenever competition is a limiting factor thinning or sprout selection could be considered as well as pruning for small dimension timber products, to increase quality [1, 117]. But this approach has the disadvantage of having lower densities and longer rotations [1, 19], resins, and other undesirable chemical components for the use of biomass for energy [19]. The energy plantations with higher densities and trees of small dimensions have the advantage of maximising the conversion of solar energy in biomass, which results in a yield of biomass oriented to bioenergy, but with less flexibility in terms of woody products [1, 107]. Other advantages are reducing the spontaneous vegetation [120] and not needing thinning, sprout selection, or pruning [1].

The *harvest cycle, i.e.,* the number of harvests until the end of the production cycle, when there is the need to regenerate the stands, is constrained by stump vigour, stump mortality, and rotation. *Stump vigour* influences the stump's ability to resprout as well as stool survival. The higher the stump vigour the higher the resprout ability and the stool survival. Thus, the higher the stump vigour the higher the potential yield. *Stump mortality* influences density and productivity. The lower mortality rates enhance higher productions [121]. Productivity is also affected by the successive *rotations* with a trend towards the increase from the first to second or third rotation, and a tendency towards yield decrease more or less accentuated, from the fourth rotation onwards. Yet, it also depends on the stump vigour, stump mortality, species, and site. In general, the harvest cycle's length is determined by productivity. When productivity between successive harvests decrease it is considered that the end of the production cycle has been reached [122]. Several authors refer to cutting cycles between 10 and 30 years with 3 to 10 rotations [12, 17, 26, 56, 63, 123].

Site Selection and Management Practices

In the establishment of any forest stand, and in particular of energy plantations, *site selection* (Fig. 4), which is related to the soil and climate [1, 3, 17], has a strong influence on the survival, growth, and yield. Overall, there is a trend toward higher yields on better quality sites [1, 26, 104]. But it is also dependent on the ecological traits of the species or clones. Thus, considering that a high yield is to be attained, soils should have adequate physical and chemical properties. The soil characteristics that enhance biomass production are soil moisture availability during the yearly growing season, nutrient availability, and aeration. The soils that should be avoided are those with drainage problems (either poorly or excessively drained), with pH too acid or too alkaline, degraded through erosion, saline, shallow, or infertile. Climate should also be considered in particular, the mean annual temperature, annual precipitation and precipitation during the growing season, frosts, and snow. Climate conditions should be within the ecological range of the species or clones, preferably close to their optimum for their growth. Steep slopes should be avoided if mechanisation is foreseen [1, 12, 19, 26].

One issue related to biomass for energy is the identification of the areas available for energy plantations. These areas can be determined following a methodology in four steps [17]: (1) selection of the species to be used and assessment of their ecological and cultural characteristics; (2) determination of the suitability of the sites, which refers to the selection of a set of data, frequently in a geographical

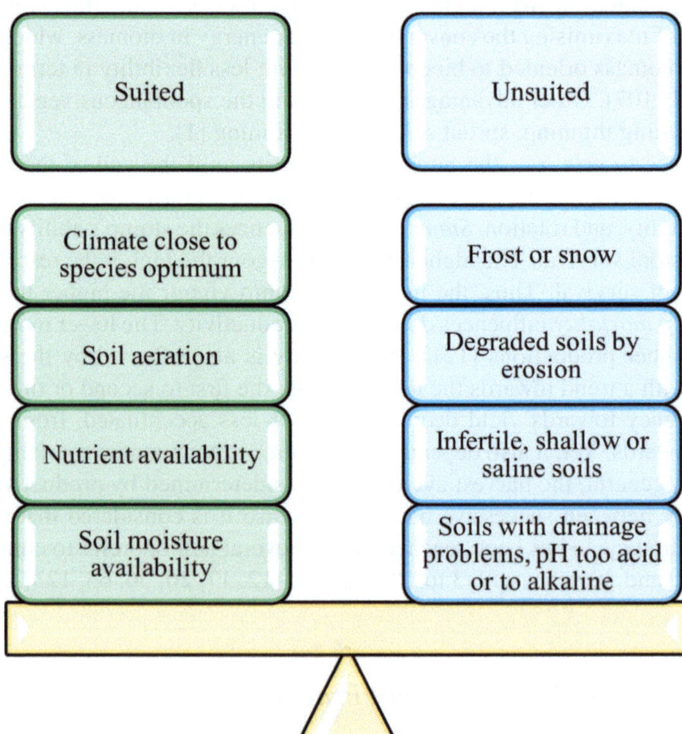

Fig. 4 Site characteristics suited and unsuited for energy plantations

information systems (GIS) environment, including soils, land morphology, climate, protected areas, administrative boundaries, a suite of assumptions and a subsequent set of operations that enable the identification of the areas where the selected species can be grown; (3) availability of land, which refers to the identification of the potential areas available, considering the existing restrictions, whether economic or social; (4) assignment of the land, which refers to the definition of a decision process that enables the determination of the areas where the energy plantations can be installed. The areas identified are dependent on the initial assumptions made. If only the optimal conditions that potentially originate the higher yields are considered, then the area estimation could be rather conservative [17].

Management practices include the control of spontaneous vegetation, planting techniques, initial development, control of pathogens, and irrigation (Fig. 5). The *control of spontaneous vegetation* enables the reduction of competition between the tree species or clones and other vegetation. It is especially relevant in the competition for light, water, and nutrients [19, 124]. It should be done during site preparation, as the control of herbaceous and shrub species is simpler and makes plantation operations easier. Several methods of control of spontaneous vegetation can be considered. Their selection should take into account a suite of factors that include the type of spontaneous vegetation, site, climate, and tree species or clones to be planted. It can be mechanical or chemical or even a combination of both [1, 26, 86, 124–126]. The control of spontaneous vegetation after each harvest might [127] or not [128] be necessary and is frequently chemical [129]. It is recommended when competition with spontaneous vegetation and/or production losses are expected, though care should be taken not to affect either the stumps or the sprouts [6, 26].

Regarding the selection of *planting techniques*, two main choices can be pointed out, namely plantation of cuttings or plantation of seedlings. Cuttings, *i.e.*, unrooted hardwood cuttings for species that have a good ability to develop the root system and the aerial part, are frequently used with *Salix* spp. [1, 26, 130–132]. Seedlings, most of the times, with plants produced in containers whether from seed origin or

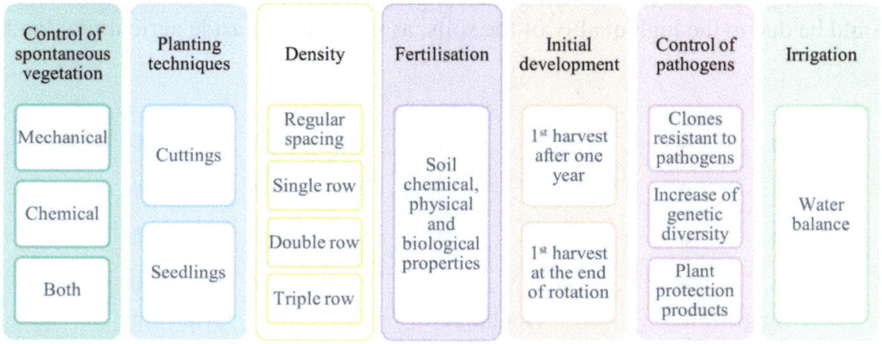

Fig. 5 Management practices

vegetative propagation [1], are used with for example *Populus* spp. or *Eucalyptus* spp. [1, 130–132],

Energy plantations establishment should consider the spatial arrangement of the individuals, which is related to spacing and density. Regular spacing design is frequently used to promote better use of the growing space while at the same time enhancing its mechanical harvest. The spacing can be in single, double, or triple rows (Fig. 6) depending on the density. Typically, the distance between rows ranges from 2.0 m to 3.0 m in the higher density and shorter rotation plantations and between 4.0 m to 6.0 m in the lower density and longer rotation ones [12, 16, 75, 133]. In the single row design, the distance within the rows ranges from 0.5 m to 3.0 m [12, 16, 75, 133]. In the double row design, the spacing between the double rows is between 0.75 m and 1.50 m, and the distance within the rows ranges from 0.45 m to 0.80 m [12, 133, 134]. For the triple row design, the spacing between the triple rows is about 0.6 m, and the distance within the rows of 0.6 m [135]. According to some authors [12, 136], long lines increase the efficiency of harvest. Also, the head and boundaries of the energy plantations should be wide enough for the machinery manoeuvres and to reduce to a minimum its turns [12]. Density and spatial arrangement affect competition between individuals. There is a trend towards lower competition in the square spacing when compared with the rectangular one [103].

Regarding the *initial development* of the plantation, two approaches can be followed [26, 107, 108]: (i) the plantation is harvested at about 1 year old to promote coppicing, which increases the density. The sprouts take advantage of the existing stump root system that promotes the growth rate and the increase of biomass production; and (ii) the first harvest is done at the end of the rotation and the coppice derives from this first harvest. Before choosing one or the other, both costs and biomass production should be considered to increase the economic viability [6].

The need for *fertilisation* is determined by the site, especially by the site's productive potential as it plays a key role in the intensive forest systems of biomass production [6]. Because it is an expensive operation its necessity should be evaluated [1]. Some energy plantations' studies reported that fertilisation did not increase yield when compared with not fertilised ones (e.g., [12, 130–132, 137–141]). This effect could be due to the high quality of the soils, as many are set aside agricultural lands

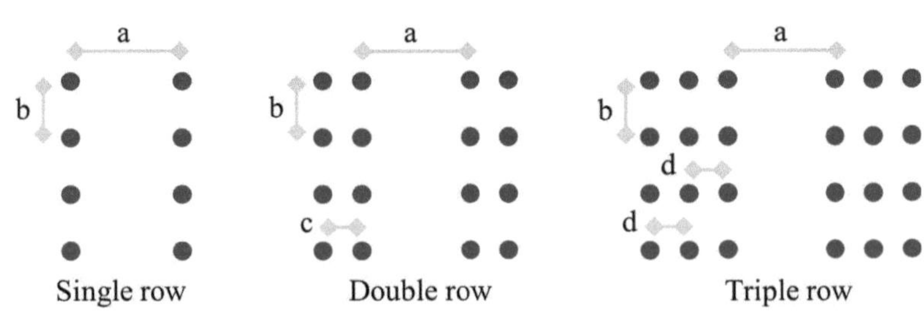

Fig. 6 Schematic representation of single, double and triple row energy plantation design (where a is the distance between rows, b distance within the row, c the distance between double rows, and d the distance between triple rows)

[6, 132]. Conversely, in other studies, fertilisation originated the increase in yield, due to the increase of the nutrients' availability (*e.g.*, [12, 131, 139, 141–143]). In others still, a negative effect of fertilisation was described, which was related to polluting mineral elements (*e.g.*, salts) and/or antibiotics (*e.g.*, [144–146]). A thorough revision of the effects of fertilisation on energy plantations can be found in Marron [147].

The assessment of soil and foliar nutrient levels should be considered to determine the need and quantity of fertiliser. If the nutrient's concentrations are below their critical level then fertilisation should be done [1]. It can either be done by inorganic fertilisers [26, 86, 148], or organic fertilisers, residual waters, or intensive cattle grazing muds [98, 149–153]. In any case, a thorough evaluation of the soil's physical, chemical, and biological (*e.g.*, organic matter and amount of seeds of spontaneous vegetation) characteristics should be carried out before considering the application of fertilisers [26] as well as their application costs [6]. The distinction between the first and the other successive harvests should also be made. In successive harvests, the export of large amounts of nutrients from the site is expectable. Yet, leaf fall and its decomposition incorporate nutrients in the soil, though the reallocation of nutrients from leaves to woody organs varies between species. Thus, the decomposition of leaves incorporates larger or smaller amounts of nitrogen, phosphorous, potassium, calcium, and magnesium promoting the maintenance of soil fertility and, potentially, compensating for the removal of woody biomass [154]. When the exports are not compensated forest energy plantations might need to be fertilised [1, 22, 155].

The *control of pathogens* should be considered whenever outbreaks of pests or diseases occur. One of the ways to minimise the effect of pathogens is by using clones resistant to pests and diseases [156, 157] or increasing the genetic diversity [1, 158]. Alternatively, plant protection products (phytopharmaceuticals) can be used, in which case legislation should be followed [6] and the economic and environmental viability should be justified [1, 86, 159]. These products should only be applied as an answer to a specific problem when large damages are to be expected and not as a prophylactic treatment [6].

Irrigation in forestry is not frequent [10, 116–118, 160]. In many forest energy plantations, annual precipitation and its annual distribution along with soil water holding capacity are sufficient to cover the trees' water needs. Irrigation should be considered when water stress is expected to occur, to avoid the reduction of biomass production or mortality [161–163]. The quantity of water to be used should be calculated as a function of the plantation evapotranspiration and cultural coefficient (*i.e.*, water balance) to promote the best possible use of water [6, 164].

Harvesting

The optimisation of harvesting is of the utmost importance [165], due to its share of costs and inputs. The harvesting costs correspond to about 45% of the total energy plantation costs [134]. Also, the energy input corresponds to up to 33% of the total input of energy [166], being the second largest (the first is fertilisation) fossil energy inputs in the system [167].

Several machinery existing on the market has been used to harvest energy plantations. Moreover, to improve harvest efficiency other machinery has been developed (for detailed revision see [168]). In literature four main harvesting techniques are referred [136]: single pass cut-and-chip, double pass cut-and-store, single pass cut-and-bale, and single pass cut-and-billet. The single pass cut-and-chip being the most flexible, can be used with different stand structures (species, ages, diameter, density, and stocking). The harvesting is done with a single pass making the operations simpler and reducing labour and machinery costs [169] because other cultural practices can be done with these machinery [170]. Furthermore, single tree harvesting productivity was improved by multiple tree harvesting with a system based on software [171]. In the double pass cut-and-store harvest system, the stems are cut and left to dry in a specific location after which are chipped, corresponding to two passes. When compared with the other systems its advantages are related to not needing biomass to be stored in a covered place; to the reduction of the losses due to microbial activity and emission of undesired gases during the storage of the chips; the reduction of the costs of transport because of the lower moisture content of the chips; forest chipper provides a high material effective capacity as well as a favourable particle size distribution [172]. The two latter harvesting systems are much less representative than the former two. The cut-and-bale and cut-and-billet derive in different biomass formats than the single pass cut-and-chip and double pass cut-and-store, resulting in biomass bales, billets, and chips, respectively [136]. The most used and improved mechanical harvesting technique is single pass cut-and chip, followed by double pass cut-and store (for details see [136]).

Species dormant season (winter in the northern hemisphere) is the best one for harvesting. The advantages are related to the recycling of leaf nutrients [154]. Additionally, cutting should leave stumps between 10 cm and 20 cm to preserve the buds and to maintain resprouting stump ability [173].

3 Biomass Yields in Energy Plantations

Estimation of Biomass in Energy Plantations

Biomass can be estimated by destructive or non-destructive methods (*cf. chapter* "Modelling Biomass"). The former can be done either through sampling, frequently used for modelling; or at the end of the rotation when trees in a certain area are harvested. The disadvantage of the latter is that it does not allow to make predictions. The non-destructive methods use allometric biomass functions and enable to predict yields. However, due to the specificities of these forest systems the numerous existing allometric equations, many developed for high forest systems, originate bias in the estimation of biomass. Thus, several authors developed allometric equations specific to energy plantations for tree species and/or clones. In literature was found a set of allometric equations for *Populus* spp. [39, 47, 59, 88, 91, 93, 108, 109, 113, 133, 174], for *Salix* spp. ([26, 59, 62, 112, 175–178], and for *Eucalyptus* spp. [37, 74, 75, 114, 179, 180].

Estimations of biomass of energy plantations have been done at the local or regional levels. Frequently growth and production models (which include allometric biomass functions) are used to generate several scenarios of management [130, 181–185]. At local level, the models are frequently based on the biomass allometric functions per species. Conversely, at broader scales, the models used in the estimation of biomass include usually several soil and climatic data variables, along with plant growth principles and management options, and also the interaction between the four factors. Bandaru et al. [186] classified climatic data sets in two categories: (i) collected from meteorological stations; and (ii) gridded weather data sets. The first is predominantly used at a local scale while the latter are used at a regional scale [187]. The gridded weather data can be obtained by (i) interpolation techniques of weather data and topographic characteristics or (ii) modelling and assimilation techniques [188]. A modified version of 3-PG for energy plantations with coppice management, 3-PG-Coppice model [183] was used by Bandaru et al. [186]. It requires four types of variables, namely weather, soil characteristics, plant growth parameters, and management regime. The main goal of the study was to analyse the effect of different weather data sets in the estimation of biomass from short rotation woody crops of hybrid *Populus* spp., using flux towers and four different high resolution gridded weather data at five different locations. The same authors refer that high resolution gridded weather data has some bias when compared with that of the flux towers [186]. This can be, at least partially, explained as modelling and assimilation techniques are not able to characterise in detail the climate that is affected by the topography and land use [186, 187, 189, 190]. Moreover, there seem to be smaller biases for the higher spatial resolutions [186, 188, 191, 192]. Bandaru et al. [186] also stress the importance of the bias determination on the weather as influences as well the biomass estimations. Other authors [28, 193] estimated the biomass for a short rotation coppice in a geographical information systems environment allowing the inclusion of climate and soil variables and the analysis of biomass spatial variability. The use of average yields to estimate biomass over a region results in bias on biomass potential, which might affect the planning of its use due to the variability of local conditions, species, and growth rates. Moreover, in the case of need, short term biomass potential yield can be increased by reducing coppices rotation lengths [28].

Biomass Yield of Energy Plantations

Biomass yield is related to initial density, regime, rotation length, and cultural practices. The analysis will be focused on three *genera Populus* spp., *Salix* spp., and *Eucalyptus* spp. It was based on 33 literature references, corresponding to a total of 415 trials (Table 2). Overall, there seems to be a trend towards higher densities for

Salix spp., when compared to *Populus* spp. and *Eucalyptus* spp., whereas yield tends to be higher for *Eucalyptus* spp. than for the other two *genera*. Rotation length shows similar trends for all three *genera*. Yet, the variability is rather large (Fig. 7). This variability results from the interactions between species and/or clone traits, site, and management practices. The yield of the *Eucalyptus* spp., *Populus* spp., and *Salix* spp. varies between 1–63.8 t·ha^{-1}·y^{-1},0.3–66 t·ha^{-1}·y^{-1} and 0.3–27.5 t·ha^{-1}·y^{-1}; density varies between 2000–7142 stems·ha^{-1}, 278–33,333 stemsha^{-1} and 6666–107,600 stems·ha^{-1}; and rotation length among 2–6 y, 1–12 y and 2–19 y, respectively.

For all *genera*, there is a yield increase from the first to the second rotation, due to the increase of density, *i.e.*, each stump had more than one stool, for *Eucalyptus* [74], *Populus* [101, 109, 113, 133, 165, 174] and *Salix* [26]. However, other studies report a decrease in yield from the first to the second rotation for *Populus* [21, 108]. From the second to the third rotation some studies report an increase in yield for *Eucalyptus* [74] and *Populus* [101, 109, 113] while others account for its reduction for *Populus* [133, 165] and for *Salix* [26]. In the fourth rotation, it is observed a reduction of yield for all species [26, 74, 113].

There was no clear trend between density and production. This is probably related to the site quality and climate as well as the management practices. Yet, considering studies where several densities have been analysed it can be seen an increase in production with the increase of density. For example, for *Salix*, Schweier and Becker [178] reported for an initial density of 12,000 stems·ha^{-1} a yield of 6.8 t·ha^{-1}·y^{-1} and 9.7 t·ha^{-1}·y^{-1}, while for a density of 13,200 stems·ha^{-1} a yield of 11.7 t·ha^{-1}·y^{-1}. This corresponds to an increase of 10% in the number of stems and an increase in yield of 72% and 20%, respectively. For *Populus*, in Italy, Di Matteo et al. [111] reported that an increase in initial density from 7140 stems·ha^{-1} to 10,360 stems·ha^{-1} (an increase of *circa* 45%) resulted in an increase of yield, from 12.2 t·ha^{-1}·y^{-1} to 13.9 t·ha^{-1}·y^{-1} (*circa* plus 14%). For *Populus*, in Germany, an increase of 10% in density (from 10,000 stems·ha^{-1} to 11,000 stems·ha^{-1}) [199] attained an increase in yield between 27% and 86% (from 4.4 t·ha^{-1}·y^{-1} and 5.9 t·ha^{-1}·y^{-1}to 5.6 t·ha^{-1}·y^{-1}, and 8.2 t·ha^{-1}·y^{-1}, respectively). Yet, the same authors for the same initial density increase observed also a reduction of yield of—6.2% (from 5.9 t·ha^{-1}·y^{-1}to 5.6 t·ha^{-1}·y^{-1}). In another study, for *Populus*, Oliveira at al. [39] tested eight different initial densities (6666 stems·ha^{-1}, 10,000 stems·ha^{-1}, 13,333 stems·ha^{-1}, 15,000 stems·ha^{-1}, 17,316 stems·ha^{-1}, 20,000 stems·ha^{-1}, 25,000 stems·ha^{-1}, and 33,333 stems·ha^{-1}) resulting in the increase of yield from the lowest initial density (6666 stems·ha^{-1}) to the fourth lowest (15,000 stems·ha^{-1}) when compared with the highest one (33,333 stems·ha^{-1}) of 179.6%, 54.3%, 76.0%, and 24.7%, respectively. The fifth and the seventh initial densities (17,316 stems·ha^{-1} and 25,000 stems·ha^{-1}) resulted in a reduction of yield of −38.8% and −13.2%, respectively, while for the sixth (20,000 stems·ha^{-1}) a small increase of 8.3% was observed. These results underpin the variability in the yields, which are probably related to the site conditions, climate, and competition between individuals.

It is known that some broadleaved species show considerable variability in their ability to coppice which is associated with their ability to produce sprouts from dormant or adventitious buds or ligno-tubers [117, 200, 201]. Sims et al. [74] observed

Table 2 Energy plantations density, rotation, and cultural practices referred on literature (where C is control of spontaneous vegetation, F fertilisation, and I irrigation)

Genus	Country	Density	Rotation length	Rotation number	Cultural practices	Number of trials	Reference
Eucalyptus	Brazil	2380, 7142	2	1	F	4	[114]
Eucalyptus	Spain	2000	6	1	C,F	1	[37]
Eucalyptus	New Zeeland	2200	3	1	C	19	[75]
Eucalyptus	New Zeeland	2200	3	1,2,3,4,5	C	62	[74]
Eucalyptus	Australia	4000	3	1	F	2	[180]
Populus	UK	10,000	2,4	1,2	C	9	[101]
Populus	UK	10,000	3	1,2	C	32	[194]
Populus	Czech Republic	2222	4	1,2,3,4	C,F,I	48	[113]
Populus	Italy	6666	3	2	–	7	[179]
Populus	Italy	7140, 10,360	3	1	C,F,I	2	[111]
Populus	Belgium	10,000	4	1–4	C	1	[195]
Populus	Italy	14,100	2	1,2,3	–	5	[165]
Populus	Canada	18,000	4	1	None	4	[98]
Populus	Belgium	10,000	3	1,2	C,F,I	3	[21]
Populus	Belgium	10,000	4	1	C,F,I	2	[108]
Populus	Belgium	10,000	3	2	None	17	[108]
Populus	Belgium	10,000	3	2	C	2	[107]
Populus	Belgium	10,000	4	2	None	6	[107]
Populus	Italy	5000	3	2	F	1	[196]
Populus	Italy	5000	3	2	F	1	[196]
Populus	Spain	6666, 10,000, 13,333, 15,000, 17,316, 20,000, 25,000, 33,333	3	1	F,I	8	[39]
Populus	Italy	5900	3, 10	1,2	C,F,I	52	[109]
Populus	France	15,625	2	1	C,F,I	2	[197]
Populus	UK	4444, 10,000	5	1	None	3	[63]
Populus	UK	10,000	5	1	–	1	[92]

(continued)

Table 2 (continued)

Genus	Country	Density	Rotation length	Rotation number	Cultural practices	Number of trials	Reference
Populus	China	278, 4000	12	1	F	12	[103]
Populus	Spain	33,333	3	1	C,F,I	1	[198]
Populus	Italy	10,000	1, 2	1,2,3	–	10	[133]
Populus	Czech Republic	2222, 7407	3	5,6	None	16	[123]
Populus	France	7272	3	1	C	4	[174]
Populus	Belgium	20,000	4	1	None	1	[58]
Salix	USA	107,600	3	–	F, F,I	10	[175]
Salix	UK	10,000	3	1,2	C	32	[194]
Salix	Canada	17,000	3	1,2,3,4,5	C,F	5	[26])
Salix	Finland	20,000	6, 11, 19	1	F	6	[112]
Salix	Canada	18,000	4	1	None	10	[98])
Salix	UK	10,000	5	1	None	1	[63]
Salix	UK	10,000	5	1	–	1	[92]
Salix	Germany	14,800	3	1	–	5	[199]
Salix	Germany	13,200	2,3	1	C,I	6	[178]
Salix	Belgium	20,000	4	1	None	1	[58]

a wide variation in the ability to sprout of 19 *Eucalyptus* species and Dillen at al. [195] of 17 *Populus* clones. Furthermore, survival rates had considerable variations in both studies. As yield in energy plantations depends on density [75] the *Eucalyptus* species with higher densities were those that reached the higher yields [74], regardless of the rotation. The higher yields in the coppice regime can also be explained by the faster growth of the sprouts as their initial development takes advantage of the existing stump root system, thus not experiencing the plantation stress that the seedlings have to surpass [117, 200–202]. It is also known that the ability of a stump to resprout after successive harvests tends to decline. One or several factors can contribute to this decline: stump mortality due to competition, root mortality, disease infection of the cut surfaces, nutrient depletion of the soil, and variation of the tree ratios root/shoot [74, 117, 200, 201]. Thus, the better suited species for energy plantations, under the coppice regime, are those that are able to maintain stumps and root systems with high vigour, to resprout vigorously, and have sprouts with high growth rates, enabling in this way to have stands of high densities and yields [74].

Overall, *circa* 19% of the studies had no information (7%) or was not made (12%) the control of spontaneous vegetation, fertilisation, and irrigation. From the remaining 81% of studies, control of spontaneous vegetation was used in about 85% of the studies, fertilisation with or without control of spontaneous vegetation in 51%, and irrigation with or without spontaneous vegetation and fertilisation in

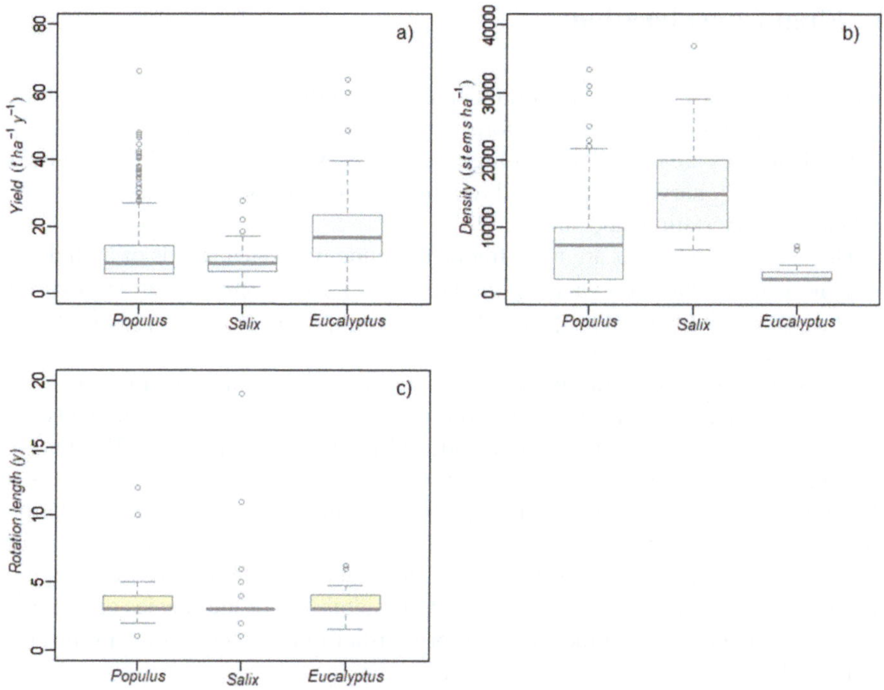

Fig. 7 Boxplots of yield, density, and rotation length

32%. The analysis per species revealed that for *Eucalyptus* spp., none of the trials was irrigated, in 60% control of spontaneous vegetation was done, in 36% control of spontaneous vegetation and fertilisation were used, and in 4% were fertilised. For *Populus* spp., 35% had control of spontaneous vegetation, 54% had control of spontaneous vegetation, fertilisation, and irrigation, 7% were fertilised and 4% were fertilised and irrigated. For *Salix* spp., in 61% control of spontaneous vegetation was made, 12% were fertilised, 12% were fertilised, and irrigated, 7% control of spontaneous vegetation, and fertilisation was made, and 7% control of spontaneous vegetation, and irrigation were used.

The large variability of yields of the energy plantations seems to be related to soil fertility (physical, and chemical characteristics) [58, 59]. The decrease in yields was also associated with the decrease in rainfall (the lower the site quality, and rainfall the lower the yield) [108, 113]. Another source of variability in yields is related to different mortality, growth rates, and patterns of the species, and clones [74, 75, 107, 108, 174].

4 Final Considerations

Energy plantations have an important role in the biomass for energy availability and may release pressure on other forest systems to supply bioenergy (*e.g.*, [2, 7]). Their advantages are related to their high yields [26], short rotations [12], worldwide availability [31], harvest flexibility through anticipated or delayed harvest [28, 33], and easy transportation, and storage [28].

The energy plantations are most frequently pure even-aged stands of high densities, and usually under coppice regime [1, 12, 21]. Many species can be used (*cf.* Table 1) although the most frequent, at least in Europe, are *Populus* spp., *Salix* spp., and *Eucalyptus* spp. [51, 53].

The selection of the site and management should be suited to the species or clones [1]. Soil and climate are of primordial importance to achieve high yields, and should be near the optimum of the ecological range of the species [12]. Management practices range from planting to harvest and include the selection of density, [12], rotation [109], harvest cycle [110], spatial arrangement [133, 135], plantations techniques [1], control of spontaneous vegetation [86], fertilisation [147], irrigation [163], control of pests, and diseases [86], and harvesting [136]. A wide range of species and management options exist, and the suitability of the species or clones to the site and management practices is of primordial importance to the optimisation of the yield.

Biomass estimation is frequently assessed with allometric functions. Due to the energy plantations' specificities, the existing functions resulted in biased estimations, especially the functions developed for high forest, and long production cycles [113]. Thus, allometric functions were developed for energy plantations (*e.g.*, [26, 174, 179]).

Moreover, yield has a trend towards the increase from the first to the second rotation, due to density increase; and a decrease from the third to the fourth rotation. Yet, the variability is high and contrasting results are found in the literature (*cf.* Sect. 3), which were related to site productivity and climate.

Funding This work is funded by National Funds through FCT–Foundation for Science and Technology under the Project UIDB/05183/2020.

References

1. Dickmann D (2006) Silviculture and biology of short-rotation woody crops in temperate regions: then and now. Biomass Bioenerg 30:696–705. https://doi.org/10.1016/j.biombioe.2005.02.008
2. Bruckman VJ, Terada T, Fukuda K et al (2016) Overmature periurban Quercus-Carpinus coppice forests in Austria and Japan: a comparison of carbon stocks, stand characteristics and conversion to high forest. Eur J Forest Res 135:857–869. https://doi.org/10.1007/s10342-016-0979-2

3. McKendry P (2002) Energy production from biomass (part 1): overview of biomass. Biores Technol 83:37–46. https://doi.org/10.1016/S0960-8524(01)00118-3
4. Pietras J, Stojanović M, Knott R, Pokorný R (2016) Oak sprouts grow better than seedlings under drought stress. iForest–Biogeosci For 9:529–535. https://doi.org/10.3832/ifor1823-009
5. Salomón R, Rodríguez-Calcerrada J, Zafra E et al (2016) Unearthing the roots of degradation of Quercus pyrenaica coppices: a root-to-shoot imbalance caused by historical management? For Ecol Manage 363:200–211. https://doi.org/10.1016/j.foreco.2015.12.040
6. Sixto H, Hernández MJ, Barrio M et al (2007) Plantaciones del género Populus para la producción de biomasa con fines energéticos: revisión. Invest Agrar Sist Recursos For 16:277–294
7. Stojanović M, Sánchez-Salguero R, Levanič T et al (2017) Forecasting tree growth in coppiced and high forests in the Czech Republic. The legacy of management drives the coming Quercus petraea climate responses. For Ecol Manage 405:56–68. https://doi.org/10.1016/j.foreco.2017.09.021
8. Suchomel C, Pyttel P, Becker G, Bauhus J (2012) Biomass equations for sessile oak (*Quercus petraea* (Matt.) Liebl.) and hornbeam (*Carpinus betulus* L.) in aged coppiced forests in southwest Germany. Biomass Bioenerg 46:722–730. https://doi.org/10.1016/j.biombioe.2012.06.021
9. Kirby KJ, Buckley GP, Mills J (2017) Biodiversity implications of coppice decline, transformations to high forest and coppice restoration in British woodland. Folia Geobot 52:5–13. https://doi.org/10.1007/s12224-016-9252-1
10. Matthews JD (1989) Silvicultural systems. Claredon Press, Oxford
11. Szabó P, Müllerová J, Suchánková S, Kotačka M (2015) Intensive woodland management in the Middle Ages: spatial modelling based on archival data. J Hist Geogr 48:1–10. https://doi.org/10.1016/j.jhg.2015.01.005
12. Dimitriou I, Rutz D (2015) Sustainable short rotation coppice a handbook. WIP Renewable Energies, Munich
13. Bisoffi S, Minotta G, Paris P (2009) Indirizzi colturali e valorizzazione delle produzioni legnose fuori foresta. In: Atti del Terzo Congresso Nazionale di Selvicoltura. Accademia Italiana di Scienze Forestali, pp 729–736
14. Cañellas I, Huelin P, Hernández MJ et al (2012) The effect of density on short rotation Populus sp. plantations in the Mediterranean area. Biomass Bioenerg 46:645–652. https://doi.org/10.1016/j.biombioe.2012.06.032
15. Deraedt W, Ceulemans R (1998) Clonal variability in biomass production and conversion efficiency of poplar during the establishment year of a short rotation coppice plantation. Biomass Bioenerg 15:391–398. https://doi.org/10.1016/S0961-9534(98)00045-2
16. Fiala M, Bacenetti J, Scaravonati A, Bergonzi A (2010) Short rotation coppice in northern Italy: comprehensive sustainability. In: 18th European Biomass Conference and Exhibition. Lyon, pp 342–348
17. Fiorese G, Guariso G (2010) A GIS-based approach to evaluate biomass potential from energy crops at regional scale. Environ Model Softw 25:702–711. https://doi.org/10.1016/j.envsoft.2009.11.008
18. Karacic A, Weih M (2006) Variation in growth and resource utilisation among eight poplar clones grown under different irrigation and fertilisation regimes in Sweden. Biomass Bioenerg 30:115–124. https://doi.org/10.1016/j.biombioe.2005.11.007
19. Kauter D, Lewandowski I, Claupein W (2003) Quantity and quality of harvestable biomass from Populus short rotation coppice for solid fuel use—a review of the physiological basis and management in uences. Biomass Bioenerg 24:411–427. https://doi.org/10.1016/S0961-9534(02)00177-0
20. Klasnja B, Orlovic S, Drekic M, Markovic M (2003) Energy production from short rotation poplar plantations. In: Interdisciplinary regional research–ISIRR 2003. Hunedoara, Romania, pp 161–166
21. Laureysens I, Deraedt W, Indeherberge T, Ceulemans R (2003) Population dynamics in a 6-year old coppice culture of poplar. I. Clonal di erences in stool mortality, shoot dynamics and

shoot diameter distribution in relation to biomass production. Biomass Bioenerg 24:81–95. https://doi.org/10.1016/S0961-9534(02)00105-8
22. Mitchell CP, Stevens EA, Watters MP (1999) Short-rotation forestry—operations, productivity and costs based on experience gained in the UK. For Ecol Manage 121:123–136. https://doi.org/10.1016/S0378-1127(98)00561-1
23. Pérez-Cruzado C, Sanchez-Ron D, Rodríguez-Soalleiro R et al (2014) Biomass production assessment from *Populus* spp. short-rotation irrigated crops in Spain. GCB Bioenergy 6:312–326. https://doi.org/10.1111/gcbb.12061
24. Volk T, Abrahamson L, Nowak C et al (2006) The development of short-rotation willow in the northeastern United States for bioenergy and bioproducts, agroforestry and phytoremediation. Biomass Bioenerg 30:715–727. https://doi.org/10.1016/j.biombioe.2006.03.001
25. Berhongaray G, El Kasmioui O, Ceulemans R (2013) Comparative analysis of harvesting machines on an operational high-density short rotation woody crop (SRWC) culture: one-process versus two-process harvest operation. Biomass Bioenerg 58:333–342. https://doi.org/10.1016/j.biombioe.2013.07.003
26. Guidi W, Pitre EF, Labrecque M (2013) Short-rotation coppice of willows for the production of biomass in Eastern Canada. In: Matovic MD (ed) Biomass now–sustainable growth and use. InTech, Rijeka, Croatia, pp 421–448
27. Vega-Nieva DJ, Tomé M, Tomé J et al (2013) Developing a general method for the estimation of the fertility rating parameter of the 3-PG model: application in *Eucalyptus globulus* plantations in northwestern Spain. Can J For Res 43:627–636. https://doi.org/10.1139/cjfr-2012-0491
28. Vávrová K, Knápek J, Weger J (2017) Short-term boosting of biomass energy sources—determination of biomass potential for prevention of regional crisis situations. Renew Sustain Energy Rev 67:426–436. https://doi.org/10.1016/j.rser.2016.09.015
29. Gonçalves AC, Malico I, Sousa AMO (2019) Solid biomass from forest trees to energy: a review. In: Jacob-Lopes E, Queiroz Zepka L (eds) Renewable resources and biorefineries. IntechOpen
30. Malico I, Nepomuceno Pereira R, Gonçalves AC, Sousa AMO (2019) Current status and future perspectives for energy production from solid biomass in the European industry. Renew Sustain Energy Rev 112:960–977. https://doi.org/10.1016/j.rser.2019.06.022
31. Verwijst T, Lundkvist A, Edelfeldt S, Albertsso J (2013) Development of sustainable willow short rotation forestry in Northern Europe. In: Matovic MD (ed) Biomass now–sustainable growth and use. InTech
32. Yemshanov D, McKenney D (2008) Fast-growing poplar plantations as a bioenergy supply source for Canada. Biomass Bioenerg 32:185–197. https://doi.org/10.1016/j.biombioe.2007.09.010
33. Hauk S, Gandorfer M, Wittkopf S et al (2017) Ecological diversification is risk reducing and economically profitable—the case of biomass production with short rotation woody crops in south German land-use portfolios. Biomass Bioenerg 98:142–152. https://doi.org/10.1016/j.biombioe.2017.01.018
34. Rothe A, Moroni M, Neyland M, Wilnhammer M (2015) Current and potential use of forest biomass for energy in Tasmania. Biomass Bioenerg 80:162–172. https://doi.org/10.1016/j.biombioe.2015.04.021
35. Ciancio O, Corona P, Lamonaca A et al (2006) Conversion of clearcut beech coppices into high forests with continuous cover: a case study in central Italy. For Ecol Manage 224:235–240. https://doi.org/10.1016/j.foreco.2005.12.045
36. Kopecký M, Hédl R, Szabó P (2013) Non-random extinctions dominate plant community changes in abandoned coppices. J Appl Ecol 50:79–87. https://doi.org/10.1111/1365-2664.12010
37. Pérez S, Renedo CJ, Ortiz A et al (2011) Energetic density of different forest species of energy crops in Cantabria (Spain). Biomass Bioenerg 35:4657–4664. https://doi.org/10.1016/j.biombioe.2011.09.008
38. White EM (2010) Woody biomass for bioenergy and biofuels in the United States: a briefing paper. USDA Forest Service, Portland, Oregon, USA

39. Oliveira N, Sixto H, Cañellas I et al (2015) Productivity model and reference diagram for short rotation biomass crops of poplar grown in Mediterranean environments. Biomass Bioenerg 72:309–320. https://doi.org/10.1016/j.biombioe.2014.09.019
40. Van Calster H, Baeten L, De Schrijver A et al (2007) Management driven changes (1967–2005) in soil acidity and the understorey plant community following conversion of a coppice-with-standards forest. For Ecol Manage 241:258–271. https://doi.org/10.1016/j.foreco.2007.01.007
41. Guidi W, Kadri H, Labrecque M (2012) Establishment techniques to using willow for phytoremediation on a former oil refinery in southern Quebec: achievements and constraints. Chem Ecol 28:49–64. https://doi.org/10.1080/02757540.2011.627857
42. Kuzovkina YA, Volk TA (2009) The characterization of willow (Salix L.) varieties for use in ecological engineering applications: co-ordination of structure, function and autecology. Ecol Eng 35:1178–1189. https://doi.org/10.1016/j.ecoleng.2009.03.010
43. Licht LA, Isebrands JG (2005) Linking phytoremediated pollutant removal to biomass economic opportunities. Biomass Bioenerg 28:203–218. https://doi.org/10.1016/j.biombioe.2004.08.015
44. Moreno-Jiménez E, Peñalosa JM, Manzano R et al (2009) Heavy metals distribution in soils surrounding an abandoned mine in NW Madrid (Spain) and their transference to wild flora. J Hazard Mater 162:854–859. https://doi.org/10.1016/j.jhazmat.2008.05.109
45. Nagendran R, Selvam A, Joseph K, Chiemchaisri C (2006) Phytoremediation and rehabilitation of municipal solid waste landfills and dumpsites: a brief review. Waste Manage 26:1357–1369. https://doi.org/10.1016/j.wasman.2006.05.003
46. Weetman GF (2000) Silvicultural systems for biomass production in Canada. NZ J Forest Sci 30:5–15
47. Liang W, Hu H, Liu F, Zhang D (2006) Research advance of biomass and carbon storage of poplar in China. J For Res 17:75–79. https://doi.org/10.1007/s11676-006-0018-0
48. DeBell DS, Harrington CA (1993) Deploying genotypes in short-rotation plantations: mixtures and pure cultures of clones and species. For Chron 69:705–713. https://doi.org/10.5558/tfc69705-6
49. Updegraff K, Baughman MJ, Taff SJ (2004) Environmental benefits of cropland conversion to hybrid poplar: economic and policy considerations. Biomass Bioenerg 27:411–428. https://doi.org/10.1016/j.biombioe.2004.05.002
50. Ceulemans R, McDonald AJS, Pereira JS (1996) A comparison among eucalypt, poplar and willow characteristics with particular reference to a coppice, growth-modelling approach. Biomass Bioenerg 11:215–231. https://doi.org/10.1016/0961-9534(96)00035-9
51. Hinchee M, Rottmann W, Mullinax L et al (2009) Short-rotation woody crops for bioenergy and biofuels applications. In Vitro Cell Dev Biol Plant 45:619–629. https://doi.org/10.1007/s11627-009-9235-5
52. Rockwood DL, Naidu CV, Carter DR et al (2004) Short-rotation woody crops and phytoremediation: opportunities for agroforestry? Agrofor Syst 61:51–63. https://doi.org/10.1023/B:AGFO.0000028989.72186.e6
53. Sims REH, Venturi P (2004) All-year-round harvesting of short rotation coppice eucalyptus compared with the delivered costs of biomass from more conventional short season, harvesting systems. Biomass Bioenerg 26:27–37. https://doi.org/10.1016/S0961-9534(03)00081-3
54. Goel VL, Behl HM (2004) Productivity assessment of three leguminous species under high-density plantations on degraded soil sites. Biomass Bioenerg 27:403–409. https://doi.org/10.1016/j.biombioe.2004.04.004
55. Kumar BM, George SJ, Jamaludheen V, Suresh TK (1998) Comparison of biomass production, tree allometry and nutrient use eficiency of multipurpose trees grown in woodlot and silvopastoral experiments in Kerala, India. For Ecol Manage 112:145–163. https://doi.org/10.1016/S0378-1127(98)00325-9
56. Geyer W (2006) Biomass production in the Central Great Plains USA under various coppice regimes. Biomass Bioenerg 30:778–783. https://doi.org/10.1016/j.biombioe.2005.08.002

57. Tuskan GA (1998) Short-rotation woody crop supply systems in the United States: what do we know and what we need to know? Biomass Bioenerg 14:307–315. https://doi.org/10.1016/S0961-9534(97)10065-4
58. Vande Walle I, Van Camp N, Van de Casteele L et al (2007) Short-rotation forestry of birch, maple, poplar and willow in Flanders (Belgium) I—biomass production after 4 years of tree growth. Biomass Bioenerg 31:267–275. https://doi.org/10.1016/j.biombioe.2007.01.019
59. Vande Walle I, Van Camp N, Van de Casteele L et al (2007) Short-rotation forestry of birch, maple, poplar and willow in Flanders (Belgium) II. Energy production and CO2 emission reduction potential. Biomass Bioenerg 31:276–283. https://doi.org/10.1016/j.biombioe.2007.01.002
60. Aosaar J, Varik M, Uri V (2012) Biomass production potential of grey alder (Alnus incana (L.) Moench.) in Scandinavia and Eastern Europe: a review. Biomass Bioenerg 45:11–26. https://doi.org/10.1016/j.biombioe.2012.05.013
61. Johansson T (2000) Biomass equations for determining fractions of common and grey alders growing on abandoned farmland and some practical implications. Biomass Bioenerg 18:147–159. https://doi.org/10.1016/S0961-9534(99)00078-1
62. Proe MF (2002) E ects of spacing, species and coppicing on leaf area, light interception and photosynthesis in short rotation forestry. Biomass Bioenergy 12
63. Proe MF, Craig J, Gri J (1999) Comparison of biomass production in coppice and single stem woodland management systems on an imperfectly drained gley soil in central Scotland. Biomass Bioenerg 17:141–151. https://doi.org/10.1016/S0961-9534(99)00029-X
64. Uri V, Aosaar J, Varik M et al (2014) The dynamics of biomass production, carbon and nitrogen accumulation in grey alder (Alnus incana (L.) Moench) chronosequence stands in Estonia. For Ecol Manage 327:106–117. https://doi.org/10.1016/j.foreco.2014.04.040
65. Uri V, Lõhmus K, Mander Ü et al (2011) Long-term effects on the nitrogen budget of a short-rotation grey alder (Alnus incana (L.) Moench) forest on abandoned agricultural land. Ecol Eng 37:920–930. https://doi.org/10.1016/j.ecoleng.2011.01.016
66. Uri V, Vares A, Tullus H, Kanal A (2007) Above-ground biomass production and nutrient accumulation in young stands of silver birch on abandoned agricultural land. Biomass Bioenerg 31:195–204. https://doi.org/10.1016/j.biombioe.2006.08.003
67. Uri V, Tullus H, Lõhmus K (2002) Biomass production and nutrient accumulation in short-rotation grey alder (Alnus incana (L.) Moench) plantation on abandoned agricultural land. For Ecol Manage 161:169–179. https://doi.org/10.1016/S0378-1127(01)00478-9
68. Embaye K (2001) The potential of bamboo as an interceptor and converter of solar energy into essential goods and services: focus on Ethiopia. Int J Sust Dev World 8:346–355. https://doi.org/10.1080/13504500109470092
69. Scurlock JMO, Dayton DC, Hames B (2000) Bamboo: an overlooked biomass resource? Biomass Bioenerg 19:229–244. https://doi.org/10.1016/S0961-9534(00)00038-6
70. Srivastava AK (1995) Biomass and energy production in Casuarina equisetifolia plantation stands in the degraded dry tropics of the Vindhyan plateau, India. Biomass Bioenerg 9:465–471. https://doi.org/10.1016/0961-9534(95)00048-8
71. Pérez-Cruzado C, Merino A, Rodríguez-Soalleiro R (2011) A management tool for estimating bioenergy production and carbon sequestration in Eucalyptus globulus and Eucalyptus nitens grown as short rotation woody crops in north-west Spain. Biomass Bioenerg 35:2839–2851. https://doi.org/10.1016/j.biombioe.2011.03.020
72. Rockwood DL, Rudie AW, Ralph SA et al (2008) Energy product options for Eucalyptus species grown as short rotation woody crops. Int J Mol Sci 9:1361–1378. https://doi.org/10.3390/ijms9081361
73. Senelwa K, Sims REH (1997) Tree biomass equations for short rotation eucalypts grown in New Zealand. Biomass Bioenerg 13:133–140. https://doi.org/10.1016/S0961-9534(97)00026-3
74. Sims REH, Senelwa K, Maiava T, Bullock BT (1999) Eucalyptus species for biomass energy in New Zealand—II: coppice performance. Biomass Bioenerg 17:333–343. https://doi.org/10.1016/S0961-9534(99)00043-4

75. Sims REH, Senelwa K, Maiava T, Bullock BT (1999) Eucalyptus species for biomass energy in New Zealand—I: growth screening trials at first harvest. Biomass Bioenerg 16:199–205. https://doi.org/10.1016/S0961-9534(98)00078-6
76. Khamzina A, Lamers JPA, Worbes M et al (2006) Assessing the potential of trees for afforestation of degraded landscapes in the aral sea basin of Uzbekistan. Agrofor Syst 66:129–141. https://doi.org/10.1007/s10457-005-4677-1
77. Fuwape JA, Akindele SO (1997) Biomass yield and anergy value of some fast-growing multipurpose trees in Nigeria. Biomass Bioenerg 12:101–106. https://doi.org/10.1016/S0961-9534(96)00061-X
78. Onyekwelu JC (2004) Above-ground biomass production and biomass equations for even-aged Gmelina arborea (ROXB) plantations in south-western Nigeria. Biomass Bioenerg 26:39–46. https://doi.org/10.1016/S0961-9534(03)00100-4
79. Tenorio C, Moya R, Arias-Aguilar D, Briceño-Elizondo E (2016) Biomass yield and energy potential of short-rotation energy plantations of Gmelina arborea one year old in Costa Rica. Ind Crops Prod 82:63–73. https://doi.org/10.1016/j.indcrop.2015.12.005
80. Coyle DR, Aubrey DP, Coleman MD (2016) Growth responses of narrow or broad site adapted tree species to a range of resource availability treatments after a full harvest rotation. For Ecol Manage 362:107–119. https://doi.org/10.1016/j.foreco.2015.11.047
81. Davis A, Trettin C (2006) Sycamore and sweetgum plantation productivity on former agricultural land in South Carolina. Biomass Bioenerg 30:769–777. https://doi.org/10.1016/j.biombioe.2005.08.001
82. López F, Pérez A, Zamudio MAM et al (2012) Paulownia as raw material for solid biofuel and cellulose pulp. Biomass Bioenerg 45:77–86. https://doi.org/10.1016/j.biombioe.2012.05.010
83. Zuazo VHD, Bocanegra JAJ, Torres FP et al (2013) Biomass yield potential of paulownia trees in a semi-arid Mediterranean environment (S Spain). Int J Renew Energy Res (IJRER) 3:789–793
84. Coleman MD, Coyle DR, Blake J, et al (2003) Production of short-rotation woody crops grown with a range of nutrient and water availability: establishment report and first-year responses. Forest Service, USDA
85. Devine WD, Tyler DD, Mullen MD et al (2006) Conversion from an American sycamore (Platanus occidentalis L.) biomass crop to a no-till corn (Zea mays L.) system: crop yields and management implications. Soil Tillage Res 87:101–111. https://doi.org/10.1016/j.still.2005.03.006
86. Fischer M, Kelley AM, Ward EJ et al (2017) A critical analysis of species selection and high vs. low-input silviculture on establishment success and early productivity of model short-rotation wood-energy cropping systems. Biomass Bioenerg 98:214–227. https://doi.org/10.1016/j.biombioe.2017.01.027
87. Tang Z, Land SB Jr (1996) Early growth, leaf development, and dry-weight production of sycamore rooted cuttings. Biomass Bioenerg 10:221–229
88. Fang S, Xue J, Tang L (2007) Biomass production and carbon sequestration potential in poplar plantations with different management patterns. J Environ Manage 85:672–679. https://doi.org/10.1016/j.jenvman.2006.09.014
89. Fang S, Xu X, Lu S, Tang L (1999) Growth dynamics and biomass production in short-rotation poplar plantations: 6-year results for three clones at four spacings. Biomass Bioenerg 17:415–425. https://doi.org/10.1016/S0961-9534(99)00060-4
90. Liesebach M, von Wuehlisch G, Muhs H-J (1999) Aspen for short-rotation coppice plantations on agricultural sites in Germany: effects of spacing and rotation time on growth and biomass production of aspen progenies. For Ecol Manage 121:25–39. https://doi.org/10.1016/S0378-1127(98)00554-4
91. Pontailler J, Ceulemans R, Guittet J, Mau F (1997) Linear and non-linear functions of volume index to estimate woody biomass in high density young poplar stands. Ann Sci For 54:335–345. https://doi.org/10.1051/forest:19970402
92. Proe MF, Griffths JH, Craig J (2002) Effects of spacing, species and coppicing on leaf area, light interception and photosynthesis in short rotation forestry. Biomass Bioenerg 23:315–326. https://doi.org/10.1016/S0961-9534(02)00060-0

93. Zabek LM, Prescott CE (2006) Biomass equations and carbon content of aboveground leafless biomass of hybrid poplar in Coastal British Columbia. For Ecol Manage 223:291–302. https://doi.org/10.1016/j.foreco.2005.11.009
94. Felker P (1981) Uses of tree legumes in semiarid regions. Econ Bot 35:174–186. https://doi.org/10.1007/BF02858684
95. Lemus R, Lal R (2005) Bioenergy crops and carbon sequestration. Crit Rev Plant Sci 24:1–21. https://doi.org/10.1080/07352680590910393
96. Gasol CM, Brun F, Mosso A et al (2010) Economic assessment and comparison of acacia energy crop with annual traditional crops in Southern Europe. Energy Policy 38:592–597. https://doi.org/10.1016/j.enpol.2009.10.011
97. Manzone M, Bergante S, Facciotto G (2015) Energy and economic sustainability of woodchip production by black locust (Robinia pseudoacacia L.) plantations in Italy. Fuel 140:555–560. https://doi.org/10.1016/j.fuel.2014.09.122
98. Labrecque M, Teodorescu TI (2005) Field performance and biomass production of 12 willow and poplar clones in short-rotation coppice in southern Quebec (Canada). Biomass Bioenerg 29:1–9. https://doi.org/10.1016/j.biombioe.2004.12.004
99. Sudha P, Ravindranath NH (1999) Land availability and biomass production potential in India. Biomass Bioenerg 16:207–221. https://doi.org/10.1016/S0961-9534(98)00083-X
100. Duku MH, Gu S, Hagan EB (2011) A comprehensive review of biomass resources and biofuels potential in Ghana. Renew Sustain Energy Rev 15:404–415. https://doi.org/10.1016/j.rser.2010.09.033
101. Armstrong A, Johns C, Tubby I (1999) Effects of spacing and cutting cycle on the yield of poplar grown as an energy crop. Biomass Bioenerg 17:305–314. https://doi.org/10.1016/S0961-9534(99)00054-9
102. Marchadier H, Sigaud P (2005) Los álamos en la investigación biotecnológica. Unasylva 221:38–39
103. Qian Z, Ge X, Bai Y et al (2022) Effects of different planting configurations and clones on biomass and carbon storage of a 12-year-old poplar ecosystem in southern China. Can J For Res 52:70–78. https://doi.org/10.1139/cjfr-2021-0041
104. Mosseler A, Major JE (2022) Clonal variation in coppiced and uncoppiced growth, root sprout stem formation, and biomass partitioning in *Salix interior* on two highly disturbed site types. Can J For Res 52:148–157. https://doi.org/10.1139/cjfr-2021-0025
105. Tharakan PJ, Robison DJ, Abrahamson LP, Nowak CA (2001) Multivariate approach for integrated evaluation of clonal biomass production potential. Biomass Bioenerg 21:237–247. https://doi.org/10.1016/S0961-9534(01)00038-1
106. Willebrand E, Verwijst T (1993) Population dynamics of willow coppice systems and their implications for management of short-rotation forests. For Chron 69:699–704. https://doi.org/10.5558/tfc69699-6
107. Laureysens I, Pellis A, Willems J, Ceulemans R (2005) Growth and production of a short rotation coppice culture of poplar. III. Second rotation results. Biomass Bioenergy 29:10–21. https://doi.org/10.1016/j.biombioe.2005.02.005
108. Laureysens I, Bogaert J, Blust R, Ceulemans R (2004) Biomass production of 17 poplar clones in a short-rotation coppice culture on a waste disposal site and its relation to soil characteristics. For Ecol Manage 187:295–309. https://doi.org/10.1016/j.foreco.2003.07.005
109. Paris P, Mareschi L, Sabatti M et al (2011) Comparing hybrid Populus clones for SRF across northern Italy after two biennial rotations: survival, growth and yield. Biomass Bioenerg 35:1524–1532. https://doi.org/10.1016/j.biombioe.2010.12.050
110. Resquin F, Bentancor L, Carrasco-Letelier L et al (2022) Rotation length of intensive Eucalyptus plantations: how it impacts on productive and energy sustainability. Biomass Bioenerg 166:106607. https://doi.org/10.1016/j.biombioe.2022.106607
111. Di Matteo G, Sperandio G, Verani S (2012) Field performance of poplar for bioenergy in southern Europe after two coppicing rotations: effects of clone and planting density. iForest Biogeosci For 5:224–229. https://doi.org/10.3832/ifor0628-005

112. Hytönen J, Saarsalmi A (2009) Long-term biomass production and nutrient uptake of birch, alder and willow plantations on cut-away peatland. Biomass Bioenerg 33:1197–1211. https://doi.org/10.1016/j.biombioe.2009.05.014
113. Benetka V, Novotná K, Štochlová P (2014) Biomass production of Populus nigra L. clones grown in short rotation coppice systems in three different environments over four rotations. iForest Biogeosci For 7:233–239. https://doi.org/10.3832/ifor1162-007
114. Eufrade Junior HJ, de Melo RX, Sartori MMP et al (2016) Sustainable use of eucalypt biomass grown on short rotation coppice for bioenergy. Biomass Bioenerg 90:15–21. https://doi.org/10.1016/j.biombioe.2016.03.037
115. Giannini V, Silvestri N, Dragoni F et al (2017) Growth and nutrient uptake of perennial crops in a paludicultural approach in a drained Mediterranean peatland. Ecol Eng 103:478–487. https://doi.org/10.1016/j.ecoleng.2015.11.049
116. Assmann E (1970) The principles of forest yield study. Pergamon Press, Oxford, UK
117. Boudru M (1989) Forêt et sylviculture. Presses Agronomiques de Gembloux, Gembloux, Le Traitement des Forêts
118. Smith DM, Larson BC, Kelty MJ, Ashton PMS (1997) The practice of silviculture. Applied forest ecology, 9th edn. John Wiley & Sons, Inc, New York
119. Bullard MJ, Mustill SJ, McMillan SD et al (2002) Yield improvements through modiÿcation of planting density and harvest frequency in short rotation coppice Salix spp.—1. Yield response in two morphologically diverse varieties. Biomass Bioenerg 22:15–25. https://doi.org/10.1016/S0961-9534(01)00054-X
120. Bergkvist P, Ledin S (1998) Stem biomass yields at different planting designs and spacings in willow coppice systems. Biomass Bioenerg 14:149–156. https://doi.org/10.1016/S0961-9534(97)10021-6
121. Nordh N, Verwijst T (2004) Above-ground biomass assessments and first cutting cycle production in willow (Salix sp.) coppice—a comparison between destructive and non-destructive methods. Biomass Bioenerg 27:1–8. https://doi.org/10.1016/j.biombioe.2003.10.007
122. Dillen M, Vanhellemont M, Verdonckt P et al (2016) Productivity, stand dynamics and the selection effect in a mixed willow clone short rotation coppice plantation. Biomass Bioenerg 87:46–54. https://doi.org/10.1016/j.biombioe.2016.02.013
123. Štochlová P, Novotná K, Costa M, Rodrigues A (2019) Biomass production of poplar short rotation coppice over five and six rotations and its aptitude as a fuel. Biomass Bioenerg 122:183–192. https://doi.org/10.1016/j.biombioe.2019.01.011
124. Willoughby I, Clay D (1996) Herbicides for farm woodlands and short rotation coppice. Forestry Comission, London
125. Stanturf JA, Van Oosten C, Netzer DA, et al (2001) Ecology and silviculture of poplar plantations. Poplar Culture in North America 153–206
126. Tubby I, Armstrong A (2002) Establishment and management of short rotation coppice. Forestry Comission, Edinburgh
127. Buhler DD, Netzer DA, Riemenschneider DE, Hartzler RG (1998) Weed management in short rotation poplar and herbaceous perennial crops grown for biofuel production. Biomass Bioenerg 14:385–394. https://doi.org/10.1016/S0961-9534(97)10075-7
128. Sixto H, Grau JM, Garc JM (2001) Assessment of the e!ect of broad-spectrum pre-emergence herbicides in poplar nurseries. Crop Prot 20:121–126. https://doi.org/10.1016/S0261-2194(00)00064-8
129. Karačić A (2005) Production and ecological aspects of short rotation poplars in Sweden. PhD Thesis, Deprtment of Short Rotation Forestry, Swedish University of Agricultural Sciences
130. Deckmyn G, Laureysens I, Garcia J et al (2004) Poplar growth and yield in short rotation coppice: model simulations using the process model SECRETS. Biomass Bioenerg 26:221–227. https://doi.org/10.1016/S0961-9534(03)00121-1
131. Kopp RF, Abrahamson LP, White EH et al (2001) Willow biomass production during ten successive annual harvests. Biomass Bioenerg 20:1–7. https://doi.org/10.1016/S0961-9534(00)00063-5

132. Telenius B (1999) Stand growth of deciduous pioneer tree species on fertile agricultural land in southern Sweden. Biomass Bioenerg 16:13–23. https://doi.org/10.1016/S0961-9534(98)00073-7
133. Spinelli R, Nati C, Magagnotti N (2008) Harvesting short-rotation poplar plantations for biomass production. Croatian J Forest Eng 29:129–139
134. El Kasmioui O, Ceulemans R (2013) Financial analysis of the cultivation of short rotation woody crops for bioenergy in belgium: barriers and opportunities. BioEnergy Res 6:336–350. https://doi.org/10.1007/s12155-012-9262-7
135. Savoie P, Hébert P-L, Robert F-S, Sidders D (2013) Harvest of short-rotation woody crops in plantations with a biobaler. Energy Power Eng 05:39–47. https://doi.org/10.4236/epe.2013.52A006
136. Vanbeveren SPP, Spinelli R, Eisenbies M et al (2017) Mechanised harvesting of short-rotation coppices. Renew Sustain Energy Rev 76:90–104. https://doi.org/10.1016/j.rser.2017.02.059
137. Ceulemans R, Deraedt W (1999) Production physiology and growth potential of poplars under short-rotation forestry culture. For Ecol Manage 121:9–23. https://doi.org/10.1016/S0378-1127(98)00564-7
138. Fillion M, Brisson J, Teodorescu TI et al (2009) Performance of Salix viminalis and Populus nigra × Populus maximowiczii in short rotation intensive culture under high irrigation. Biomass Bioenerg 33:1271–1277. https://doi.org/10.1016/j.biombioe.2009.05.011
139. Hytönen J (1998) Effect of peat ash fertilization on the nutrient status and biomass production of short-rotation willow on cut-away peatland area. Biomass Bioenerg 15:83–92. https://doi.org/10.1016/S0961-9534(97)10050-2
140. Laidlaw WS, Baker AJM, Gregory D, Arndt SK (2015) Irrigation water quality influences heavy metal uptake by willows in biosolids. J Environ Manage 155:31–39. https://doi.org/10.1016/j.jenvman.2015.03.005
141. Scholz V, Ellerbrock R (2002) The growth productivity, and environmental impact of the cultivation of energy crops on sandy soil in Germany. Biomass Bioenerg 23:81–92. https://doi.org/10.1016/S0961-9534(02)00036-3
142. Coleman M, Tolsted D, Nichols T et al (2006) Post-establishment fertilization of Minnesota hybrid poplar plantations. Biomass Bioenerg 30:740–749. https://doi.org/10.1016/j.biombioe.2006.01.001
143. Dimitriou I, Rosenqvist H (2011) Sewage sludge and wastewater fertilisation of Short Rotation Coppice (SRC) for increased bioenergy production—biological and economic potential. Biomass Bioenerg 35:835–842. https://doi.org/10.1016/j.biombioe.2010.11.010
144. Park BB, Yanai RD, Sahm JM et al (2005) Wood ash effects on plant and soil in a willow bioenergy plantation. Biomass Bioenerg 28:355–365. https://doi.org/10.1016/j.biombioe.2004.09.001
145. Quaye AK, Volk TA, Hafner S et al (2011) Impacts of paper sludge and manure on soil and biomass production of willow. Biomass Bioenerg 35:2796–2806. https://doi.org/10.1016/j.biombioe.2011.03.008
146. Sevel L, Nord-Larsen T, Ingerslev M et al (2014) Fertilization of SRC Willow, I: biomass production response. BioEnergy Res 7:319–328. https://doi.org/10.1007/s12155-013-9371-y
147. Marron N (2015) Agronomic and environmental effects of land application of residues in short-rotation tree plantations: a literature review. Biomass Bioenerg 81:378–400. https://doi.org/10.1016/j.biombioe.2015.07.025
148. Alriksson B, Ledin S, Seeger P (1997) Effect of nitrogen fertilization on growth in a *Salix viminalis* stand using a response surface experimental design. Scand J For Res 12:321–327. https://doi.org/10.1080/02827589709355418
149. Cavanagh A, Gasser MO, Labrecque M (2011) Pig slurry as fertilizer on willow plantation. Biomass Bioenerg 35:4165–4173. https://doi.org/10.1016/j.biombioe.2011.06.037
150. Heaton RJ, Randerson, Slater FM (2000) The silviculture, nutrition & economics of short rotation willow coppice in the uplands of mid-Wales. Cardiff University
151. Jonsson M, Dimitriou I, Aronsson P, Elowson T (2006) Treatment of log yard run-off by irrigation of grass and willows. Environ Pollut 139:157–166. https://doi.org/10.1016/j.envpol.2005.04.026

152. Martin PJ, Stephens W (2006) Willow growth in response to nutrients and moisture on a clay landfill cap soil. I. Growth and biomass production. Bioresource Technol 97:437–448. https://doi.org/10.1016/j.biortech.2005.03.003
153. Perttu KL (1999) Environmental and hygienic aspects of willow coppice in Sweden. Biomass Bioenerg 16:291–297. https://doi.org/10.1016/S0961-9534(98)00012-9
154. González I, Sixto H, Rodríguez-Soalleiro R et al (2022) How can leaf-litter from different species growing in short rotation coppice contribute to the soil nutrient pool? For Ecol Manage 520:120405. https://doi.org/10.1016/j.foreco.2022.120405
155. Bungart R, Hüttl RF (2001) Production of biomass for energy in post-mining landscapes and nutrient dynamics. Biomass Bioenerg 20:181–187. https://doi.org/10.1016/S0961-9534(00)00078-7
156. Dowkiw A, Husson C, Frey P et al (2003) Partial Resistance to *Melampsora larici-populina* leaf rust in hybrid poplars: genetic variability in inoculated excised leaf disk bioassay and relationship with complete resistance. Phytopathology 93:421–427. https://doi.org/10.1094/PHYTO.2003.93.4.421
157. Pei MH, Ruiz C, Bayon C et al (2005) Pathogenic variation in poplar rust Melampsora larici-populina from England. Eur J Plant Pathol 111:147–155. https://doi.org/10.1007/s10658-004-1920-y
158. Lonsdale D, Tabbush P (2002) Poplar rust and its recent impact in Great Britain. For Commis Inform Note 7:1–4
159. Kline KL, Coleman MD (2010) Woody energy crops in the southeastern United States: two centuries of practitioner experience. Biomass Bioenerg 34:1655–1666. https://doi.org/10.1016/j.biombioe.2010.05.005
160. Schütz JP (1990) Sylviculture 1. Presses Polytechniques et Universitaires Romandes, Lausanne, Principes d'Éducation des Forêts
161. Allen SJ, Hall RL, Rosier PT (1999) Transpiration by two poplar varieties grown as coppice for biomass production. Tree Physiol 19:493–501. https://doi.org/10.1093/treephys/19.8.493
162. Perry CH, Miller RC, Brooks KN (2001) Impacts of short-rotation hybrid poplar plantations on regional water yield. For Ecol Manage 143:143–151. https://doi.org/10.1016/S0378-1127(00)00513-2
163. Petzold R, Schwärzel K, Feger K-H (2011) Transpiration of a hybrid poplar plantation in Saxony (Germany) in response to climate and soil conditions. Eur J Forest Res 130:695–706. https://doi.org/10.1007/s10342-010-0459-z
164. Yin C, Wang X, Duan B et al (2005) Early growth, dry matter allocation and water use efficiency of two sympatric species as affected by water stress. Environ Exp Bot 53:315–322. https://doi.org/10.1016/j.envexpbot.2004.04.007
165. Fiala M, Bacenetti J (2012) Economic, energetic and environmental impact in short rotation coppice harvesting operations. Biomass Bioenerg 42:107–113. https://doi.org/10.1016/j.biombioe.2011.07.004
166. Djomo SN, Ac A, Zenone T et al (2015) Energy performances of intensive and extensive short rotation cropping systems for woody biomass production in the EU. Renew Sustain Energy Rev 41:845–854. https://doi.org/10.1016/j.rser.2014.08.058
167. Eisenbies MH, Volk TA, Posselius J et al (2014) Evaluation of a single-pass, cut and chip harvest system on commercial-scale, short-rotation shrub willow biomass crops. Bioenerg Res 7:1506–1518. https://doi.org/10.1007/s12155-014-9482-0
168. Santangelo E, Scarfone A, Giudice AD et al (2015) Harvesting systems for poplar short rotation coppice. Ind Crops Prod 75:85–92. https://doi.org/10.1016/j.indcrop.2015.07.013
169. Culshaw D, Stokes B (1995) Mechanisation of short rotation forestry. Biomass Bioenerg 9:127–140. https://doi.org/10.1016/0961-9534(95)00085-2
170. Pecenka R, Hoffmann T (2015) Harvest technology for short rotation coppices and costs of harvest, transport and storage. Agron Res 13:361–371
171. Magagnotti N, Spinelli R, Kärhä K, Mederski PS (2021) Multi-tree cut-to-length harvesting of short-rotation poplar plantations. Eur J Forest Res 140:345–354. https://doi.org/10.1007/s10342-020-01335-y

172. Pari L, Civitarese V, del Giudice A et al (2013) Influence of chipping device and storage method on the quality of SRC poplar biomass. Biomass Bioenerg 51:169–176. https://doi.org/10.1016/j.biombioe.2013.01.019
173. Telenius B, Verwijst T (1995) The influnece of allometric variation, vertical, biomass distribution and sapmpling procedure on biomass estimates in commercial short-rotation forests. Biores Technol 51:24–253
174. Toillon J, Dallé E, Bodineau G et al (2016) Plasticity of yield and nitrogen removal in 56 Populus deltoides × P. nigra genotypes over two rotations of short-rotation coppice. For Ecol Manage 375:55–65. https://doi.org/10.1016/j.foreco.2016.05.023
175. Adegbidi HG, Volk TA, White EH et al (2001) Biomass and nutrient removal by willow clones in experimental bioenergy plantations in New York State. Biomass Bioenerg 20:399–411. https://doi.org/10.1016/S0961-9534(01)00009-5
176. Bergante S, Manzone M, Facciotto G (2016) Alternative planting method for short rotation coppice with poplar and willow. Biomass Bioenerg 87:39–45. https://doi.org/10.1016/j.biombioe.2016.02.016
177. Mosseler A, Major JE, Labrecque M, Larocque GR (2014) Allometric relationships in coppice biomass production for two North American willows (*Salix* spp.) across three different sites. For Ecol Manage 320:190–196. https://doi.org/10.1016/j.foreco.2014.02.027
178. Schweier J, Becker G (2013) Economics of poplar short rotation coppice plantations on marginal land in Germany. Biomass Bioenerg 59:494–502. https://doi.org/10.1016/j.biombioe.2013.10.020
179. Civitarese V, Faugno S, Pindozzi S et al (2015) Effect of short rotation coppice plantation on the performance and chips quality of a self-propelled harvester. Biosys Eng 129:370–377. https://doi.org/10.1016/j.biosystemseng.2014.11.009
180. Sochacki SJ, Harper RJ, Smettem KRJ et al (2013) Evaluating a sustainability index for nutrients in a short rotation energy cropping system. GCB Bioenergy 5:315–326. https://doi.org/10.1111/j.1757-1707.2012.01202.x
181. Amichev BY, Hangs RD, Van Rees KCJ (2011) A novel approach to simulate growth of multi-stem willow in bioenergy production systems with a simple process-based model (3PG). Biomass Bioenerg 35:473–488. https://doi.org/10.1016/j.biombioe.2010.09.007
182. Bandaru V, Parker NC, Hart Q et al (2015) Economic sustainability modeling provides decision support for assessing hybrid poplar-based biofuel development in California. Calif Agric 69:171–176. https://doi.org/10.3733/ca.v069n03p171
183. Hart QJ, Tittmann PW, Bandaru V, Jenkins BM (2015) Modeling poplar growth as a short rotation woody crop for biofuels in the Pacific Northwest. Biomass Bioenerg 79:12–27. https://doi.org/10.1016/j.biombioe.2015.05.004
184. Malico I, Gonçalves AC (2021) Eucalyptus globulus coppices in Portugal: influence of site and percentage of residues collected for energy. Sustainability 13:5775. https://doi.org/10.3390/su13115775
185. Tallis MJ, Casella E, Henshall PA et al (2013) Development and evaluation of ForestGrowth-SRC a process-based model for short rotation coppice yield and spatial supply reveals poplar uses water more efficiently than willow. GCB Bioenergy 5:53–66. https://doi.org/10.1111/j.1757-1707.2012.01191.x
186. Bandaru V, Pei Y, Hart Q, Jenkins BM (2017) Impact of biases in gridded weather datasets on biomass estimates of short rotation woody cropping systems. Agric For Meteorol 233:71–79. https://doi.org/10.1016/j.agrformet.2016.11.008
187. van Wart J, Grassini P, Cassman KG (2013) Impact of derived global weather data on simulated crop yields. Glob Change Biol 19:3822–3834. https://doi.org/10.1111/gcb.12302
188. Eum H-I, Dibike Y, Prowse T, Bonsal B (2014) Inter-comparison of high-resolution gridded climate data sets and their implication on hydrological model simulation over the Athabasca Watershed, Canada: inter-comparison of gridded climate data and hydrologic simulations. Hydrol Process 28:4250–4271. https://doi.org/10.1002/hyp.10236
189. Miao Z, Shastri Y, Grift TE et al (2012) Lignocellulosic biomass feedstock transportation alternatives, logistics, equipment configurations, and modeling. Biofuels Bioprod Biorefin 6:351–362. https://doi.org/10.1002/bbb.1322

190. Zhong S, Fast J (2003) An evaluation of the MM5, RAMS, and Meso-Eta models at subkilometer resolution using VTMX field campaign data in the Salt Lake Valley. Mon Weather Rev 131:1301–1322. https://doi.org/10.1175/1520-0493(2003)131%3c1301:AEOTMR%3e2.0.CO;2
191. Abatzoglou JT (2013) Development of gridded surface meteorological data for ecological applications and modelling. Int J Climatol 33:121–131. https://doi.org/10.1002/joc.3413
192. Daly C, Halbleib M, Smith JI et al (2008) Physiographically sensitive mapping of climatological temperature and precipitation across the conterminous United States. Int J Climatol 28:2031–2064. https://doi.org/10.1002/joc.1688
193. Aust C, Schweier J, Brodbeck F et al (2014) Land availability and potential biomass production with poplar and willow short rotation coppices in Germany. GCB Bioenergy 6:521–533. https://doi.org/10.1111/gcbb.12083
194. Aylott MJ, Casella E, Tubby I et al (2008) Yield and spatial supply of bioenergy poplar and willow short-rotation coppice in the UK. New Phytol 178:358–370
195. Dillen SY, Djomo SN, Al Afas N et al (2013) Biomass yield and energy balance of a short-rotation poplar coppice with multiple clones on degraded land during 16 years. Biomass Bioenerg 56:157–165. https://doi.org/10.1016/j.biombioe.2013.04.019
196. Liberloo M, Calfapietra C, Lukac M et al (2006) Woody biomass production during the second rotation of a bio-energy Populus plantation increases in a future high CO2 world. Glob Change Biol 12:1094–1106. https://doi.org/10.1111/j.1365-2486.2006.01118.x
197. Pontailler JY, Ceulemans R, Guittet J (1999) Biomass yield of poplar after five 2-year coppice rotations. Forestry 72:157–163
198. Sixto H, Gil P, Ciria P, Camps F (2013) Performance of hybrid poplar clones in short rotation coppice in Mediterranean environments: analysis of genotypic stability. GCB Bioenergy 6:661–671. https://doi.org/10.1111/gcbb.12079
199. Schweier J, Becker G (2012) Harvesting of short rotation coppice—harvesting trials with a cut and storage system in Germany. Silva Fennica 46:287–299. https://doi.org/10.14214/sf.61
200. Boudru M (1992) Forêt et Sylviculture. Presses Agronomiques de Gembloux, Gembloux, Boisements te reboisements artificieles
201. Florence RG (1996) Ecology and silviculture of eucalyptus forests. Csiro Publishing, Collingwood, Victoria
202. Gonçalves AC, Fonseca TF (2023) Influence management and disturbances on the regeneration of forest stands. Front Forests Glob Change 6:1123215. https://doi.org/10.3389/ffgc.2023.1123215

Modelling Biomass

Ana Cristina Gonçalves

Abstract Models are abstractions that enable to assess and predict forest stands variables. Two broad methods to estimate biomass were defined. The direct method, the most accurate, has the disadvantage of resulting from destructive sampling. Inversely, the indirect method uses a variety of mathematical methods, with forest inventory, remote sensing, and ancillary data as explanatory variables. The accuracy of the biomass models is dependent on data acquisition precision and accuracy as well as on the model's uncertainties. Moreover, model accuracy is also dependent on species, individual tree biomass partitioning, stand structure, region, and spatial and temporal scales. This chapter overviews the data sets and mathematical methods used for modelling biomass and their uncertainties. Overall, the performance of the forest biomass functions is linked to its ability to accommodate the variability inherent to forest data and to make biomass assessments, monitoring, and predictions with the best possible precision and accuracy and the smallest bias.

Keywords Data sets · Forest inventory · Remote sensing · Regression · Uncertainties

1 Introduction

The long span of production cycles and the variability in species, site, and stand structure [1] is reflected in the tree allometry as well as in the variability within and between stands (*cf*. chapter "Stand Structure and Biomass"). This brings about several challenges when estimating biomass in stands and forests whether in the selection of predictive variables, of the modelling approach, and the target uncertainties.

A. C. Gonçalves (✉)
MED—Mediterranean Institute for Agriculture, Environment and Development & CHANGE—Global Change and Sustainability Institute, Instituto de Investigação e Formação Avançada, Departamento de Engenharia Rural, Escola de Ciências e Tecnologia, Universidade de Évora, Apartado 94, 7002-544 Évora, Portugal
e-mail: acag@uevora.pt

A. C. Gonçalves and I. Malico (eds.), *Forest Bioenergy*, Green Energy and Technology,
https://doi.org/10.1007/978-3-031-48224-3_5

Models are abstractions, that apply mathematical methods and techniques to develop equations (or systems of equations) to assess and predict dependent variables. Yet, they are much more than the application of the aforementioned methods as they can be used not only to assess a certain variable in a determined moment in time but also to predict its dynamics in time and to explore management alternatives [2]. Models are also determinant in the assessment and monitoring, in space and time, of biomass due to the dynamics of stand structure, disturbances whether natural (biotic or abiotic) and/or artificial (forest management). In space, small, medium, and/or large scales can be used whereas in time short, medium, and/or long term may be used. The space and time levels influence the data sets used for modelling and the mathematical models [3, 4].

The choice of the mathematical modelling approach is dependent on the available data sets and their quality, which in turn are linked to the species, individual tree biomass partitioning, stand structure, geographic region, and spatial and temporal scales. The aforementioned originated the use of many mathematical methods less or more complex, from parametric to non-parametric methods as well as the selection of the variables with the best predictive ability that have low correlation among them. The vast number of biomass models reflects the variability and complexity of the stands and forests and the continuous search for improving predictions.

This chapter aims to provide an overview of the data sets, modelling methods, and techniques and their uncertainties for biomass estimation. The chapter is divided into three sections: data sets (Sect. 2), methods for estimating biomass, including at tree level and per area basis (Sect. 3), and model uncertainties (Sect. 4).

2 Data Sets

Data sets for biomass modelling can be broadly grouped into those derived from (Table 1): destructive sampling, and non-destructive sampling, the latter including forest inventory, remote sensing, and ancillary data.

The ***direct method*** implies a destructive sampling of a set of trees with an appropriate sampling design. The trees are cut and separated by component (Fig. 1), frequently stem, bark, branches, and leaves (or alternatively, crown including branches and leaves), and if below ground biomass is to be considered, roots

Table 1 Data sources for biomass modelling as function of the spatial and temporal scales (where DS is destructive sampling, RS remote sensing, FI forest inventory, AD ancillary data)

Scale		Temporal		
		Short (<5 y)	Medium (5–10 y)	Long (>10 y)
Spatial	Small	RS	FI, RS	DS, FI, RS, AD
	Medium	RS	FI, RS	FI, RS, AD
	Large	RS	FI, RS	FI, RS, AD

[3, 5, 6]. The dry weight of each component is determined and the tree biomass is the sum of the biomass of all components. Frequently, the different components are aggregated in two broad classes: above ground biomass (stem, bark, branches, and leaves) and below ground biomass (roots). The advantage of this method is its accuracy, but the disadvantages are that it is labour demanding, expensive, and destructive. Moreover, it is not suitable for forests with low accessibility or under protection and/or conservation status [4, 5] because of the disturbances on habitat, flora, and fauna [7]. One alternative to sampling design for the destructive sampling data acquisition was developed by Xu et al. [8], in which biomass data is collected from recently wind fallen trees, using the same procedures to collect data as those of the direct method. The advantages are the collection of data in areas under conservation status where tree cutting is seldom allowed. The disadvantages are that data collection is dependent on the wind, not enabling a sampling design (*e.g.*, random), and does not guarantee that all tree dimension classes are adequately represented. Despite the limitations, with appropriate methods of regression, the disadvantages can be overcome [8]. The detailed data sets obtained through direct methods are of the utmost importance for the development of the indirect methods of biomass estimation [3].

The ***indirect method*** uses data from forest inventory, remote sensing and ancillary data as explanatory variables with different mathematical formulations.

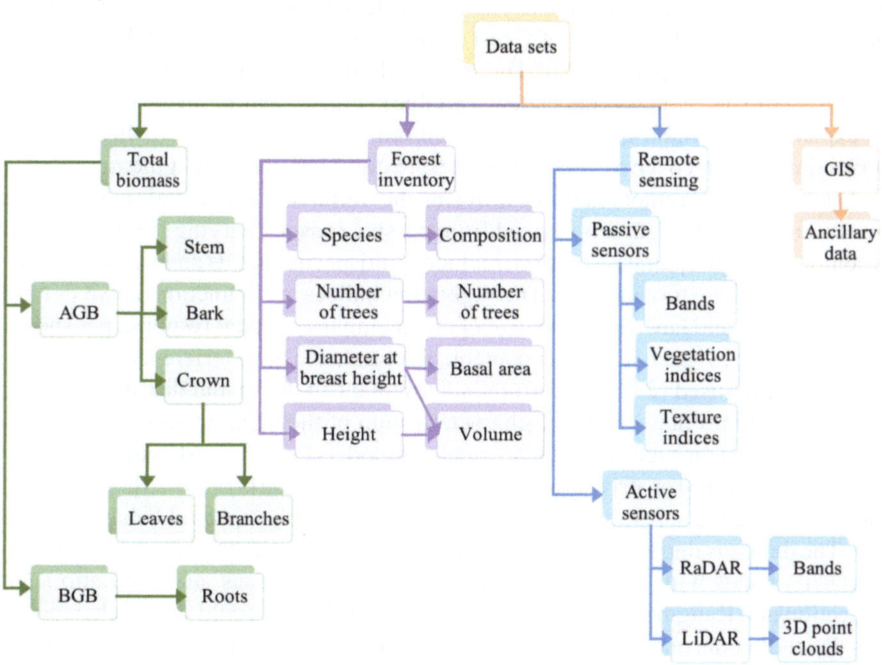

Fig. 1 Data sets used for biomass modelling

Based on the objectives and target area, ***forest inventory*** can be divided in national forest inventories (NFI) and forest management inventories (FMI). National forest inventories' main goal is to provide regional, national, and international statistics while forest management inventories' main goal is to the support planning of management and silvicultural practices [9].

Forest inventories are based on a sampling design, for a given threshold error, and field plots [5, 6, 10]. The variables most frequently measured are forest area, crown cover, tree species, number of trees, diameter at breast height (*dbh*), and height (*h*) [3, 5, 6, 11–14]. Forest area and crown cover are frequently assessed with remote sensing techniques [5, 15–18], while tree species, number of trees, diameter at breast height and height are usually obtained in field inventory plots [6, 10]. Field plots also enable the evaluation of basal area, volume, and biomass [3, 5, 6, 11–14]. The latter two with functions in which the most frequent explanatory variables are diameter at breast height and height (*cf*. Sect. 3). Similarly to the other absolute density measures, such as number of trees, basal area, and volume, biomass can be calculated with allometric functions at tree level or relations between volume and wood density that allow the estimation of the biomass per plot (sum of the biomass of all the trees in the plot) and scale it to a measure per area (typically, the hectare).

Remote sensing data is an alternative to obtain biomass estimates at different scales (from local to global), at short or long time periods, and at low cost [9, 19, 20]. The availability of remote sensing data with different spectral, spatial, radiometric, and temporal resolutions, obtained with diverse technologies enables a wide range of monitoring scales [15, 19, 21] of forest area distribution, species, and their physical and biochemical properties [22]. The tree level allometric functions are frequently used to estimate biomass (dependent variable) (Fig. 2) and biomass models have, as explanatory variables, data derived from remote sensing (*e.g.*, crown horizontal projection, crown cover, vegetation indices, textural indices, height metrics).

Data sets from remote sensing differ in the variables and spatial and temporal resolution [15]. The data sets derived from passive sensors, varying from low spatial resolution to high spatial resolution, include multispectral bands and auxiliary bands (vegetation and texture indices). The vegetation indices are a function of two or more spectral bands and have the advantage of enhancing the spectral response per species or its traits (*e.g.*, vigour, leaf water content, biomass), and the interference of the atmosphere is minimised [23]. The texture indices quantify the spatial distribution of the pixels' grey tones that enable the characterisation of the tree crowns (*e.g.*, shadows, size, and shape) through several mathematical models (*e.g.*, nearest neighbours, grey level co-occurrence matrix, grey level co-occurrence matrix) [24, 25]. The texture indices enable the separation of forest classes with similar spectral behaviours [26, 27]. The data sets derived from active sensors comprise Radio Detection and Ranging (RaDAR) bands with different wavelengths and polarisations, and for Ratio Light Detection and Ranging (LiDAR) a three-dimensional point could whose characteristics depend on the platform (spaceborne, airborne, and terrestrial), signal return (full waveform *vs* discrete) and footprint (small, medium or large) resolution [20, 28, 29].

When forest inventory and remote sensing techniques are compared several advantages can be pointed out to the latter: (i) ability to obtain measurements in the entire

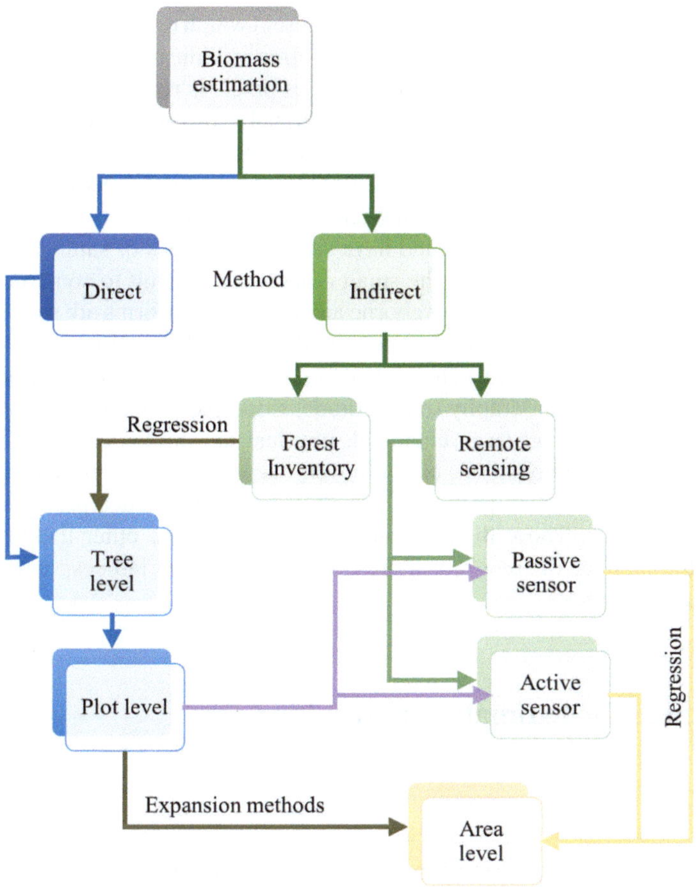

Fig. 2 Biomass direct and indirect estimation methods

study area [9, 30], (ii) relatively low cost for many types of remote sensing data and short time to obtain and process them [20]; (iii) possibility of estimating biomass for small and large areas [15, 20], which is of special importance in forest systems where inventory field work is hard to accomplished due to topography, climate and/or stand structure [5, 20] or when the number of field plots needed make forest inventory costs prohibitive [15]; (iv) possibility of collecting data in short time cycles, enabling time series analysis [9, 18], for ground plot data, cycles shorter than those of the forest inventories (5 or 10 years) are not feasible, due to labour and costs [5]; (v) possibility of collecting data, at different scales, as a function of the imagery spatial resolution [15, 18]; (v) calculation of biomass for all the area without the need of an extrapolation methods [9, 30–33].

Remote sensing data characteristics are related to the target scale. Passive sensors' images of low spatial resolution are better suited for large scales (regional and global) and homogeneous forest areas [4], due to its pixel size, especially in heterogeneous forest areas several land uses/classes may be present which may decrease the accuracy of the models [34]; the medium spatial resolution is indicated for regional or local scales and the effect of mixed land uses/classes pixels is reduced [35, 36]; the high spatial resolution is better suited for local or regional scales and mixed pixels are much less frequent than in the two former spatial resolutions, being suited for heterogeneous and complex forest areas [31, 37]. The RaDAR derived data sets are better suited for large areas where stands are homogeneous, have low density and/or are at the initial stages of development [4] and have the disadvantages of saturating in stands with high biomass [29] and also the small objects are difficult to construct [4]. The LiDAR derived data sets from spaceborne and airborne platforms are suited for large areas whereas the terrestrial ones are for smaller areas [4, 15, 38].

Ancillary data correspond, usually, to thematic maps (frequently in raster format) that characterise the topography (*e.g.*, altitude, slope, aspect), climate (*e.g.*, temperature, precipitation), soil (*e.g.*, chemical, physical and biological characteristics), forests (*e.g.*, area, composition, crown cover, harvested areas, burned areas, areas affected by storms and/or pest and diseases), and areas under protection and/or conservation status. These themes can be used to generate other themes, such as maps of the number of trees, basal area, volume, climatic indices, and soil water holding capacity [5, 9].

3 Methods for Estimating Forest Biomass

Tree Level

Biomass (and carbon stocks) at tree level can be estimated with forest inventory data by a ***ratio between volume and wood density***, which is the relation between dry weight and the volume of water equal to wood volume [3, 39, 40]. This method is used either for stem [3] or merchantable [40] volume. The biomass of crown (or branches and leaves), bark, and roots can be determined with component ratios (proportion of the component in relation to above ground or total biomass) [41]. The proportions of biomass per component present high variability as function of the species, stand development stage, stand structure, site, and silvicultural practices (*cf*. chapter "Stand Structure and Biomass"). This indirect method is easier and more cost effective than the destructive one, yet bias can occur due to the variability of wood density.

Wood density (or specific gravity) varies per species, site, genetics, and silvicultural treatments. It is dependent on trees' growth patterns which in turn are dependent on the available growing space, thus, dependent on site (both edaphic and climatic conditions) and silvicultural practices [42–47]. Moreover, genetics and age are also pivotal to wood density. At early ages, trees develop juvenile wood, which has different characteristics than that of mature wood, and frequently lower density. With ageing trees make the transition, more or less abruptly, from juvenile to mature wood. Mature wood, apart from having better technological characteristics for timber and pulp, has also higher wood density [3, 42, 44]. Additionally, wood density in a tree varies along its vertical profile [43, 48]. A comprehensive list of wood density per species can be found in Chave et al. [43] and at ICRAF (http://db.worldagroforestry.org/).

The evaluation of the biomass components' proportions per species, stand development stage, stand structure, site, and silvicultural practices can improve accuracy and precision, and reduce the bias in the biomass estimations [39]. Furthermore, uncertainties are introduced by using volume equations [49].

The allometric functions at tree level result in mathematical functions that are derived from the relation between biomass (total or per component) obtained from the direct method (dependent variable) and one or several dendrometric variables, usually easy to measure, of standing trees (explanatory variables) [3]. The relation between these two data sets is frequently done by regression analysis. The most used models are the linear with additive error ($y = \beta_0 + \sum_{i=1}^{n} (\beta_i \times x_i) + \varepsilon$, where y is the component or total biomass, x_i the ith explanatory variable, β_i the ith model parameter and ε the error term), nonlinear with additive error ($y = \beta_0 \times x_1^{\beta_1} \times x_2^{\beta_2} \times \cdots \times x_n^{\beta_n} + \varepsilon$, where n = 1,…,N is the number of explanatory variables) and nonlinear with multiplicative error ($y = \beta_0 \times x_1^{\beta_1} \times x_2^{\beta_2} \times \cdots \times x_n^{\beta_n} \times \varepsilon$) [3].

Model fitting techniques with least squares or maximum likelihood assume the independence of residuals, homoscedasticity (or constant variance), and normal distribution of the residuals. Yet, forest data may not meet the abovementioned assumptions, due to data spatial, temporal, and/or hierarchical structure [50, 51]. Mixed effects models; combining fixed effects (*i.e.*, related to all the population or with factors that have repeatable levels), and random effects (*i.e.*, linked to the experimental entities that for a population are taken at random) are an alternative. These models can be considered an extension of linear models (with only fixed effects) by the incorporation of random effects. This approach enables to accommodate the variability within and between the different variables of a model. Mixed effects models can have a linear formulation ($y_i = X_i\beta + Z_i b_i + \epsilon_i$, i = 1,…,M, where X_i and Z_i are fixed and random effects regressor matrices, β vector of fixed effects, b_i vector of random effects and ε_i error vector) or nonlinear ($y_{ij} = f(\phi_{ij}, v_{ij}) + \epsilon_{ij}$, i = 1,…,M, j = 1,…., n_i, where φ_{ij} is a vector of a specific group of parameters, v_{ij} a covariate vector, ε_{ij} error vector, M is the number of groups and n_i number of observations of the ith group) (for details see [50]).

Two issues related to biomass modelling are the functions transformation and the additivity of the per component and total biomass models. Biomass models can be fitted in their nonlinear formulation or can be transformed into a linear one (frequently the logarithm of both equation sides), which is simpler and might reduce the variance heterogeneity. However, these functions have to be re-transformed in nonlinear ones to enable to have the predictions in their assessment units [3]. Yet, the back transformation to nonlinear functions results in bias, thus a correction factor is used [52, 53], which can be an exponential function of for example the estimated mean of a variable and regression variance [3], an average of the mean square error of site and regression [53], or half of the standard error of the residuals [52].

The wide variability in species traits, site, stand structure, and silvicultural systems and practices is not only reflected in total biomass but also in the biomass per component. The per component biomass allometric equations, to minimise the prediction errors, should have compatible additivity, *i.e.*, the sum of all components should be equal to the total biomass [54]. When fitting functions per component or total biomass the compatible additivity is not always achieved. Moreover, fitting the functions per component does not take into account the correlation between them [54, 55]. This can be overcome with the simultaneous fitting of a system of equations. Three alternative approaches were developed for linear regression [56] and two for nonlinear regression [54] to force the sum of per component biomass predictions to be equal to total biomass one. From those, the most general and with more flexibility are for linear regression and nonlinear regression, the seemingly unrelated regression (SUR) and nonlinear seemingly unrelated regression (NSUR), respectively. These two techniques to force additivity have been widely used (*e.g.*, [8, 55, 57–60]). In linear (not transformed) functions additivity is relatively easy to obtain [40, 61, 62] whereas for logarithm transformed variables is not possible [60]. Furthermore, the additive models seem to have a higher predictive ability for stem and bark than for branches and leaves [8, 55] which can be explained by the interactions between the trees in the stand (*cf*. chapter "Stand Structure and Biomass").

The selection of the explanatory variables of the models should ensure that they will provide predictions that are biologically realistic, which can be done by imposing restrictions (*e.g.*, model parameters values, value at the origin, maximum, inflection point, asymptote) [63, 64] to enable the function to have a biological significance.

Explanatory variables can be continuous (*e.g.*, diameter at breast height, *dbh*, and/or height, *h*), binary (also called dummy, which is useful to separate a variable in two classes, *e.g.*, between sites, composition or structure of the stand) or qualitative (*e.g.*, different types of soils) [65]. Typically, it is assumed that the explanatory variables in a model are independent. Yet, in multiple regression, the criteria of independence of the variables are not always met [3, 65], due to collinearity (or multicollinearity), *i.e.*, amongst explanatory variables exist strong dependencies [3]. Multicollinearity results in imprecise regression coefficients, *i.e.*, include more explanatory variables than needed, which results in predictions with irregular precision [3, 65]. To tackle collinearity several alternatives, exist [3, 36, 65, 66]: (i) select the explanatory variables so that those that are correlated are not included in the model (*e.g.*, using correlation analysis); (ii) remove explanatory variables from the model that have strong correlations (*e.g.*, with variance inflation factor, VIF); (iii) a priori combining explanatory variables (*e.g.*, with principal component analysis, PCA); or iv) combine explanatory variables in the model equation (*e.g.*, dbh^2h).

The most frequently used explanatory variables for biomass modelling at tree level are diameter at breast height and height. Several studies [53, 66–69] demonstrated that the inclusion of height as an explanatory variable improved the predictive ability of the biomass functions. This can be explained by the interaction between site and height. It is known that in better quality sites trees tend to reach higher total heights than in poorer quality sites [53]. Inversely, diameter at breast height seems to be more related to competition between trees in a stand. Thus, allometric biomass functions for sites of similar quality, the inclusion of height as an explanatory variable might not improve model performance. Yet, when developing biomass allometric functions of sites of different quality, height mitigates their specificity to the site [53, 66, 70], but has the disadvantage of including and propagating the errors associated with the measurements of heights [66].

The formulation of biomass allometric functions has to be considered, as some explanatory variables might have collinearity. One example is the diameter at breast height and height. Thus, integrating these two variables in one (*e.g.*, dbh^2h) overcomes the collinearity between variables [53, 66, 70]. Other variables that are also used are related to crown dimensions, namely crown diameter, crown length, crown width, crown area, crown volume, and live crown [45, 57, 67, 71, 72]; wood density [48, 66, 67, 73]; age [52]; and base diameter [72, 74].

The allometric functions at tree level are considered the most accurate indirect method [15, 75–83]. Nonetheless, as many other forest estimation functions, they have a species and site specific character, which results in a very large number of available functions. Many allometric functions are also regime, stand development, or composition specific; some were developed for high forest stands (*e.g.*, [41, 84, 85]), while others were fitted for coppice stands (*e.g.*, [86–88]), per stand development stage (*e.g.*, [52, 74]), and per composition (*e.g.*, [72]). For more examples of the influence of species, stand structure, and site on biomass allometric functions at tree level see chapter "Overview of the Biomass Models".

The shortcoming that can be pointed out to allometric functions at tree level is that the function for a stand has to be chosen according to the species, stand structure, and site. This is frequently done for boreal and temperate forests. When more than one allometric function exists, the most suitable one has to be chosen. This can be a difficult task because more than one equation may exist for a given species, stand structure, and ecological zone, or, on the contrary, there might be no allometric equation or those that exist might not be reliable [89].

Conversely, in tropical forests, characterised by mixed stands with high species richness, it is difficult to acquire data per species, through the destructive method, to fit the allometric functions. Alternatively, many biomass equations were developed as a function of a group of species, forest types, and/or bioclimatic zones [90–92], frequently with the diameter at breast height or diameter at breast height and height as explanatory variables. The disadvantage of these functions is that they should only be applied in forests similar to those where the data sets were collected [93–97] or for species with similar wood density [48, 91, 98], otherwise, bias may arise. Additionally, environmental variables may also be included in the models, which makes the use of allometric functions in other regions or forests more difficult [76, 99].

There is still the case where no allometric equation exists for biomass for a certain species or of a specific forest. In this case, the biomass allometric functions of the web platform GlobAllomeTree [100] can be used. GlobAllomeTree is a free software (http://www.globallometree.org/), where more than 4600 allometric equations were compiled for Europe, North America, and Africa, whenever possible with the georreferencing of the data used to develop the functions [100]. Nonetheless, due to their general character, they are not as accurate as the functions that use diameter at breast height, height, and/or wood density as explanatory variables per species and/or per site [91, 97, 101]. Another compilation of data and allometric equations is Biomass and Allometric database (BAAD) which contains a very large database of dendrometric variables that were structured, organised, and standardised in one database that can be used to develop biomass functions for 678 species [102].

Area Level

Forest Inventory Data

Biomass for an area can be calculated with conversion and expansion factors [78, 84, 103–105]. While allometric functions are used at the individual tree level, these factors are used at the stand level (Fig. 3). They are frequently function of one stand absolute density measure (*e.g.*, stand development stage or age, number of trees, biomass or, most frequently, volume per hectare). Conversion factors can have simple (*e.g.*, coefficient between biomass and volume) or complex (*e.g.*, exponential or other non-linear functions) formulations [78, 103, 105]. Their advantage is that they require less labour and are used for large areas, while their disadvantages are related to their lower accuracy, due to the variability of stand structure. For example, in young stands, volume is frequently not calculated in forest inventories and the variability between volume and biomass is high [78, 103, 104].

Expansion factors (Fig. 3) can be based only on the forest area attributes (*e.g.*, stand structure and/or soil type) or on remote sensing data. The former approach uses the target area as the multiplier to attain the estimate for a certain absolute density measure, such as biomass, derived from inventory plots [5]. With this method, the accuracy decreases with the increase of the forest area and with the variability of stand structure, topography, soil, and climate [106]. These estimations have acceptable errors, between 15–40%, being 25% the reference value [39, 107]. Yet, they are frequently attained for time periods corresponding to those of the forest inventories (5–10 years). Therefore, the biomass mapping derived from field work is not continuous in space and time. For the latter, a set of thematic maps and K-nearest neighbour methods are used (*cf*. Sect. 3). The expansion factor is derived from a constellation of neighbour pixels, with similar characteristics to the pixel corresponding to the plot centre [5]. The use of thematic raster maps leads to the improvement of the absolute density measure estimates (*e.g.*, biomass). This is the result of the inclusion of the ancillary data, which enables more detail for the target area (by using several forest, soil, and topographic variables), subdividing it in more homogeneous subareas [5, 9].

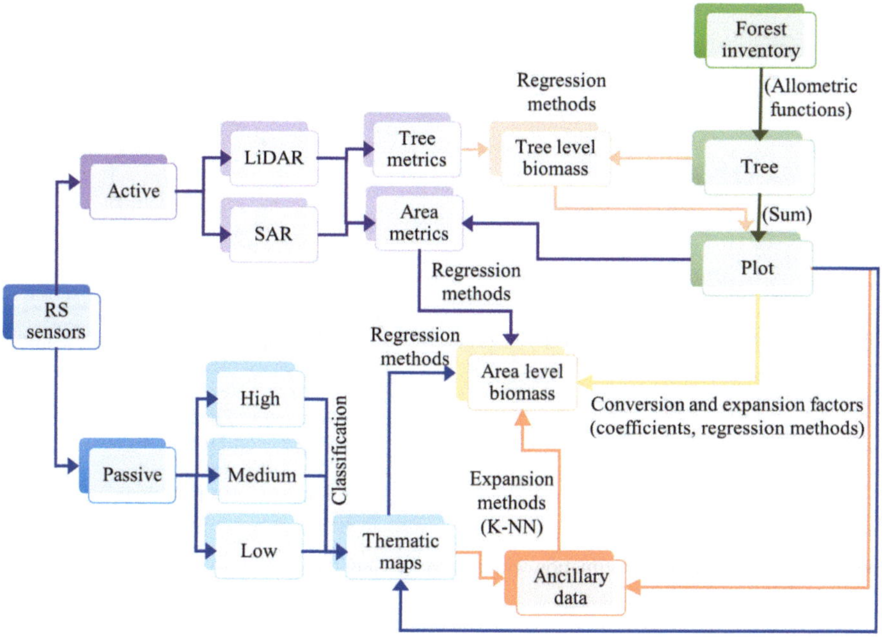

Fig. 3 Biomass indirect estimation methods

Remote Sensing Data

Biomass functions with data from remote sensing have been developed with parametric and non-parametric methods (Fig. 4). The parametric models include linear regression, both single and multiple; and non-linear models, power. The non-parametric models include k-nearest neighbour, artificial neural network, regression tree, random forest, support vector machine, and maximum entropy. The single and multiple linear regression and the power regression were already described (*cf.* Sect. 3). The difference is that the explanatory variables are derived from remote sensing imagery, such as bands, vegetation indices, texture indices, and/or crown cover. The following paragraphs describe briefly the mostly used non-parametric models.

Geographically weighted regression (GWR) is a regression method that takes into account the geographical location of the observations, whether in the plane (x, y) or space (x, y, z). It uses a fixed or moving window (kernel), depending on the data distribution function. The weighting is inversely proportional to the distance between the observations, and for the weighting function, Gaussian or bi-square methods and fixed or adaptive kernels can be used. For continuous data, the Gaussian method is the most frequently used. The fixed kernel is used when data presents a random distribution while the adaptive is used for non-random distributions. The advantages of this regression method are related to the inclusion of the spatial location while the

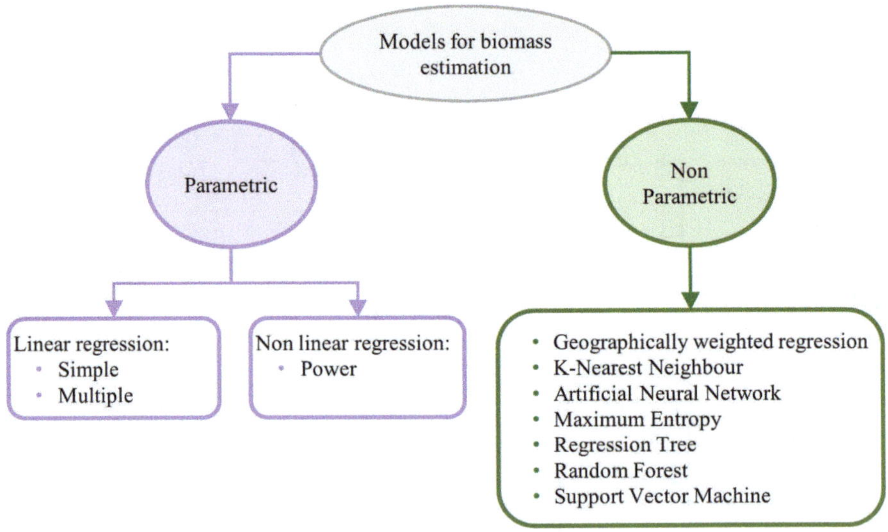

Fig. 4 Models for biomass estimation with remote sensing data

disadvantages are linked to the number of samples and the distance between them [108–111]

In the K-Nearest Neighbours' method (K-NN), the estimation of the value of a certain variable (*e.g.*, above ground biomass) is predicted considering a constellation of the k nearest neighbours, as a function of the distance (inverse) between the target pixel and their neighbours. For the weighted distance, several functions can be used, *e.g.*, Euclidean, weighted Euclidean, Mahalanobis, canonical correlation analysis based metrics, and canonical correspondence analysis. Accuracy is determined by the weighted distance function and location between the inventory plots and the satellite image pixel. Additionally, primordial to the accuracy is the number of explanatory variables. Though one variable can be used, frequently more than two variables are used to improve accuracy. The advantages that can be pointed out to this method are the avoidance of uncertainties related to unbalanced samples and that several predictor variables can be used. The disadvantages are related to the time needed to select the most appropriate variables [112–117].

Artificial Neural Network (ANN) is a black box model, that uses several complex functions that connect the output variable with the input ones through network training. It is constituted by three groups of layers: input (one), hidden (one or more), and output (one). Several advantages are referred to this method: explanatory variables from remote sensing and ancillary (continuous or categorical) can be used; it is independent of data distribution (*e.g.*, normal distribution); regardless of whether data is incomplete or imprecise or not, the solutions are robust; provides efficient and accurate solutions for both linear and nonlinear complex patterns; and physical limitations of knowledge or representativeness of the biophysical traits are analysed [118, 119]. Their disadvantages are related to their black box model, *i.e.*, the internal mechanism of the relationships between the dependent and explanatory variables is not disclosed straightforwardly; training and learning the iterative procedures are required; the number of plots required is large; and when the optimisation of the parameters is not appropriately done by the algorithm the resulting accuracy may be poor [15].

Maximum Entropy (MaxEnt) is a black box model belonging to the general purpose machine learning methods. This method either predicts or infers a target probability distribution of maximum entropy. Two data sets are needed, one is the set of sample points and the other is the set of variables such as those of remote sensing and ancillary. The variables can be continuous or categorical. The advantages of this method are not requiring any assumptions to the input data; can be efficient with small sample size data; and can be used with presence only data. The disadvantages are the need of a priori information; and the overfitting that occurs when constraints are based on the sample data empirical averages, particularly when the set of environmental variables is very large [120, 121].

Regression Tree (RT) is a tree based model, using a recursive partitioning algorithm, that assembles the input data in clusters with a set of hierarchical criteria. The clusters are function of the homogeneity of the dependent variable. Overall, three hierarchical nodes can be identified, *i.e.*, the root at the highest level (the highest homogeneity), several internal nodes at an intermediate level, and the terminal nodes at the lowest level. The terminal nodes correspond to the class with the highest probability for discrete variables, named classification tree, or to the average for continuous variables, called regression tree. The advantages that can be pointed out to this method are related to the selection of the variables; the fitting being iterative; not requiring any assumptions to the input data, can be efficient in the presence of outliers, incomplete data, and correlation and collinearity between the explanatory variables; and the output is easy to understand. The disadvantages are twofold; due to the wide range of variance, when data suffers small fluctuations the output derived is different; and it tends to result in over estimations for forest stands with low biomass and under estimations for those with high biomass [15, 122–127]. The improvement of the accuracy of this method can be done with a data resampling technique, either bagging (the probability of reaching the following bootstrap has the same probability for all the observations) or boosting (the probability of reaching the following bootstrap has a higher probability for those observations that are often mistyped) [122, 123].

Random Forest (RF) is a tree based model, where a wide range of not correlated regression trees are constructed iteratively through random bootstrap sampling. Every tree is created independently without pruning, deriving in a set of nodes. For each node, a subset of explanatory variables is selected and used to determine the best split. The definition of the latter is achieved when the largest reduction of the residual sum of squares among the observations and the average of the node is attained. When the extent of the trees reaches a maximum, the variance and the bias of trees produced by the algorithm are high and low, respectively. All the regression trees contribute to the final output by their average value [128]. It can be used with continuous or categorical data. The advantages of this approach are not requiring any assumptions to the input data; its ability to work with data with high variability; enables working with a small and large number of input variables; allows using large data sets; and is robust to overfitting. The disadvantage is related to the possibility of overfitting when data has a very wide variability (noise) [15, 19, 108, 128–130].

Support Vector Machine (SVM) is a statistical learning algorithm. Input data is transformed, by support vector regression (SVR) in a n-dimensional feature. The latter is attained with a nonlinear kernel function (*e.g.*, Gaussian radial, linear, polynomial), which minimises both the error associated with training and the model complexity. In this approach is of primordial importance the identification of the meta parameters (kernel, precision, and penalty). The advantages of this method are the performance optimisation, the overfitting minimisation attained using the minimisation of the structural risk, and the higher accuracies attained when small data sets are used for training. The disadvantages are related to the selection of the better suited kernel and to the difficulty to develop the model if the number of training samples is large [131–134].

Forest Inventory, Remote Sensing and Geographical Information Systems

The improvement of remote sensing and geographical information systems allowed the estimation of some variables, especially forest areas, crown cover (distinguishing forest from other land use classes based on a minimum crown cover threshold, frequently of 10%), and stand structures [17, 135, 136]. It enabled the forest inventory field work rationalisation, through the increase of sample efficiency and the reduction of error, labour, and costs [5, 6]. During the last two decades, a set of functions were developed to estimate several stand and forest absolute density parameters, such as the number of trees, basal area, volume, and biomass per area (*e.g.*, [30–33, 112, 137]).

In the last years, raster maps of forest resources have been produced with national forest inventory field plots and remote sensing data of passive or active sensors [138]. The accuracy of the maps depends on the target area, less accurate for larger areas (*e.g.*, country) than for small areas (*e.g.*, management units), and on the type of remote sensing data, namely 2D or 3D, and on pixel size. Typically, 3D data and smaller pixel sizes enable higher accuracy but are mainly used in small areas because of processing time and remote sensing data availability [9]. Although with less accuracy, national forest inventory data associated with auxiliary data enables the production of forest resource maps at country level [9]. In this approach, the features of the forest areas (*e.g.*, number of trees, basal area, volume, and biomass) are predicted on an area basis (grid cells or pixels), through statistical modelling, most frequently with K-Nearest Neighbours [139]. The advantages of this approach are: (i) it enables the estimation of forest variables for cells with no data from field inventory, based on remote sensing and ancillary data; (ii) it is possible to collect data in short time periods, which enables obtaining time series data sets and thematic maps with a suite of images from the same year; (iii) it enables automatic mapping; (iv) it allows working at both high and low resolution scales. Their disadvantages are: (i) the low spatial resolution, large pixels enable a nationwide analysis, but do not have enough precision for decision making at the management unit level, which requires high spatial resolution data (frequently not freely available); (ii) poor correlations of data from remote sensing from passive sensors with forest variables, yet this constraint can be circumvented by the use of data from active sensors; (iii) the different dimensions that field plots and pixel size many times have, which is reflected in the accuracy of the models that in turn is also dependent on the structure and homogeneity of the stands and forests [9].

Especially for large areas, but also for small ones, the variability and heterogeneity of the stand structures, crown allometry and ancillary variables (*e.g.*, climate, altitude, latitude) can result in significant differences in the estimation of the forest variables [138]. For the enhancement of the model accuracy, three approaches can be used [9]: (i) when local data is available, it is preferable to use models that can be calibrated with those data with kriging or mixed models; (ii) to use geographically weighted regression; (iii) to select data for a determined spatial range using non-parametric models such as the k nearest neighbours.

4 Models Uncertainties

Uncertainties are inherent to data acquisition and model development (Table 2). The errors in data acquisition are related to [15, 35, 140–145]: (i) forest inventory, in what regards sampling design and its implementation, tree measuring and data recording; (ii) tree level biomass indirect methods (allometric functions) where two sources of errors can be pointed out; one is the error of the model and the other is related to the inadequate selection of the model; (iii) upscaling biomass from tree level to plot level and to forest areas; iv) remote sensing data, which is related with the errors derived from the atmospheric conditions, topography (*e.g.*, slope), image corrections (geometrical, radiometric and atmospheric), data analysis and technical aspects (*e.g.*, platforms, scanner motions), level of saturation, effects of the understorey and soil; v) plot geographical location, linked with errors associated to the determination of the coordinates of the plots in the field with Global Navigation Satellite System (GNSS) devices, geometric correction (especially in areas with steep slopes) and discrepancy between the inventory plots and the image pixels; (vi) size of the inventory plots and image pixel, associated to field plot areas not being multiple of the image pixel area or due to trees in the inventory plots boundaries; (vii) temporal displacement, related to the difference in time between field measurements and remote sensing data.

The second source of error is associated with model development, inherent to the mathematical representation of biophysical parameters. In forestry it is well known that most tree features and stand structures have high variability, reason why the accuracy of the models is evaluated. Typically, one of three methods is used to fit and validate the models [3, 146, 147]: (i) one data set is divided randomly in two subsets, one for model fitting and another for validation; (ii) cross validation, where the

Table 2 Error sources on biomass models (where FI is forest inventory, RS remote sensing, and GIS Geographical Information Systems)

Error source	Data source / method				
	Direct tree level	Indirect tree level	FI	FI + RS	FI + RS + GIS
Forest inventory sampling design	√	√	√	√	√
Tree measurement	√	√	√	√	√
Conversion methods		√	√	√	√
Allometric functions		√	√	√	√
Remote sensing data				√	√
Plot geographical location				√	√
Size of the plots and pixels				√	√
Temporal difference between data sets				√	√

population is divided in subsets with an equal number of observations and the model is fitted for each subset, deriving the statistic prediction sum of squares (PRESS); (iii) two data sets are used, one for fitting and the other for validation. The model validation with an independent data set from that of the fitting is the best suited method as it meets the criterion of independent sampling of the two data sets. Yet, it is frequently not used due to cost and labour. From the other two methods, cross validation as it ensures the independence of the data to fit the model and to computed residuals reduces the accuracy overestimation when compared with a method that splits the data in two data sets, one for fitting and another for validation as both subsets derive from one sampling design [146, 147].

The most frequently used statistical measures to evaluate accuracy are the coefficient of determination (R^2), the adjusted coefficient of determination ($R^2{}_{aj}$), and the root mean square error (RMSE), model bias, model precision, model efficiency, as well as the graphical analysis of the plots of observed and predicted values, normal quantile–quantile plots and normality tests [3, 65, 146, 147]. The adjusted coefficient of determination ($R^2{}_{aj}$) and the relative root mean square error ($RMSE_r$) enable the comparison among studies, as they are a relative measure. Inversely, RMSE is an absolute measure (*e.g.*, $t{\cdot}ha^{-1}$), which does not allow to compare studies directly. For example, two biomass models with an RMSE of 1 $t{\cdot}ha^{-1}$, for two stands one with 10 $t{\cdot}ha^{-1}$ and another with 50 $t{\cdot}ha^{-1}$, the error corresponds likely to different accuracy. Alternatively, the relative root mean square error ($RMSE_r$) could provide a straightforward measure to compare different models.

Apart from the aforementioned methods, several others have been used. For example, four were defined to deal with the uncertainties at the regional and national levels. In the first approach, error propagation, the variance of the different sources of uncertainties are summed, with a first order Taylor series [140]. As no direct interaction exists between different sources of error, this approach may lead to inaccuracies [148]. The second approach is based on the analysis of the model and is founded on sampling theory and the principles of the series of Taylor. It enables to calculate the sampling errors independently of those of the model [148, 149]. Its disadvantage is that it is dependent on dataset characteristics [150]. The third approach is based on simulation techniques (*e.g.*, the Monte Carlo method) by attaining the distribution of probability as well as the model's parameters estimates [151]. The four approach is based on simulation techniques (*e.g.*, Monte Carlo) in which the principles of Taylor's series are associated. It enables the separation of the sampling error from that of the model and reduces the model parameters' effect variability [150]. Another method to tackle uncertainties is via covariance matrix for parameters estimates [67].

In literature, many statistics, apart from the abovementioned, have been used to evaluate the accuracy and uncertainties of the models (*e.g.*, [143, 152]). The major drawbacks that can be pointed out are the number of different statistical measures used; the same terms are sometimes used for different statistical formulations; and formulas are not always provided in the studies. The inclusion of the formulas and statistical measures providing relative measures for accuracy are suggested as they enable the comparison between studies.

Errors vary depending on input data and their variability. Relative errors ranging from 5 to 30% have been reported in the literature [140]. Additionally, the level of accuracy is linked with the target area. While at regional scale the threshold is set at 90%, and at national or global scale is set at 80% [15].

5 Final Considerations

Biomass modelling has a key role in the assessment, monitoring, and prediction of the total or per component biomass (*e.g.*, [3, 60]). It can also be used to fit biomass residues functions, that would substitute the estimations per utilisation rates between different biomass components. For example, [153] developed nonlinear models with good performance of recoverable biomass at tree level species-dependent with diameter at breast height as explanatory variable.

Data sets are of primordial importance in modelling to enable selecting the best suited explanatory variables and at the same time tackle collinearity amongst them (*e.g.*, [65, 66]) and uncertainties related to their measurement and acquisition (*e.g.*, [35, 143]). Moreover, the specificity of each dataset origin (destructive sampling, forest inventory, remote sensing, and/or ancillary) has also its own acquisition specificities and associated uncertainties that should be taken into consideration when selecting the modelling methods (*e.g.*, [15, 141]).

The wide range of variables available along with the different spatial scales resulted in a large variety of models. The modelling approach is also linked with the stand structure, site (edaphic and climatic conditions), objectives of modelling (including use and model uncertainties), and available data (some variables acquisition may be constrained by site, atmospheric conditions, stand structure or costs) (*e.g.*, [6, 15]). Additionally, the modelling approach is also related to biomass use (*e.g.*, timber, pulp, and paper, bioenergy). Thus, there is a continuous development of models for estimating biomass that have the primary goal of accommodating the variability and dynamics of biomass in space and time. In general, there is an increase in complexity of the models, which reflects the heterogeneity of the forest stands and enables the reduction of the uncertainties of the models (*cf*. Sect. 3 and Sect. 4).

The direct method can be classified small spatial scale as it is mainly used to estimate individual tree biomass and long term temporal scale (Table 1) due to its cost and labour. Yet, it is of the utmost importance to develop tree level indirect methods. The indirect methods can be used at the three spatial and temporal scales (Table 1). However, due to the stand structure heterogeneity bias increase with the increase of the spatial scale (*cf*. Sect. 3).

Funding This work is funded by National Funds through FCT - Foundation for Science and Technology under the Project UIDB/05183/2020.

References

1. Gonçalves AC (2022) Influence of stand structure on forest biomass sustainability. In: Jhariya MK, Meena RS, Banerjee A, Meena SN (eds) Natural resources conservation and advances for sustainability. Elsevier, Cambridge, United States, pp 327–352
2. Gonçalves AC (2022) Stand structure impacts on forest modelling. Appl Sci 12:6963. https://doi.org/10.3390/app12146963
3. Burkhart HE, Tomé M (2012) Modeling forest trees and stands. Springer, Netherlands, Dordrecht
4. Dittmann S, Thiessen E, Hartung E (2017) Applicability of different non-invasive methods for tree mass estimation: a review. For Ecol Manage 398:208–215. https://doi.org/10.1016/j.foreco.2017.05.013
5. McRoberts RE, Tomppo EO, Næsset E (2010) Advances and emerging issues in national forest inventories. Scand J For Res 25:368–381. https://doi.org/10.1080/02827581.2010.496739
6. Tomppo E, Olsson H, Ståhl G et al (2008) Combining national forest inventory field plots and remote sensing data for forest databases. Remote Sens Environ 112:1982–1999. https://doi.org/10.1016/j.rse.2007.03.032
7. Vashum KT, Jayakumar S (2012) Methods to estimate above-ground biomass and carbon stock in natural forests-a review. J Ecosyst Ecogr 02.https://doi.org/10.4172/2157-7625.1000116
8. Xu Z, Du W, Zhou G et al (2022) Aboveground biomass allocation and additive allometric models of fifteen tree species in northeast China based on improved investigation methods. For Ecol Manage 505:119918. https://doi.org/10.1016/j.foreco.2021.119918
9. Kangas A, Astrup R, Breidenbach J et al (2018) Remote sensing and forest inventories in Nordic countries–roadmap for the future. Scand J For Res 33:397–412. https://doi.org/10.1080/02827581.2017.1416666
10. Henttonen HM, Kangas A (2015) Optimal plot design in a multipurpose forest inventory. Forest Ecosyst 2:14. https://doi.org/10.1186/s40663-015-0055-2
11. Pretzsch H (2009) Forest dynamics, growth, and yield. Springer, Berlin
12. Avery TE, Burkhart HE (1994) Measurements, 4th edn. Macgraw-Hill Inc., New York

13. Vidal C, Lanz A, Tomppo E et al (2008) Establishing forest inventory reference definitions for forest and growing stock: a study towards common reporting. Silva Fennica 42:247–266. https://doi.org/10.14214/sf.255
14. Pretzsch H (2014) Canopy space filling and tree crown morphology in mixed-species stands compared with monocultures. For Ecol Manage 327:251–264. https://doi.org/10.1016/j.foreco.2014.04.027
15. Lu D, Chen Q, Wang G et al (2016) A survey of remote sensing-based aboveground biomass estimation methods in forest ecosystems. Int J Digital Earth 9:63–105. https://doi.org/10.1080/17538947.2014.990526
16. White JC, Coops NC, Wulder MA et al (2016) Remote sensing technologies for enhancing forest inventories: a review. Can J Remote Sens 42:619–641. https://doi.org/10.1080/07038992.2016.1207484
17. Ke Y, Quackenbush LJ (2011) A review of methods for automatic individual tree-crown detection and delineation from passive remote sensing. Int J Remote Sens 32:4725–4747. https://doi.org/10.1080/01431161.2010.494184
18. Eisfelder C, Kuenzer C, Dech S (2012) Derivation of biomass information for semi-arid areas using remote-sensing data. Int J Remote Sens 33:2937–2984. https://doi.org/10.1080/01431161.2011.620034
19. Ghosh SM, Behera MD (2018) Aboveground biomass estimation using multi-sensor data synergy and machine learning algorithms in a dense tropical forest. Appl Geogr 96:29–40. https://doi.org/10.1016/j.apgeog.2018.05.011
20. Kumar L, Sinha P, Taylor S, Alqurashi AF (2015) Review of the use of remote sensing for biomass estimation to support renewable energy generation. J Appl Remote Sens 9:097696. https://doi.org/10.1117/1.JRS.9.097696
21. Gail WB (2007) Remote sensing in the coming decade: the vision and the reality. J Appl Remote Sens 1:012505. https://doi.org/10.1117/12.694379
22. Boyd DS, Danson FM (2005) Satellite remote sensing of forest resources: three decades of research developmen. Progress Phys Geogr Earth Environ 29:1–26. https://doi.org/10.1191/0309133305pp432
23. Koskinen J, Leinonen U, Vollrath A et al (2019) Participatory mapping of forest plantations with open foris and google earth engine. ISPRS J Photogramm Remote Sens 148:63–74. https://doi.org/10.1016/j.isprsjprs.2018.12.011
24. Fassnacht FE, Latifi H, Stereńczak K et al (2016) Review of studies on tree species classification from remotely sensed data. Remote Sens Environ 186:64–87. https://doi.org/10.1016/j.rse.2016.08.013
25. Thomlinson JR, Bolstad PV, Cohen WB (1999) Coordinating methodologies for scaling land-cover classifications from site-specific to global: steps toward validating global map products. Remote Sens Environ 70:16–28. https://doi.org/10.1016/S0034-4257(99)00055-3
26. Li M, Zang S, Zhang B et al (2014) A review of remote sensing image classification techniques: the role of Spatio-contextual information. European Journal of Remote Sensing 47:389–411. https://doi.org/10.5721/EuJRS20144723
27. Chen Q, Gong P (2004) Automatic variogram parameter extraction for textural classification of the panchromatic IKONOS imagery. IEEE Trans Geosci Remote Sens 42:1106–1115. https://doi.org/10.1109/TGRS.2004.825591
28. Brovkina O, Novotny J, Cienciala E et al (2017) Mapping forest aboveground biomass using airborne hyperspectral and LiDAR data in the mountainous conditions of Central Europe. Ecol Eng 100:219–230. https://doi.org/10.1016/j.ecoleng.2016.12.004
29. Chao Z, Liu N, Zhang P et al (2019) Estimation methods developing with remote sensing information for energy crop biomass: a comparative review. Biomass Bioenerg 122:414–425. https://doi.org/10.1016/j.biombioe.2019.02.002
30. Gonçalves AC, Sousa AMO, Mesquita PG (2017) Estimation and dynamics of above ground biomass with very high resolution satellite images in *Pinus pinaster* stands. Biomass Bioenerg 106:146–154. https://doi.org/10.1016/j.biombioe.2017.08.026

31. Gonçalves AC, Sousa AMO, Mesquita P (2019) Functions for aboveground biomass estimation derived from satellite images data in Mediterranean agroforestry systems. Agrofor Syst 93:1485–1500. https://doi.org/10.1007/s10457-018-0252-4
32. Sousa AMO, Gonçalves AC, da Silva JRM (2017) Above-ground biomass estimation with high spatial resolution satellite images. In: Tumuluru JS (ed) Biomass volume estimation and valorization for energy. InTech
33. Sousa AMO, Gonçalves AC, Mesquita P, Marques da Silva JR (2015) Biomass estimation with high resolution satellite images: a case study of *Quercus rotundifolia*. ISPRS J Photogramm Remote Sens 101:69–79. https://doi.org/10.1016/j.isprsjprs.2014.12.004
34. Fernández-Manso O, Fernández-Manso A, Quintano C (2014) Estimation of aboveground biomass in Mediterranean forests by statistical modelling of ASTER fraction images. Int J Appl Earth Obs Geoinf 31:45–56. https://doi.org/10.1016/j.jag.2014.03.005
35. Tomppo E, Nilsson M, Rosengren M et al (2002) Simultaneous use of Landsat-TM and IRS-1C WiFS data in estimating large area tree stem volume and aboveground biomass. Remote Sens Environ 82:156–171. https://doi.org/10.1016/S0034-4257(02)00031-7
36. Zhang Q, He HS, Liang Y et al (2018) Integrating forest inventory data and MODIS data to map species-level biomass in Chinese boreal forests. Can J For Res 48:461–479. https://doi.org/10.1139/cjfr-2017-0346
37. Lu D (2006) The potential and challenge of remote sensing-based biomass estimation. Int J Remote Sens 27:1297–1328. https://doi.org/10.1080/01431160500486732
38. Ko C, Remmel TK (2017) Airborne LiDAR applications in forest landscapes. In: Remmel TK, Perera AH (eds) Mapping forest landscape patterns. Springer, New York, pp 105–185
39. Radtke P, Walker D, Frank J et al (2017) Improved accuracy of aboveground biomass and carbon estimates for live trees in forests of the eastern United States. Forestry 90:32–46. https://doi.org/10.1093/forestry/cpw047
40. Skovsgaard JP, Nord-Larsen T (2012) Biomass, basic density and biomass expansion factor functions for European beech (Fagus sylvatica L.) in Denmark. Eur J Forest Res 131:1035–1053. https://doi.org/10.1007/s10342-011-0575-4
41. Jenkins JC, Chojnacky DC, Heath LS, Birdsey RA (2003) National-scale biomass estimators for United States tree species. Forest Sci 49:12–35. https://doi.org/10.1093/forestscience/49.1.12
42. Alfaro-Sánchez R, Valdés-Correcher E, Espelta JM et al (2020) How do social status and tree architecture influence radial growth, wood density and drought response in spontaneously established oak forests? Ann For Sci 77:49. https://doi.org/10.1007/s13595-020-00949-x
43. Chave J, Coomes D, Jansen S et al (2009) Towards a worldwide wood economics spectrum. Ecol Lett 12:351–366. https://doi.org/10.1111/j.1461-0248.2009.01285.x
44. Knapic S, Louzada JL, Leal S, Pereira H (2008) Within-tree and between-tree variation of wood density components in cork oak trees in two sites in Portugal. Forestry 81:465–473. https://doi.org/10.1093/forestry/cpn012
45. Kuyah S, Muthuri C, Jamnadass R et al (2012) Crown area allometries for estimation of aboveground tree biomass in agricultural landscapes of western Kenya. Agrofor Syst 86:267–277. https://doi.org/10.1007/s10457-012-9529-1
46. Resquin F, Navarro-Cerrillo RM, Carrasco-Letelier L, Casnati CR (2019) Influence of contrasting stocking densities on the dynamics of above-ground biomass and wood density of Eucalyptus benthamii, Eucalyptus dunnii, and Eucalyptus grandis for bioenergy in Uruguay. For Ecol Manage 438:63–74. https://doi.org/10.1016/j.foreco.2019.02.007
47. Vannoppen A, Boeckx P, De Mil T et al (2018) Climate driven trends in tree biomass increment show asynchronous dependence on tree-ring width and wood density variation. Dendrochronologia 48:40–51. https://doi.org/10.1016/j.dendro.2018.02.001
48. Henry M, Besnard A, Asante WA et al (2010) Wood density, phytomass variations within and among trees, and allometric equations in a tropical rainforest of Africa. For Ecol Manage 260:1375–1388. https://doi.org/10.1016/j.foreco.2010.07.040
49. Poudel KP, Temesgen H, Radtke PJ, Gray AN (2019) Estimating individual-tree aboveground biomass of tree species in the western U.S.A. Can J For Res 49:701–714. https://doi.org/10.1139/cjfr-2018-0361

50. Pinheiro JC, Bates DM (2000) Mixed-Effects models in S and S-PLUS. Springer, New York
51. Robinson A, Hamann JD (2011) Forest analytics with R: an introduction. Springer, New York
52. Xiang W, Li L, Ouyang S et al (2021) Effects of stand age on tree biomass partitioning and allometric equations in Chinese fir (Cunninghamia lanceolata) plantations. Eur J Forest Res 140:317–332. https://doi.org/10.1007/s10342-020-01333-0
53. Zhang J, Fiddler GO, Young DH et al (2021) Allometry of tree biomass and carbon partitioning in ponderosa pine plantations grown under diverse conditions. For Ecol Manage 497:119526. https://doi.org/10.1016/j.foreco.2021.119526
54. Parresol BR (2001) Additivity of nonlinear biomass equations. Can J For Res 31:865–878. https://doi.org/10.1139/x00-202
55. Zhao D, Westfall J, Coulston JW et al (2019) Additive biomass equations for slash pine trees: comparing three modeling approaches. Can J For Res 49:27–40. https://doi.org/10.1139/cjfr-2018-0246
56. Parresol BR (1999) Assessing tree and stand biomass: a review with examples and critical comparisons. Forest Sci 45:573–593. https://doi.org/10.1093/forestscience/45.4.573
57. Carvalho JP, Parresol BR (2003) Additivity in tree biomass components of Pyrenean oak (*Quercus pyrenaica* Willd.). For Ecol Manage 179:269–276. https://doi.org/10.1016/S0378-1127(02)00549-2
58. Correia A, Faias S, Tomé M (2008) Ajustamento Simultâneo de Equações de Biomassa de Pinheiro Manso no Sul de Portugal. Silva Lusitana 16:197–205
59. Levine J, de Valpine P, Battles J (2021) Generalized additive models reveal among-stand variation in live tree biomass equations. Can J For Res 51:546–564. https://doi.org/10.1139/cjfr-2020-0219
60. Nord-Larsen T, Meilby H, Skovsgaard JP (2017) Simultaneous estimation of biomass models for 13 tree species: effects of compatible additivity requirements. Can J For Res 47:765–776. https://doi.org/10.1139/cjfr-2016-0430
61. Nord-Larsen T, Nielsen AT (2015) Biomass, stem basic density and expansion factor functions for five exotic conifers grown in Denmark. Scand J For Res 30:135–153. https://doi.org/10.1080/02827581.2014.986519
62. Skovsgaard JP, Bald C, Nord-Larsen T (2011) Functions for biomass and basic density of stem, crown and root system of Norway spruce (*Picea abies* (L.) Karst.) in Denmark. Scand J For Res 26:3–20. https://doi.org/10.1080/02827581.2011.564381
63. Fang Z, Bailey RL (1998) Height–diameter models for tropical forests on Hainan Island in southern China. For Ecol Manage 110:315–327. https://doi.org/10.1016/S0378-1127(98)00297-7
64. Yuancai L, Parresol BR (2001) Remarks on height-diameter modeling. U.S. Department of Agriculture, Forest Service, Southern Research Station, Asheville, NC
65. Vanclay JK (1994) Modelling forest growth and yield: applications to mixed tropical forests. CAB International, Wallingford, U.K.
66. Picard N, Rutishauser E, Ploton P et al (2015) Should tree biomass allometry be restricted to power models? For Ecol Manage 353:156–163. https://doi.org/10.1016/j.foreco.2015.05.035
67. Asrat Z, Eid T, Gobakken T, Negash M (2020) Aboveground tree biomass prediction options for the Dry Afromontane forests in south-central Ethiopia. For Ecol Manage 473:118335. https://doi.org/10.1016/j.foreco.2020.118335
68. Dutcă I, Mather R, Ioraş F (2018) Tree biomass allometry during the early growth of Norway spruce (*Picea abies*) varies between pure stands and mixtures with European beech (*Fagus sylvatica*). Can J For Res 48:77–84. https://doi.org/10.1139/cjfr-2017-0177
69. Elfving B, Ulvcrona KA, Egnell G (2017) Biomass equations for lodgepole pine in northern Sweden. Can J For Res 47:89–96. https://doi.org/10.1139/cjfr-2016-0131
70. Dutcă I, Mather R, Blujdea VNB et al (2018) Site-effects on biomass allometric models for early growth plantations of Norway spruce (Picea abies (L.) Karst.). Biomass Bioenerg 116:8–17. https://doi.org/10.1016/j.biombioe.2018.05.013
71. Forrester DI (2021) Does individual-tree biomass growth increase continuously with tree size? For Ecol Manage 481:118717. https://doi.org/10.1016/j.foreco.2020.118717

72. Sillett SC, Van Pelt R, Carroll AL et al (2020) Aboveground biomass dynamics and growth efficiency of Sequoia sempervirens forests. For Ecol Manage 458:117740. https://doi.org/10.1016/j.foreco.2019.117740
73. Mankou GS, Ligot G, Loubota Panzou GJ et al (2021) Tropical tree allometry and crown allocation, and their relationship with species traits in central Africa. Forest Ecol Manag 493:119262. https://doi.org/10.1016/j.foreco.2021.119262
74. Annighöfer P, Ameztegui A, Ammer C et al (2016) Species-specific and generic biomass equations for seedlings and saplings of European tree species. Eur J Forest Res 135:313–329. https://doi.org/10.1007/s10342-016-0937-z
75. Greenberg JA, Dobrowski SZ, Ustin SL (2005) Shadow allometry: estimating tree structural parameters using hyperspatial image analysis. Remote Sens Environ 97:15–25. https://doi.org/10.1016/j.rse.2005.02.015
76. Brown S, Gillespie ARJ, Lugo AE (1989) Biomass estimation methods for tropical forests with aplications to forest inventory data. Forest Science 35:881–902. https://doi.org/10.1093/forestscience/35.4.881
77. Fehrmann L, Kleinn C (2006) General considerations about the use of allometric equations for biomass estimation on the example of Norway spruce in central Europe. For Ecol Manage 236:412–421. https://doi.org/10.1016/j.foreco.2006.09.026
78. Jagodziński AM, Dyderski MK, Gęsikiewicz K et al (2018) How do tree stand parameters affect young Scots pine biomass?–Allometric equations and biomass conversion and expansion factors. For Ecol Manage 409:74–83. https://doi.org/10.1016/j.foreco.2017.11.001
79. Keith H, Barrett D, Keenan R (2000) Review of allometric relationships for estimating woody biomass for New South Wales, the Australian Capital Territory, Victoria, Tasmania and South Australia. Australian Greenhouse Office, Canberra
80. Li H, Zhao P (2013) Improving the accuracy of tree-level aboveground biomass equations with height classification at a large regional scale. For Ecol Manage 289:153–163. https://doi.org/10.1016/j.foreco.2012.10.002
81. Návar J (2009) Biomass component equations for Latin American species and groups of species. Ann For Sci 66:208–208. https://doi.org/10.1051/forest/2009001
82. Djomo AN, Ibrahima A, Saborowski J, Gravenhorst G (2010) Allometric equations for biomass estimations in Cameroon and pan moist tropical equations including biomass data from Africa. For Ecol Manage 260:1873–1885. https://doi.org/10.1016/j.foreco.2010.08.034
83. Ter-Mikaelian MT, Korzukhin MD (1997) Biomass equations for sixty-five North American tree species. For Ecol Manage 97:1–24. https://doi.org/10.1016/S0378-1127(97)00019-4
84. Zianis D, Suomen Metsätieteellinen Seura, Metsäntutkimuslaitos (2005) Biomass and stem volume equations for tree species in Europe. Finnish Society of Forest Science, Finnish Forest Research Institute, Helsinki, Finland
85. Paul KI, Roxburgh SH, England JR et al (2013) Development and testing of allometric equations for estimating above-ground biomass of mixed-species environmental plantings. For Ecol Manage 310:483–494. https://doi.org/10.1016/j.foreco.2013.08.054
86. Oliveira TS, Tomé M (2017) Improving biomass estimation for *Eucalyptus globulus* Labill at stand level in Portugal. Biomass Bioenerg 96:103–111. https://doi.org/10.1016/j.biombioe.2016.11.010
87. Paul KI, Roxburgh SH, Ritson P et al (2013) Testing allometric equations for prediction of above-ground biomass of mallee eucalypts in southern Australia. For Ecol Manage 310:1005–1015. https://doi.org/10.1016/j.foreco.2013.09.040
88. Zabek LM, Prescott CE (2006) Biomass equations and carbon content of aboveground leafless biomass of hybrid poplar in Coastal British Columbia. For Ecol Manage 223:291–302. https://doi.org/10.1016/j.foreco.2005.11.009
89. Henry M, Picard N, Trotta C et al (2011) Estimating tree biomass of sub-Saharan African forests: a review of available allometric equations. Silva Fennica 45:477–569. https://doi.org/10.14214/sf.38
90. Brown S (1997) Estimating biomass and biomass change of tropical forests: a primer. FAO, Rome

91. Chave J, Andalo C, Brown S et al (2005) Tree allometry and improved estimation of carbon stocks and balance in tropical forests. Oecologia 145:87–99. https://doi.org/10.1007/s00442-005-0100-x
92. Cole TG, Ewel JJ (2006) Allometric equations for four valuable tropical tree species. For Ecol Manage 229:351–360. https://doi.org/10.1016/j.foreco.2006.04.017
93. Hairiah K, Sitompul S (2001) Methods for sampling carbon stocks above and below ground. International Centre for Research in Agroforestry, Bogor, Indonesia
94. Terakunpisut J, Gajaseni N, Ruankawe N (2007) Carbon sequestration potential in aboveground biomass of Thong Pha Phum national forest, Thailand. Appl Ecol Environ Res 5:93–102
95. Feldpausch TR, Lloyd J, Lewis SL et al (2012) Tree height integrated into pantropical forest biomass estimates. Biogeosciences 9:3381–3403. https://doi.org/10.5194/bg-9-3381-2012
96. Vieilledent G, Vaudry R, Andriamanohisoa SF et al (2012) A universal approach to estimate biomass and carbon stock in tropical forests using generic allometric models. Ecol Appl 22:572–583. https://doi.org/10.1890/11-0039.1
97. Mattsson E, Ostwald M, Wallin G, Nissanka SP (2016) Heterogeneity and assessment uncertainties in forest characteristics and biomass carbon stocks: important considerations for climate mitigation policies. Land Use Policy 59:84–94. https://doi.org/10.1016/j.landusepol.2016.08.026
98. Chave J, Réjou-Méchain M, Búrquez A et al (2014) Improved allometric models to estimate the aboveground biomass of tropical trees. Glob Change Biol 20:3177–3190. https://doi.org/10.1111/gcb.12629
99. Holdridge LR, Tosi, Jr JA (1967) Life zone ecology. Tropical Science Center, San Jose, Costa Rica
100. Henry M, Bombelli A, Trotta C, et al (2013) GlobAllomeTree: international platform for tree allometric equations to support volume, biomass and carbon assessment. iForest Biogeosci Forest 6:326–330. https://doi.org/10.3832/ifor0901-006
101. Feldpausch TR, Banin L, Phillips OL et al (2011) Height-diameter allometry of tropical forest trees. Biogeosciences 8:1081–1106. https://doi.org/10.5194/bg-8-1081-2011
102. Falster DS, Duursma RA, Ishihara MI et al (2015) BAAD: a biomass and allometry database for woody plants: *ecological archives* E096–128. Ecology 96:1445–1445. https://doi.org/10.1890/14-1889.1
103. Jagodziński AM, Dyderski MK, Gęsikiewicz K, Horodecki P (2019) Effects of stand features on aboveground biomass and biomass conversion and expansion factors based on a *Pinus sylvestris* L. chronosequence in Western Poland. Eur J Forest Res 138:673–683. https://doi.org/10.1007/s10342-019-01197-z
104. Neumann M, Moreno A, Mues V et al (2016) Comparison of carbon estimation methods for European forests. For Ecol Manage 361:397–420. https://doi.org/10.1016/j.foreco.2015.11.016
105. Teobaldelli M, Somogyi Z, Migliavacca M, Usoltsev VA (2009) Generalized functions of biomass expansion factors for conifers and broadleaved by stand age, growing stock and site index. For Ecol Manage 257:1004–1013. https://doi.org/10.1016/j.foreco.2008.11.002
106. Somogyi Z, Cienciala E, Mäkipää R et al (2007) Indirect methods of large-scale forest biomass estimation. Eur J Forest Res 126:197–207. https://doi.org/10.1007/s10342-006-0125-7
107. Kangas A, Maltamo M (2006) Forest inventory: methodology and applications. Springer, Dordrecht
108. Chen L, Wang Y, Ren C et al (2019) Optimal combination of predictors and algorithms for forest above-ground biomass mapping from sentinel and SRTM data. Remote Sensing 11:414. https://doi.org/10.3390/rs11040414
109. Chen L, Ren C, Zhang B et al (2018) Estimation of forest above-ground biomass by geographically weighted regression and machine learning with sentinel imagery. Forests 9:582. https://doi.org/10.3390/f9100582
110. Propastin P (2012) Modifying geographically weighted regression for estimating aboveground biomass in tropical rainforests by multispectral remote sensing data. Int J Appl Earth Obs Geoinf 18:82–90. https://doi.org/10.1016/j.jag.2011.12.013

111. Qi YJ, Zhang YC, Wang K et al (2020) Application of spatial regression models for forest biomass estimation in Guizhou province, southwest China. Appl Ecol Environ Res 18:7215–7232. https://doi.org/10.15666/aeer/1805_72157232
112. Chirici G, Giannetti F, McRoberts RE et al (2020) Wall-to-wall spatial prediction of growing stock volume based on Italian national forest inventory plots and remotely sensed data. Int J Appl Earth Obs Geoinf 84:101959. https://doi.org/10.1016/j.jag.2019.101959
113. Chirici G, Mura M, McInerney D et al (2016) A meta-analysis and review of the literature on the k-nearest neighbors technique for forestry applications that use remotely sensed data. Remote Sens Environ 176:282–294. https://doi.org/10.1016/j.rse.2016.02.001
114. Chirici G, Barbati A, Corona P et al (2008) Non-parametric and parametric methods using satellite images for estimating growing stock volume in alpine and Mediterranean forest ecosystems. Remote Sens Environ 112:2686–2700. https://doi.org/10.1016/j.rse.2008.01.002
115. McRoberts RE (2009) Diagnostic tools for nearest neighbors techniques when used with satellite imagery. Remote Sens Environ 113:489–499. https://doi.org/10.1016/j.rse.2008.06.015
116. McRoberts RE, Gobakken T, Næsset E (2012) Post-stratified estimation of forest area and growing stock volume using lidar-based stratifications. Remote Sens Environ 125:157–166. https://doi.org/10.1016/j.rse.2012.07.002
117. Tomppo EO, Gagliano C, Natale FD et al (2009) Predicting categorical forest variables using an improved k-Nearest Neighbour estimator and Landsat imagery. Remote Sensing Environ 18
118. Foody GM, Cutler ME, McMorrow J et al (2001) Mapping the biomass of Bornean tropical rain forest from remotely sensed data. Glob Ecol Biogeogr 10:379–387. https://doi.org/10.1046/j.1466-822X.2001.00248.x
119. Mas JF, Flores JJ (2008) The application of artificial neural networks to the analysis of remotely sensed data. Int J Remote Sens 29:617–663. https://doi.org/10.1080/01431160701352154
120. Phillips SJ, Dudík M (2008) Modeling of species distributions with Maxent: new extensions and a comprehensive evaluation. Ecography 31:161–175. https://doi.org/10.1111/j.0906-7590.2008.5203.x
121. Saatchi SS, Harris NL, Brown S et al (2011) Benchmark map of forest carbon stocks in tropical regions across three continents. Proc Natl Acad Sci 108:9899–9904
122. Blackard J, Finco M, Helmer E et al (2008) Mapping U.S. forest biomass using nationwide forest inventory data and moderate resolution information. Remote Sens Environ 112:1658–1677. https://doi.org/10.1016/j.rse.2007.08.021
123. Carreiras JMB, Vasconcelos MJ, Lucas RM (2012) Understanding the relationship between aboveground biomass and ALOS PALSAR data in the forests of Guinea-Bissau (West Africa). Remote Sens Environ 121:426–442. https://doi.org/10.1016/j.rse.2012.02.012
124. Ford SE, Keeton WS (2017) Enhanced carbon storage through management for old-growth characteristics in northern hardwood-conifer forests. Ecosphere 8:e01721. https://doi.org/10.1002/ecs2.1721
125. Lin D, Anderson-Teixeira KJ, Lai J et al (2016) Traits of dominant tree species predict local scale variation in forest aboveground and topsoil carbon stocks. Plant Soil 409:435–446. https://doi.org/10.1007/s11104-016-2976-0
126. Powell SL, Cohen WB, Healey SP et al (2010) Quantification of live aboveground forest biomass dynamics with Landsat time-series and field inventory data: a comparison of empirical modeling approaches. Remote Sens Environ 114:1053–1068. https://doi.org/10.1016/j.rse.2009.12.018
127. Saatchi SS, Houghton RA, Dos Santos Alvalá RC et al (2007) Distribution of aboveground live biomass in the Amazon basin: AGLB IN THE AMAZON BASIN. Glob Change Biol 13:816–837. https://doi.org/10.1111/j.1365-2486.2007.01323.x
128. Breiman L (2001) Random forests. Mach Learn 45:5–32. https://doi.org/10.1023/A:1010933404324

129. Chi H, Sun G, Huang J et al (2017) Estimation of forest aboveground biomass in changbai mountain region using ICESat/GLAS and Landsat/TM Data. Remote Sens 9:707. https://doi.org/10.3390/rs9070707
130. Waser L, Ginzler C, Rehush N (2017) Wall-to-wall tree type mapping from countrywide airborne remote sensing surveys. Remote Sens 9:766. https://doi.org/10.3390/rs9080766
131. Axelsson C, Skidmore AK, Schlerf M et al (2013) Hyperspectral analysis of mangrove foliar chemistry using PLSR and support vector regression. Int J Remote Sens 34:1724–1743. https://doi.org/10.1080/01431161.2012.725958
132. Bishop CM (2006) Pattern recognition and machine learning. Springer, New York
133. Mountrakis G, Im J, Ogole C (2011) Support vector machines in remote sensing: a review. ISPRS J Photogramm Remote Sens 66:247–259. https://doi.org/10.1016/j.isprsjprs.2010.11.001
134. Smola AJ, Schölkopf B (2004) A tutorial on support vector regression. Stat Comput 14:199–222. https://doi.org/10.1023/B:STCO.0000035301.49549.88
135. Giannetti F, Barbati A, Mancini LD et al (2018) European forest types: toward an automated classification. Ann For Sci 75:6. https://doi.org/10.1007/s13595-017-0674-6
136. Waser L, Fischer C, Wang Z, Ginzler C (2015) Wall-to-wall forest mapping based on digital surface models from image-based point clouds and a NFI forest definition. Forests 6:4510–4528. https://doi.org/10.3390/f6124386
137. Irulappa-Pillai-Vijayakumar DB, Renaud J-P, Morneau F et al (2019) Increasing precision for French forest inventory estimates using the k-NN technique with optical and photogrammetric data and model-assisted estimators. Remote Sensing 11:991. https://doi.org/10.3390/rs11080991
138. Nilsson M, Nordkvist K, Jonzén J et al (2017) A nationwide forest attribute map of Sweden predicted using airborne laser scanning data and field data from the National Forest Inventory. Remote Sens Environ 194:447–454. https://doi.org/10.1016/j.rse.2016.10.022
139. Tomppo E, Halme M (2004) Using coarse scale forest variables as ancillary information and weighting of variables in k-NN estimation: a genetic algorithm approach. Remote Sens Environ 92:1–20. https://doi.org/10.1016/j.rse.2004.04.003
140. Chave J, Condit R, Aguilar S et al (2004) Error propagation and scaling for tropical forest biomass estimates. Phil Trans R Soc Lond B 359:409–420. https://doi.org/10.1098/rstb.2003.1425
141. Cohen R, Kaino J, Okello JA et al (2013) Propagating uncertainty to estimates of above-ground biomass for Kenyan mangroves: a scaling procedure from tree to landscape level. For Ecol Manage 310:968–982. https://doi.org/10.1016/j.foreco.2013.09.047
142. Lu D, Chen Q, Wang G et al (2012) Aboveground forest biomass estimation with landsat and LiDAR data and uncertainty analysis of the estimates. Int J Forest Res 2012:1–16. https://doi.org/10.1155/2012/436537
143. McRoberts RE, Westfall JA (2014) Effects of uncertainty in model predictions of individual tree volume on large area volume estimates. Forest Sci 60:34–42. https://doi.org/10.5849/forsci.12-141
144. Shao Z, Zhang L (2016) Estimating forest aboveground biomass by combining optical and SAR data: a case study in Genhe, Inner Mongolia, China. Sensors 16:834. https://doi.org/10.3390/s16060834
145. Wang G, Oyana T, Zhang M et al (2009) Mapping and spatial uncertainty analysis of forest vegetation carbon by combining national forest inventory data and satellite images. For Ecol Manage 258:1275–1283
146. Clutter JL, Fortson JC, Pienaar LV et al (1983) Timber managment: a quantitative approach. John Wiley & Sons, New York
147. Myers RH (1986) Classical and modern regression with applications. Duxbury Press, Boston
148. Ståhl G, Holm S, Gregoire TG et al (2011) Model-based inference for biomass estimation in a LiDAR sample survey in Hedmark County, Norway. Can J For Res 41:96–107. https://doi.org/10.1139/X10-161

149. Ståhl G, Heikkinen J, Petersson H et al (2014) Sample-based estimation of greenhouse gas emissions from forests—a new approach to account for both sampling and model errors. Forest Sci 60:3–13. https://doi.org/10.5849/forsci.13-005
150. Fu Y, Lei Y, Zeng W et al (2017) Uncertainty assessment in aboveground biomass estimation at the regional scale using a new method considering both sampling error and model error. Can J For Res 47:1095–1103. https://doi.org/10.1139/cjfr-2016-0436
151. McRoberts RE, Næsset E, Gobakken T (2013) Inference for lidar-assisted estimation of forest growing stock volume. Remote Sens Environ 128:268–275. https://doi.org/10.1016/j.rse.2012.10.007
152. McRoberts RE, Chen Q, Domke GM et al (2016) Hybrid estimators for mean aboveground carbon per unit area. Ecol Manage 378:44–56. https://doi.org/10.1016/j.foreco.2016.07.007
153. Thiel M, Basiliko N, Caspersen J et al (2015) Operational biomass recovery of small trees: equations for six central Ontario tree species. Can J For Res 45:372–377. https://doi.org/10.1139/cjfr-2014-0429

Overview of the Biomass Models

Ana Cristina Gonçalves and Adélia M. O. Sousa

Abstract The diversity of species, tree allometry, and stand structure makes modelling forest biomass a challenge. At tree level diameter at breast height, and height are the most frequently used explanatory variables. Yet, other variables that encompass the variability in tree allometry due to species, stand structure, competition between trees, and site allow better performances of the biomass models. Similarly, at area level, the biomass functions have large variability in the data and explanatory variables used for modelling. This is due to the differences in species, stand structure, and their correlation with the remote sensing data. The combination of different remote sensing data sets from passive and/or active sensors linked with ancillary data enabled to improve models' performance. Furthermore, a wide set of mathematical methods have been used to capture the stands and forests diversity and variability and accommodate it in the models to improve predictions. Overall, the wide range of biomass models corresponds to a continuous need to develop biomass functions that enable assessing, monitoring and predicting total or per component biomass.

Keywords Data sets · Explanatory variables · Mathematical models · Accuracy · Uncertainties

A. C. Gonçalves (✉)
MED—Mediterranean Institute for Agriculture, Environment and Development & CHANGE—Global Change and Sustainability Institute, Instituto de Investigação e Formação Avançada, Departamento de Engenharia Rural, Escola de Ciências e Tecnologia, Universidade de Évora, Apartado 94, 7002-544 Évora, Portugal
e-mail: acag@uevora.pt

A. M. O. Sousa
MED—Mediterranean Institute for Agriculture, Environment and Development & Change—Global Change and Sustainability Institute, Laboratório de Deteção Remota—EaRSLab, Instituto de Investigação e Formação Avançada, Departamento de Engenharia Rural, Escola de Ciências e Tecnologia, Universidade de Évora, Apartado 94, 7002-544 Évora, Portugal
e-mail: asousa@uevora.pt

A. C. Gonçalves and I. Malico (eds.), *Forest Bioenergy*, Green Energy and Technology,
https://doi.org/10.1007/978-3-031-48224-3_6

1 Introduction

The biomass estimation functions reflect the variability of the stand structure (*cf*. chapter "Stand Structure and Biomass"), the data sets available and the mathematical methods and techniques (*cf*. chapter "Modelling Biomass"). The biomass models can be, according to the data sets used, broadly grouped in: tree level and area level. Tree level functions have as explanatory variables dendrometric, tree classification systems and, sometimes, absolute density variables. These allometric functions are most frequently derived from destructive sampling. The area level functions are based on the former to obtain the biomass (dependent variable) with data from forest inventory, whereas the explanatory variables are from forest inventory, remote sensing and ancillary. The remote sensing data is dependent on the sensors (passive or active), and its spatial resolution and temporal resolution. The passive sensors data includes spectral bands, vegetation indices and texture indices while the active sensors data comprise bands, point clouds and indices. Many ancillary variables have also been used in biomass modelling, including topographic (altitude, slope, aspect), climatic (precipitation, temperature) and soil (pH, texture, depth).

The main difference between volume and biomass is that volume evaluates the mass that can be used for timber while biomass is of more general application as it evaluates not only stem but also crown and root system mass. Volume modelling preceded biomass modelling due to its importance in the evaluation of timber yield of stands and forests. Biomass has been gaining importance as a parameter to evaluate biomass and carbon sequestration and sustainability of forest areas. Furthermore, in forest systems where the main production is not timber, the volume has little importance, conversely to biomass that can be used in all forest types.

Reflecting the diversity of stand structures, data sets and mathematical methods, there has been a wide range of studies that developed biomass estimation functions. In this chapter some examples will be given of the models developed at tree level and at area level with data from destructive sampling, forest inventory, and remote sensing, for both passive and active sensors and ancillary data. The goal of this chapter is to present the diversity of biomass models, their potential use, and their advantages and disadvantages.

2 Functions at Tree Level

Biomass has been having the attention of researchers as a way to evaluate forest sustainability, yield, biomass and carbon stocks [1, 2]. The most frequently used explanatory variables in the biomass functions at tree level are diameter at breast height and total height, though other variables such as wood density and crown area have also been used [3–5]. At tree level, single and multiple linear regression and nonlinear regression (exponential) are the most frequent functions [6].

Many biomass allometric functions at tree level have been developed for different species and stand structures. Most are linear or exponential functions. Their number is related to uncertainties encountered when calculating biomass. The biomass allometric functions have been usually fitted per species [7–13]. This is a consequence of the growth patterns of each specie (*cf*. chapter "Stand Structure and Biomass"). The regime also introduces differences in the growth and allometry of the individual trees. As a consequence, biomass functions were developed for high forest individuals (e.g., [7–10, 12–16]) and coppice individuals (e.g., [17–24]). For high forest stands several authors analysed and compiled biomass allometric functions at tree level, with the best accuracy, for 39 forest species in Europe ([13]; 65 [12] and 85 [9] in North America; 40 in Australia [8], 102 in South America [11] and 8 for mixed tropical American, African and Asian forests [7]. For Australia, Keith et al. [10] revised more than 100 equations, improved some of them and stressed the need to improve their accuracy.

For some forest areas, especially in tropical forests, is not always possible to fit biomass functions per species. Inversely, functions have been developed per group of species, or even with groups more broad, like the forest type or the bioclimatic zones. The groups of species are defined according to the species traits, such as patterns of growth, density of wood and the species' habits. The drawback of this method is that accuracy might decrease and bias increase when compared with the allometric functions per species [25–32].

The influence of the variability in tree allometry, due to species, stand structure, social status, stand development stage (age), regime and site were studied. Tabacchi et al. [33] modelled above ground biomass, at national level, for the 25 most important species in Italy, with destructive sampling, using linear regression with a weighted function. The authors referred that above ground biomass estimates were unbiased. Yet, the authors pinpointed two problems: (i) individual tree allometry variability, caused by the geographical gradients coupled with the low number of sampled trees for some species, might explain the error level; and (ii) though the use of weighted regression overcome the heteroscedasticity of the residuals, error distribution did not meet the normality criterion, thus, in general, the intervals of confidence are not valid. Mankou et al. [4] also found that species traits affected tree biomass allometry in a study of 54 species in 8 sites in Africa. Mixed effects models were able to accommodate diameter at breast height, height, shade tolerance (shade intolerant species had lower crown ratios than shade tolerant ones), and wood density (which seems to be linked to the direct relation between crown and stem diameters) variability [4]. The inclusion of crown and/or height as explanatory variables in biomass models (mixed effects) for 10 *Eucalyptus* spp. plantations in 24 sites in Australia, increased the accuracy of the allometric equations at tree level, in particular for the foliage component. Crown dimensions or height may express, in biomass models, the competition effects between trees, which in turn are dependent on stand structure (e.g., density, composition, stand development stage) or site (edaphic and climatic conditions) [34]. For *Pinus contorta* var *latifolia* plantations, with different ages (20–87 years), variability among 37 sites was detected and minimised with mixed effects models, by the inclusion of height as explanatory variable [35].

Apart from diameter at breast height, other variables amongst which stand development stage, are determinant in the model accuracy. In mixed stands with three different development stages (young, mature and old) the inclusion of the species and the stand development stage as explanatory variables in biomass models attained better performances than just with diameter at breast height and height. Moreover, mixed models fixed and random effects accommodated better the variability in biomass than the linear models, reaching better performances [36]. Silvicultural practices such as control of spontaneous vegetation and fertilisation were tested for their effect on estimating biomass with mixed effects models for *Pinus ponderosa* plantations in young stands in three sites. Both control of spontaneous vegetation and/or fertilisation resulted in the increase of biomass per stand, especially in poor quality sites. Yet, in biomass modelling, while the site had a significant effect, silvicultural practices did not, implying that the site was a better predictor of biomass than silvicultural practices [37]. Jorge et al. [38] developed biomass allometric functions for *Quercus suber* with a joined data set from Portugal, Spain and Tunisia, using seemingly unrelated regression (SUR) with diameter at breast height, with and without height and a dummy variable for the country as explanatory variables. The models with the best performance were those with all the explanatory variables. The increase of accuracy by the inclusion of height in the model was that it enabled to accommodate the variability of the diameter at breast height/height relationship, due to competition whereas the dummy variable for the country accounted for the variability of tree allometry due to stand structure, silvicultural practices as well as site variability.

In Europe, Annighöfer et al. [39], considering the areas with young stands, identified the need to have biomass allometric functions for seedlings and saplings. The authors collected data from 19 European forest species and developed specific and generic allometric functions, with root collar diameter or height or the combination of both variables. Similarly, Jagodziński et al. [16] identified bias in the estimation of biomass with the existing allometric functions for young *Pinus sylvestris* stands and developed new functions with improved accuracy. Sillett et al. [40] developed and compiled a large set of allometric functions at tree level for several species. The authors identified the need to develop functions considering climatic areas, stand composition, tree social status and stand development stage, to improve the models' accuracy, due to the difference in biomass increments [40]. Li and Zhao [41] for *Cunninghamia lanceolate* in China improved the existing allometric functions with diameter at breast height as explanatory variable, considering the vertical stand profile divided into 2 to 5 height classes. The best performance was attained with three height classes. Paul et al. [15] developed, for Australian forest stands, allometric functions considering or not the specificity of species, *genus* and growth habit. They concluded that, as species and site introduce variability in the tree allometry, at a local level the best accuracy was attained by the allometric functions species and site specific while at regional level the generic functions were better suited.

The coppice regime corresponds to tree habits different from those of the high forest, reason why specific coppice biomass allometric functions were developed. Reed and Tomé [17] fitted models for coppice *Eucalyptus globulus* stands in Portugal with and without irrigation and referred that the development of models for irrigated and non-irrigated stands improved the performance of the models, due to the different growth patterns of the poles. Oliveira and Tomé [24] also for *Eucalyptus* spp. stands in Portugal stressed that dominant height, age and regime improved model accuracy. The different pole growth habits occurred not only for different species but also for different clones. Vande Walle et al. [19] developed allometric functions for four species while Zabek and Prescott [18] and Dillen at el. [42] did it for clones of *Populus* spp. and *Salix* spp. As referred, high forest and coppice regime derived in different tree biomass partitioning, which resulted in the development of biomass allometric functions for coppices of *Castanea sativa*, *Quercus frainetto*, *Quercus cerris* and *Quercus petraea* [20, 43, 44].

The differences in biomass partition and their effects on biomass modelling were studied for high forest stands of 13 species and 39 sites with nonlinear models. Though similarities were observed between species, there were differences in biomass partitioning between broadleaved and conifer species (with higher variability for the former than for the latter) along with the variability of wood density per species. The species specific allometric equations were better suited than group-species aggregations. The biomass models with diameter at breast height and height as explanatory variables attained better performances than those with nonlinear seemingly unrelated regression, due to the increase of bias [45].

Modelling 4–15 years old *Picea abies* plantations with mixed effects models indicated that 2/3 of the model variance was due to interactions between the trees in the stands whereas the remaining 1/3 was due to variability among stands. The allocation patterns were considered the main factors in the variability of biomass between trees in a stand. The proportion of crown biomass was interlinked with competition within the stand, *hd* (the ratio between height and diameter at breast height) was site related (better quality sites had higher *hd* at early stages of development) and wood density. As a result, the *formulae* that better described the aforementioned relation was a compound explanatory variable with both diameter at breast height and height (dbh^2h). Moreover, it seemed that for small young trees (in contrast to large trees) the primordial factor of specificity to the site was the patterns of allocation of photoassimilates to roots, needles and branches [46].

Many other examples could be given that emphasise the specificity of the biomass functions at tree level. Overall, there seems to be a trend towards linear and exponential biomass equations; the functions' accuracy improves when they are specific to the species, clones, regime, composition, structure, stand development stage or age, site and tree social status; and the generic functions are better suited for the regional level for a similar performance than the local level functions.

3 Biomass Functions with Remote Sensing Data

The diversity of models to estimate biomass with satellite image data reflects the spatial and temporal scales, the species, the stand structure, the forest inventory, the remote sensing sensors and the mathematical methods (*cf*. chapter "Modelling Biomass"). The range of variables derived from satellite images (including the combination of data with different spatial resolutions and/or passive and/or active sensors) and the variety of mathematical models to fit the functions are related to a constant search to improve the accuracy of the models.

Biomass Functions with Passive Sensors Data

Passive sensors can be classified by their spatial resolution in three broad classes, low spatial resolution (pixel larger than 500 × 500 m), corresponding to the sensors National Oceanic Atmospheric Administration-Advanced Very High Resolution Radiometer (NOAA-AVHRR), and Moderate Resolution Imaging Spectroradiometer (MODIS); medium spatial resolution (pixel size between 10 × 10 m and 30 × 30 m), corresponding to Landsat, Advanced Spaceborne Thermal Emission and Reflection Radiometer (ASTER), Sentinel-2 and *Satellite pour l'Observation de la Terre* (SPOT) sensors; and high spatial resolution (pixel smaller than 5 × 5 m), corresponding to Geoeye, IKONOS, Quickbird, WorldView and Pleiades sensors.

The passive sensors with ***low spatial resolution*** have been used to derive biomass models. Linear regression was used by Al-Bakri et al. [47] to developed a biomass function for arid and semi-arid climate in Jordan (Africa) for herbaceous, shrub and arboreal strata with NOAA-AVHRR sensor data. The best performing model had vegetation indices as explanatory variables. Li et al. [48] used random forest (RF) to model the biomass of a bamboo forest in subtropical China, using MODIS (spatial resolution of 1000 m) with a time series, including spectral bands and Leaf Area Index (LAI) product. The authors referred that the inclusion of LAI as explanatory variable greatly improved the model accuracy, due to LAI being able to reflect the bamboo growth. MODIS, climate, topographic, soil data, central coordinates of the raster cells and distance metrics with K-Nearest Neighbours (K-NN) and random forest, for boreal forests in China were used by Zhang et al. [49]. The performance of the models with single month or with a time series data had similar performance. Moreover, accuracy depended on the species and the stand structure variability. K-nearest neighbours model was able to account for disturbances like fire and harvest, and was considered a useful tool for large scale monitoring. The accuracy of this model for young and early mature stands was similar to that derived from Light Detection and Ranging (LiDAR) data.

Blackard et al. [50] used both MODIS and Landsat TM along with climatic and topographic data with regression tree to model biomass at country level for United States of America and Puerto Rico. Overall, the models had a good performance being comparable to those obtained by plot estimation and conversion/expansion functions. Yet, under and over estimation of biomass occurred for stands with low and high density, which was related to satellite image data saturation. The authors stressed the importance of presenting uncertainty maps along with those that predict biomass. Propastin [51] combined MODIS and Landsat ETM+ sensors for Indonesia tropical forests with linear regression where spectral bands were the explanatory variables. The combination of the two sensors' spectral bands enabled to map with accuracy large forest areas, which was not feasible with only Landsat ETM+ data.

The need to improve accuracy led to the use of ***medium spatial resolution*** satellite image data. Steininger [52] used Landsat TM and linear and nonlinear (exponential) regression for tropical forests in Brazil and Bolivia. Contrasting results were observed among canopy reflectance/biomass for the two countries due probably to the low sun angle of the Bolivian satellite images, the age of the forest stands and the difference in the canopy structure. In Bolivia shading effect was smaller on the upper canopy than in Brazil, explaining, at least partially, the weak correlation between age and biomass. Additionally, the contrasting patterns among the stands of Brazil and Bolivia could be related to the differences in the canopy structure composition of the stands. The best performance model was the linear model with reflectance as explanatory variable. Landsat 7 ETM+ was used by Zheng et al. [53] with multiple linear regression for pure and mixed forests of conifer and broadleaved species in Wisconsin, United States of America. The above ground biomass model with medium spatial resolution aimed to discriminate the forest management practices and stand development stages (in three classes: young, intermediate and mature). The models' performance improved when conifer and broadleaved species were fitted separately. This was due to different correlations between biomass and the explanatory variables. While the broadleaved species biomass had stronger correlations with stand stand development stage (or age) and near-infrared reflectance, for the conifers the strongest correlation was with the Normalized Difference Vegetation Index (NDVI).

With Landsat 7 ETM+, Gasparri et al. [54] modelled biomass in a semi-arid region in Argentina, with linear and non-linear models and vegetation indices as the explanatory variables. The best performing model was the linear with NDVI as explanatory variable, which was related to the sparse density resulting that NDVI did not reach saturation levels, and was sensitive to traits related to the absorption of photosynthetically active radiation. Landsat TM and neural networks were used by Foody et al. [55] in Malaysia, with vegetation indices as explanatory variables. The authors concluded that the fitting method was suited to model biomass with accuracy. For tropical forests in Brazil, Thailand and Malaysia, Foody et al. [56] used Landsat 4 and 5 with multiple linear regression and neural networks algorithms, where vegetation indices were the explanatory variables. The best performing models were attained with neural networks fitted per site. The variability of the forests between sites and their influence on the spectral signal explained the variation in accuracy.

Powell et al. [57] used Landsat time series (1985–2006) with Reduced Major Axis regression (RMA), Gradient Nearest Neighbour imputation (GNN), and random forest, for forest stands in Arizona and Minnesota, United States of America. The explanatory variables considered were spectral bands, tasselled cap indices, vegetation indices, topographic variables and climate variables. It was not clear whether one regression method was superior to another. Random forest was the best model in error minimisation, while for maintaining the variation (e.g., old growth forests) reduced major axis regression or gradient nearest neighbour imputation had better performances, and for the maintenance of the covariance of structure forest attributes gradient nearest neighbour imputation should be chosen. The authors stressed the importance of plot and scene validation, and that the Landsat time series enabled a consistent evaluation of the dynamics of biomass accurately. Maps of forest disturbances and biomass made the bridge between temporal dynamics of biomass, including the regrowth, and accuracy.

Zhu and Liu [58] used Landsat time series, with simple linear, multiple-linear (SML), Partial Least Squares (PLS), Reduced Major Axis (RMA), random forest and Artificial Neural Network (ANN) regression for temperate forest stands in Ohio, United States of America. The explanatory variables considered were NDVI at the time frame (different seasons). In autumn the correlation between NDVI and above ground biomass was stronger than in the peak season. The time series highlighted the importance of the NDVI to make the bridge to the trees' phenological phase. The accuracy of the models improved with the use of time series. Different modelling methods had similar accuracies which indicated that they did not play a key role in the improvement of the models' accuracy. The different models had weaknesses and strengths, thus the selection should be dependent on the use. The advantages of passive sensors were that they were easily accessible and had a wide temporal range. Yet, the models' accuracy might be lower than those with active sensors data.

ASTER sensor was used by several authors to model biomass. Fernández-Manso et al. [59] for *Pinus pinaster* in Spain, used multiple linear regression. The best performance model included as explanatory variable the fraction images from Linear Spectral Mixture Analysis (LSMA), which much improved its performance. This model can be used at regional level for monitoring biomass, carbon sequestration and management practices. Heiskanen [60] used linear and non-linear regression for forest stands in Finland. The best models had Simple Ratio index (SR) or NDVI vegetation indices or reflectance canonical correlation analysis as explanatory variables. Linear and nonlinear models had similar performances. The authors indicated as possible sources of bias the understorey vegetation and the reflectance of the background. Poulain et al. [61] for *Nothofagus pumilio* stands in Chilean Andes used linear and nonlinear regression (exponential). The explanatory variables were simple ratio index and the vegetation cover fraction. As in the former study, the performance of the linear regression was better than the nonlinear one.

Random forests were used by Pham and Brabyn [62] with SPOT 4 and 5 (fusion with panchromatic spectral bands - 10 m spatial resolution) with a 12 year time series (2000–2011) in mangrove forests in Vietnam. Vegetation type and texture improved models' accuracy for complex forest systems like mangroves. Random forest was suited for modelling above ground biomass with a small number of plots. The increase in the image classification accuracy (e.g., with ancillary data) improved the models' performance. Plot location might originate bias, which could be reduced by placing the plots in large homogeneous patches.

Linear regression and Sentinel-2 data were used by Askar et al. [63] to model above ground biomass for tree forest species in Indonesia with vegetation indices as explanatory variables. The explanatory variables were the Normalised Difference Index attained with bands 4 and 5 (NDI45) and Enhanced Vegetation Index (EVI). The authors stated that these two vegetation indices had a better performance in linking the biophysical parameters of vegetation (NDI45) and dense vegetation (EVI, as it was able to reduce the effect of atmosphere and canopy background). Random forest was used by Pandit et al. [64] with Sentinel-2 in modelling biomass in forest stands in Nepal. The explanatory variables considered were the spectral bands and vegetation indices. Though the model had a good performance the authors stated that spectral texture indices might improve model accuracy.

High spatial resolution (Geoeye, IKONOS, Quickbird, WorldView and Pleiades) passive sensors have been used for smaller areas than low and medium spatial resolution remote sensing data, for modelling biomass with high accuracy. Quickbird sensor was used in several studies. Leboeuf et al. [65] used linear regression for *Picea mariana* stands for three sites, in Canada. The model had shadow fraction as explanatory variable, and similar performance at the three sites, which was considered efficient for mapping. The authors pointed out that the methodology was easy to apply, and as local regressions were similar among the three sites it was possible to derive a generic function. Sousa et al. [66] and Macedo et al. [67] used linear regression to model above ground biomass in *Quercus rotundifolia* stands in Portugal. The former used crown horizontal projection as explanatory variable while the latter used vegetation indices (the best model had the median simple ratio as explanatory variable). When comparing both studies it seemed that crown horizontal projection was a better predictor than the vegetation indices. This could be, at least partially, explained by stand structure; one layer, low density, with many free growth trees. Sousa et al. [68] used multiple linear regression to model above ground biomass in *Quercus rotundifolia* and *Quercus suber* stands in Portugal, with crown horizontal projection (total and per species) and dummy variables for the stand composition. Overall, the best model included the total crown horizontal projection and dummy variables for the species.

Quickbird and WorldView were used in other studies. Gonçalves et al. [69] used multiple linear regression to model above ground biomass in *Quercus rotundifolia, Quercus suber* and *Pinus pinea* stands in Portugal, with crown cover (total and per species) and dummy variables for the stand composition as explanatory variables. The best model had as explanatory variables the total crown cover and dummy variables for composition. The function reflected the variability among species and stand structure. Moreover, as modelling was based on a square grid the estimations should use the same grid dimension. Gonçalves et al. [70] used multiple linear regression for *Pinus pinaster* stands in Portugal, with crown horizontal projection as explanatory variable. A time series of satellite images (2004, 2007 and 2011) was used to evaluate the temporal dynamics of the stands. It enabled the evaluation of growth as well as of fire effects on above ground biomass. Lourenço et al. [71] for *Quercus rotundifolia*, *Quercus suber* and *Pinus pinea* modelled above ground biomass with random forest and the spectral bands, vegetation indices and texture indices (Grey-Level Co-occurrence Matrix) and crown horizontal projection (vegetation mask) as explanatory variables. The explanatory variable with the highest relative importance was Grey-Level Co-occurrence Matrix, followed by the vegetation indices and the spectral bands. Moreover, the explanatory variables calculated with the vegetation mask had a stronger relationship with above ground biomass. This was related to the stand structure (including low density) and irregular spatial distribution of the trees in the stands (both in clusters or isolated).

Biomass has been modelled for mangrove forests in China with Pléiades and single and multiple regression by Wang et al. [72]. The best performance model had as explanatory variables texture indices. The authors' main conclusions were that the variability captured by texture variables was larger than that of the spectral ones; the window size determined the sensitivity of the texture variables with accuracy increasing with the window increase; and when texture and spectral variables were associated, the accuracy of the model decreased.

Random forest was used by Ploton et al. [73] with Pléiades in tropical forests in French Guiana, Africa and India, with texture indices (Fourier and lacunity), principal component analysis, site and forest type as explanatory variables. The goal was to derive a generalised biomass model based on a canopy texture index that could be used at local and global levels. The performance of the model improved with lacunity and bioclimatic as explanatory variables, that captured the variability of *hd*. The drawbacks found were related to the variability of the tree height and slenderness, which could not be evaluated using remote sensing 2D canopy texture metrics. Jachowski et al. [74] used GeoEye-1 and ASTER sensors data with support vector machine algorithms, in mangroves in Thailand. Higher errors were observed for higher biomass ranges, indicating the need to improve the existing models. Also, it was a good tool for monitoring forest systems.

Biomass Functions with Active Sensors Data

Active remote sensing sensors have been used to acquire forest stand variables, especially for areas where forest inventory plots are scarce, due to stand structure and/or topographic features, cost and labour. The main goal is acquiring active sensor data and fitting biomass estimation functions with high accuracy. Two main active sensors, Radio Detection and Ranging (RaDAR) and Light Detection and Ranging (LiDAR) have been used either independently or a combination of both, to derive a wide range of variables. Their combination showed the potential to develop continuous detailed maps at global and regional scales [75].

Synthetic Aperture RaDAR (SAR) was used to model forest biomass in several studies. Santos [76] used logarithmic and polynomial functions for Amazon tropical forest in Brazil. Both models had a suitable performance. The Iterated Conditional Modes (ICM) classification originated accurate above ground biomass mapping.

Solberg et al. [77] utilised linear and exponential regression for *Picea abies* and *Pinus sylvestris* stands in boreal forests in Norway. The best model was the linear and the saturation effect did not occur. The higher accuracy was attained with digital terrain model of high quality derived from Airborne Laser Scanning (ALS) which outperformed the digital terrain model derived from topographic maps. Moreover, accuracy slightly increased when the boundaries of the stands were removed from the analysis.

Other studies focused on LiDAR. Linear regression was used by Lau et al. [78] with Terrestrial Laser Scanning (TLS) and destructive tree sampling, to model biomass at tree level, in tropical forests in Guyana. The best model included crown diameter as explanatory variable while the inclusion of height had a worse performance. Salum et al. [79] used linear regression, for mangroves in Guaras Island in Brazil. The model with canopy height as explanatory variable had the best performance. The model enabled to detect the differences between different species which had diverse morphologies and allowed the mapping above ground biomass. Yet, its use in other mangroves could not be extrapolated due to species morphologies variability. For mangroves in other regions, it should be carried out the calibration of the model.

Esteban et al. [80] modelled biomass and their dynamics in time with Airborne Laser Scanning and random forest, at an area level, for Spanish and Norwegian temperate forests. The authors concluded that the relationship between biomass and remote sensing data were described adequately with random forest; the biomass estimations with random forest models had a better performance than the expansion based estimations; and the bootstrapping influenced the performance of the model, with wild bootstrapping being the better suited.

Biomass and its variability in time were studied by Knapp et al. [81] with LiDAR, and a forest model (FORMIND), with linear regression, nonlinear regression (exponential) and random forest, for tropical forests in Panama. Linear regression models had bias, while the other two methods did not. The two latter models had similar performance. It was considered a tool to monitor biomass in time, but further improvements are needed, as it was not able to detect small variations in biomass. Swatantran et al. [82] developed biomass models with Laser Vegetation Imaging Sensor (LVIS) and Airborne Visible Infrared Imaging Spectrometer (AVIRIS) with linear (single and multiple) regression for Mediterranean forests in California, United States of America. The models with LVIS data had good performances, improving when species were stratified. The models with the fusion of the data of the two sensors and with each sensor alone did not had significant differences in the derived maps. LiDAR derived variables were suitable when no species stratification was made while the hyperspectral derived variables were better suited when species stratification was done. LiDAR data and ancillary variables (e.g., Gini index, wood density), with multivariate regression (nonlinear regression), for tropical forests in Panama, French Guiana, Gabon, and temperate forests in United States of America and Germany were used by Knapp et al. [83] to model biomass. The models' performance was good. It was highlighted the importance of the explanatory variables (e.g., height, density, vertical heterogeneity and wood density) in the biomass predictions as well as interactions that originated those relationships.

Biomass was also modelled with data from both RaDAR and LiDAR. Nelson et al. [84] modelled biomass with airborne BioSAR RaDAR and PALS LiDAR with linear (single and multiple) regression for *Pinus taeda* pure even-aged stands with closed canopies in North Carolina, United States of America. The models with explanatory variables derived from LiDAR were more accurate than those of RaDAR, whereas those with both sensors' data did not improve accuracy when compared with those of LiDAR. Næsset et al. [85] used LiDAR, Interferometric Synthetic Aperture RADAR (InSAR), and aerial photographs, with linear and nonlinear regression for *Picea abies*, *Pinus sylvestris* and *Betula pubecens* stands in boreal forests in Norway. The modelling was done at three levels, district, village and stand. The explanatory variables considered were the height metrics (LiDAR and InSAR), canopy density metrics (LiDAR), and strata as function of age and composition of the stands (aerial photographs). Models with LIDAR data outperformed those of InSAR. The error sources were linked to the time displacement between InSAR data and forest inventory.

Tsui et al. [86] used Advanced Land Observation Satellite and Phased Array L-band Synthetic Aperture RaDAR (ALOS PALSAR), RaDARSAT-2 and LiDAR with multiple linear regression to model biomass in *Pseudotsuga menziesii* and *Tsuga heterophyla* stands in a temperate forest in Vancouver Island, British Columbia, Canada. Stem biomass had the strongest relationship with LiDAR data while crown had the lowest; and the combination of horizontal and vertical polarisation (HH and VV, respectively) backscatter from RaDAR had the strongest correlation above ground biomass. When to the best model of LiDAR was added the C-band of RaDAR there was an improvement in the performance of the model; and the best biomass model was attained with InSAR coherence magnitude of repeated passes and the combination of the C and L bands. Montesano et al. [87] developed biomass functions with LiDAR, SAR and linear and nonlinear (power) regression for boreal forests in Maine, United States of America. The results indicated variability in accuracy and uncertainty across the biomass gradient of the studied area. The spaceborne and airborne sensors had an overall trend towards error reduction with the increase of biomass (especially from 0 to 60 t·ha^{-1}) and spaceborne data had lower accuracies than airborne. The spaceborne data did not enable the estimation of biomass with sufficient accuracy (error 50–100% for biomass <80 t·ha^{-1}). Tanase et al. [88] used Polarimetric L-band Imaging Synthetic Aperture RaDAR (PLIS) and ALS LiDAR with multiple linear regression to model biomass in *Callistris glaucophylla* and *Eucalyptus microcarpa* stands in New South Wales, Australia. The models with LiDAR explanatory variables had smaller errors than those of RaDAR; for stands with biomass between 30 and 100 t·ha^{-1} the error between both sensors was around 9%; and the models with explanatory variables of both sensors did not derived in the improvement of the model. The choice of the most appropriate sensor was dependent on the desired accuracy, forest area, time range and costs. The most suited models for forest management were those with LiDAR explanatory variables as they were the most accurate for all biomass ranges. Omar et al. [89] used SAR and Sentinel-1A with single and multiple linear regression in tropical forests in Malaysia. The best model had explanatory variables from both sensors while the models with only one sensor had worse performance. Schlund et al. [90] modelled biomass for several forest types in Indonesia with SAR and LiDAR, forest inventory and digital terrain model data with linear regression. The authors stressed the importance of SAR data to estimate the variability of the canopy height and its effects on above ground biomass modelling.

Biomass Functions with Passive and Active Sensors Data

Biomass modelling with passive sensors, due to their characteristics and constraints (*cf*. Sect. 3) might not enable to achieve the desired model performances. Similarly, active sensors (*cf*. Sect. 3) characteristics in a different way might also result in models with accuracies below the desired thresholds. The combination of variables derived from both passive and active sensors have been used to overcome the constraints and improve the model's accuracy.

Shendryk et al. [91] developed biomass functions with SPOT-5 and LiDAR, and linear regression for conifer-dominated forests in Sweden. The accuracy of the model was reasonable with the fusion of LiDAR and spectral data, with the best accuracy achieved with a grid of 3 × 3 m. This method can be used in other conifer stands and to monitor development and disturbances in time. Sun et al. [92] used LVIS, SAR ALOS PALSAR, Landsat ETM+ and ASTER with single and multiple linear regression in mixed broadleaved and conifer species forests in Maine, United States of America. The wavelength of the RaDAR constrained the accuracy of the biomass models. The best performing model was the multiple linear regression with height from LVIS as explanatory variable and was considered the reference for biomass values. The biomass models with SAR variables as explanatory variables derived from a time series had a similar performance to that of LVIS variables. Yet, the differences between the models could be due to the biomass samples not covering the total range for the existing stand structures; and to the gap of time between the forest inventory and SAR and LVIS data acquisition.

Lucas et al. [93] used Landsat TM (1988–2016), Synthetic Aperture RaDAR (SAR) (1988–2016), Interferometric Shuttle RaDAR Topographic Mission (SRTM) X/C-band (2000), TanDEM-X-band (2010–2016) and WorldView-2 (2016), in mangroves in Malaysia, with linear and nonlinear regression. The nonlinear regression was considered the best model. The time series enabled the quantification of biomass dynamics, including those resulting from silvicultural practices such as thinning and regeneration. Brovkina et al. [94] modelled biomass with high spatial resolution hyperspectral (HS), airborne laser scanning and fusion between hyperspectral and LiDAR data, for temperate forest stands in Czech Republic, with linear and nonlinear regression. The model with the highest uncertainties was the one with HS data as explanatory variables, while the best performance was attained with the explanatory variable derived from the fusion of HS and LiDAR.

Multiple linear regression and the combination of several remote sensing data were used to develop biomass models. Persson [95] used Pléiades and ALS in boreal forests in Sweden, with explanatory variables the height metrics, principal component analysis (PCA) of spectral bands and texture metrics. The most important variables were the height metrics while the latter two gave similar information. Yet, most models included all variables types. Basuki et al. [96] utilised Landsat-7 ETM+ and PALSAR in mixed tropical forest, in Indonesia. The fusion of the images improved above ground biomass estimation and discrete wavelet transform was a key variable to improve qualitatively and quantitatively the performance of the model. Phua et al. [97] used Landsat 8 OLI, airborne LiDAR for tropical forests in Malaysia. The model with the highest accuracy included explanatory variables of both sensors. The authors highlighted that: Landsat 8 OLI texture measures had a higher correlation with biomass than the vegetation indices probably due to saturation; LiDAR penetration variables identified the differences in the canopy structures of the different types of stands; and Landsat 8 OLI and LiDAR explanatory variables of the developed model improved the accuracy when compared with the models with explanatory variables of only one sensor. Berninger [98] combined data from ALOS PALSAR, ALOS-2 PALSAR-2, Sentinel-1, airborne LiDAR, Digital Elevation Model (DEM) from the Shuttle RaDAR Topography Mission (SRTM), MODIS hotspot information (product MCD14DL) and European Space Agency (ESA) Climate Change Initiative (CCI) land cover maps, for tropical forests in Indonesia. A time series of 3 years was used, deriving in one model per year, with polarisation ratios as explanatory variables. The accuracy improved by using a time series and enabled the evaluation of the dynamics of biomass.

Other studies used random forest with a fusion of different remote sensing data. Cortés et al. [99] used Landsat, ASTER, LiDAR in *Pinus radiata*, *Eucalyptus globulus* and *Nothophagus* glauca stands in Chile. The best performing model was the one with the combination of Landsat and LiDAR variables while the combination of Landsat and ASTER ones performed worse than the former. For *Pinus radiata* and *Nothophagus glauca* the best correlations and smaller errors were attained when canopy structure was included in the model. Also, the accuracy of the models seemed to be related to the accuracy of the canopy height estimations. In the case of *Eucalyptus globulus* the LiDAR point density did not seem to be dense enough which derived in under estimations of the stand height. Huang et al. [100] used Landsat TM, MODIS, PALSAR, ICEsat/GLAS (Ice, Cloud, and land Elevation Satellite, Geoscience Laser Altimeter System) in pure and mixed stands, at country level, in China. The model enabled to produce an above ground biomass map of 30 m of spatial resolution. The overall biomass estimations were similar to previous ones based on forest inventory and conversion factors. The uncertainties of the developed model were related to the biomass derived from allometric models with forest inventory data and the estimation of heights with spaceborne LiDAR. The study highlighted the use of spaceborne LiDAR to extend the forest inventory plots (of a limited number) to a wider set of GLAS data. It also evaluated, on a large scale, the importance of multi-seasonable Landsat and PALSAR data to improve the model's accuracy.

Chi et al. [101] modelled above ground biomass for high diversity forests (four types) in a temperate climate in China, with Landsat ETM+ and ICESat/GLAS. The model performance was good, enabling to detect the variability of the different types of forests. Ghosh and Behera [102] used Sentinel-1A and Sentinel-2A for *Shorea robusta* and *Tectona grandis* stands in a tropical forest in India. The best model had vegetation indices and SAR backscatter as explanatory variables. The combination of the two Sentinels' enabled a good accuracy for forests with high density. Matasci et al. [103] utilised Landsat TM and ETM+ and LiDAR for boreal forests in Canada. The model with explanatory variables of all sensors enabled deriving accurate maps both detailed and for large areas. Kashongwe et al. [104] used Landsat 8, LiDAR in tropical forests in Democratic Republic of Congo. For modelling three options were taken, using dry season image, using wet season image and both images. The model with both images had the best performance.

Guerra-Hernández et al. [105] combined data from ALS, ICESat-2, Sentinel-1 and 2, ALOS PALSAR and Shuttle RaDAR Topography Mission (STRM) to model and map the above ground biomass with random forest, for Mediterranean forests in Central-West Spain. The five forest types considered were agroforestry systems of *Quercus* spp., and stands of *Quercus suber*, *Quercus ilex*, *Pinus pinaster* and *Pinus pinea*. The above ground biomass models with explanatory variables of ICESat-2 and ICESat-2 and ALS data had good performances, though the latter was better than that of the former. The ICESat-2 model accuracy enables its use when neither field inventory nor ALS data is available, at large scale level.

A set of mathematical models have been tested with the goal of developing accurate biomass models. Kattenborn et al. [106] developed biomass models for pure and mixed stands for German temperate forests with random forest and generalised additive models, generalised boosted regression models and boosted version of the generalised additive models. Data from WorldView-2 and LiDAR and digital surface models derived from SAR were used with two field sampling designs, a cluster and a non-cluster. The best models had one or two explanatory variables (from the 26 initial ones). The random forest models were the best. Additionally, the cluster design enabled the models to detect the relationship between the remote sensing variables and field surveys. Moreover, clustering seemed to have a smooth effect in the field samples, which enabled to compensate the effect of the outliers. One drawback was the overestimation and underestimation for areas with low and high biomass, respectively. This was explained by the low availability of field samples.

Other mathematical models have been tested, such as k-nearest neighbours, maximum entropy, geographically weighted regression (GWR), artificial neural network, and support vector regression (SVR). Andersen et al. [107] used Landsat, LiDAR, and SAR with bootstrapping approach and k-nearest neighbours for forest stands in Alaska, United States of America. The biomass estimates with k-nearest neighbours method outperformed substantially that of the bootstrapping approach. Tian et al. [108] modelled biomass with SPOT-5, ALOS PALSAR, airborne LiDAR and k-nearest neighbours and multiple linear regression in a cold arid region in North-west China. Linear regression performance was very poor, except when explanatory variables included LiDAR ones. The suited model was k-nearest neighbours as it was possible to map biomass with accuracy with SPOT data which was less expensive than LiDAR. The more homogeneous the stands were the better results were attained. Carreiras et al. [109] used ALOS PALSAR data, WorldView-2 to derive crown cover and regression tree algorithms, for Miombo savannah in Mozambique. The model performance was good and it enabled the prediction of biomass per pixel. However, two drawbacks were referred to, one was related to validation, which could have underestimated the error, and the other was related to the sampling design, which did not consider samples with more than 50% crown cover.

Saatchi et al. [110] used Geoscience Laser Altimeter System (GLAS), LiDAR and MODIS with maximum entropy algorithm, for tropical forests, all over the world. A map was derived for all tropical forests with a standard methodology to enable the analysis within this forest type with accuracy, both for above and below ground biomass as well as carbon. The model uncertainties were smaller for African and Asian forests (25% and 26%, respectively) than for South American forests (49%). The uncertainties were larger at the management unit scale (5%) than at the country level (1%). Chen et al. [111] modelled above ground biomass for forests in a temperate continental monsoon climate in China, with Sentinel-1 and Sentinel-2 with geographically weighted regression, artificial neural network, support vector regression and random forest. Support vector regression model with sequential minimal optimisation outperformed the other models in what concerns spatial prediction and mapping. The variables better suited to explain the variability of biomass were the textures derived from Sentinel-1 and the vegetation biophysical variables from Sentinel-2. In addition, for forest stands with high biomass there seemed to be a saturation of Sentinel data. Chen et al. [112] for forests in a humid continental climate in China, used Sentinel-1 and Sentinel-2 and linear regression, geographically weighted regression, artificial neural network, support vector regression and random forest. The best model for prediction and mapping of above ground biomass was that derived from random forest. The authors concluded that more important for accuracy were the explanatory variables rather than the fitting algorithms, and that linear regression was more dependent on the explanatory variables than the machine learning methods. Castillo et al. [113] modelled above ground biomass for mangrove forests in the Philippines with Sentinel-1 and Sentinel-2 with linear regression and 17 non parametric models (among others, random forest, regression trees and k-nearest neighbours). Some of the non parametric models outperformed the linear regression while others did not. The explanatory variable that resulted in an overall better accuracy was LAI derived from Sentinel-2, while the spectral bands of the passive sensor had lower accuracy. The models based on SAR had higher accuracy for areas with low crown cover.

Biomass was modelled for Cerrado (grassland and savannah) in Brazil by Zimbres et al. [114] with forest inventory, LiDAR, Landsat and ALOS data using classification and regression trees (CART) and random forest. The random forest model had better accuracy for an above ground biomass range of 50–100 $t{\cdot}ha^{-1}$ whereas CART had better accuracy for lower and higher above ground biomass ranges. Moreover, data from passive sensors were able to detect differences in density that enabled to separate dense and sparse vegetation while the active sensor data allowed for the separation of the different tree habits. Su et al. [115] developed models of above ground biomass for natural forests and plantations, under tropical climate, in China. The authors used forest inventory, Landsat, ALOS PALSAR and digital elevation data with random forest and co-kriging. It was stressed the importance of texture measures in above ground biomass modelling in stands with complex composition and structure, and of vegetation indices to reduce the background influence. The above ground biomass modelling with ALOS PALSAR data had some constraints due to L-band saturation at *circa* 150 t·ha^{-1}. Random forest overestimated and under estimated above ground biomass for small and large values, respectively. The combination of random forest and co-kriging outperformed the random forest models.

4 Final Considerations

There have been developed a large set of biomass models at tree and area levels. The diversity of the models reflects the variability in species, stand structures, site and data available for modelling and modelling approaches (*cf*. chapter "Stand Structure and Biomass"). Yet, the comparison between different models is not straightforward as most studies use R^2 and RMSE. Only a few studies consider relative measures like relative RMSE (*cf*. chapter "Modelling Biomass"). For example, two models can have similar RMSE, but one has relative RMSE double the other. Another issue when analysing the accuracy of the models is that the statistics are not defined in the text, metrics with the same name have different formulations and there are no standard methods to evaluate accuracy and precision.

The set of examples of biomass models with destructive sampling, forest inventory, remote sensing and ancillary data presented reflected the need to search for the best possible accuracy. In spite of the diversity some observed trends can be pinpointed: (i) at tree level, model accuracy is dependent on tree biomass partitioning, which in turn is dependent on stand structure, thus species, regime, composition, structure, stage of development, tree social status and site specific models tend to be more accurate (e.g., [16, 24, 35, 40]); (ii) at area level, the model accuracy depends on the stand structure and their variability, the wider the gradient of the stand structure the lower the accuracy (e.g., [49, 69]); (iii) the selection of the explanatory variables has to consider the statistical algorithm, especially in what concerns data normality criterion and/or collinearity among variables (e.g., [6, 83, 116]); (iv) good accuracy can be attained with different mathematical methods, stressing that method might not be of primordial importance to achieve good accuracy (e.g., [45, 117]); (v) remote sensing passive sensors data has the advantage of their accessibility and temporal range when compared with active sensors (e.g., [118, 119]); (vi) accuracy of the models derived from remote sensing passive sensors increase from low to high spatial resolution, and when time series and/or more than one sensor are used (e.g., [51, 62]); (vii) models accuracy depends on the target area, low, medium and high spatial resolution are suited for large scale and medium and small scales, respectively (e.g., [97, 119]); (viii) models derived from remote sensing active sensors data tend to have higher accuracy than those with remote sensing passive sensors data, especially for complex stand structures (e.g., [100, 101]); (ix) models with explanatory variables from remote sensing passive and active sensors tend to have higher accuracy than those with only explanatory variables from one sensor, due to the complementarity among the variables of the two sensors (e.g., [97, 102]); (x) models accuracy frequently improves with the combination of forest inventory, remote sensing and/or ancillary data (e.g., [77, 98]).

Funding This work is funded by National Funds through FCT—Foundation for Science and Technology under the Project UIDB/05183/2020.

References

1. Kern CC, Kenefic LS, Kuehne C et al (2021) Relative influence of stand and site factors on aboveground live-tree carbon sequestration and mortality in managed and unmanaged forests. For Ecol Manage 493:119266. https://doi.org/10.1016/j.foreco.2021.119266
2. Ontl TA, Janowiak MK, Swanston CW et al (2020) Forest management for carbon sequestration and climate adaptation. J Forest 118:86–101. https://doi.org/10.1093/jofore/fvz062
3. Forrester DI (2021) Does individual-tree biomass growth increase continuously with tree size? For Ecol Manage 481:118717. https://doi.org/10.1016/j.foreco.2020.118717
4. Mankou GS, Ligot G, Loubota Panzou GJ et al (2021) Tropical tree allometry and crown allocation, and their relationship with species traits in central Africa. For Ecol Manage 493:119262. https://doi.org/10.1016/j.foreco.2021.119262
5. Xiang W, Li L, Ouyang S et al (2021) Effects of stand age on tree biomass partitioning and allometric equations in Chinese fir (Cunninghamia lanceolata) plantations. Eur J Forest Res 140:317–332. https://doi.org/10.1007/s10342-020-01333-0
6. Burkhart HE, Tomé M (2012) Modeling forest trees and stands. Springer, Netherlands, Dordrecht
7. Brown S, Gillespie ARJ, Lugo AE (1989) Biomass estimation methods for tropical forests with aplications to forest inventory data. Forest Sci 35:881–902. https://doi.org/10.1093/forestscience/35.4.881
8. Eamus D, McGuinness K, Burrows W (2000) Review of allometric relationships for estimating woody biomass for Queensland, the Northern Territory and Western Australia. Australian Greenhouse Office, Canberra
9. Jenkins JC, Chojnacky DC, Heath LS, Birdsey RA (2003) National-scale biomass estimators for United States tree species. Forest Sci 49:12–35. https://doi.org/10.1093/forestscience/49.1.12
10. Keith H, Barrett D, Keenan R (2000) Review of allometric relationships for estimating woody biomass for New South Wales, the Australian Capital Territory, Victoria, Tasmania and South Australia. Australian Greenhouse Office, Canberra
11. Návar J (2009) Biomass component equations for Latin American species and groups of species. Ann For Sci 66:208–208. https://doi.org/10.1051/forest/2009001
12. Ter-Mikaelian MT, Korzukhin MD (1997) Biomass equations for sixty-five North American tree species. For Ecol Manage 97:1–24. https://doi.org/10.1016/S0378-1127(97)00019-4
13. Zianis D, Suomen Metsätieteellinen Seura, Metsäntutkimuslaitos (2005) Biomass and stem volume equations for tree species in Europe. Finnish Society of Forest Science, Finnish Forest Research Institute, Helsinki, Finland
14. Correia A, Faias S, Tomé M (2008) Ajustamento Simultâneo de Equações de Biomassa de Pinheiro Manso no Sul de Portugal. Silva Lusitana 16:197–205

15. Paul KI, Roxburgh SH, England JR et al (2013) Development and testing of allometric equations for estimating above-ground biomass of mixed-species environmental plantings. For Ecol Manage 310:483–494. https://doi.org/10.1016/j.foreco.2013.08.054
16. Jagodziński AM, Dyderski MK, Gęsikiewicz K et al (2018) How do tree stand parameters affect young Scots pine biomass?—Allometric equations and biomass conversion and expansion factors. For Ecol Manage 409:74–83. https://doi.org/10.1016/j.foreco.2017.11.001
17. Reed D, Tomé M (1998) Total aboveground biomass and net dry matter accumulation by plant component in young *Eucalyptus globulus* in response to irrigation. For Ecol Manage 103:21–32. https://doi.org/10.1016/S0378-1127(97)00174-6
18. Zabek LM, Prescott CE (2006) Biomass equations and carbon content of aboveground leafless biomass of hybrid poplar in Coastal British Columbia. For Ecol Manage 223:291–302. https://doi.org/10.1016/j.foreco.2005.11.009
19. Vande Walle I, Van Camp N, Van de Casteele L et al (2007) Short-rotation forestry of birch, maple, poplar and willow in Flanders (Belgium) I—Biomass production after 4 years of tree growth. Biomass Bioenerg 31:267–275. https://doi.org/10.1016/j.biombioe.2007.01.019
20. Menéndez-Miguélez M, Canga E, Barrio-Anta M et al (2013) A three level system for estimating the biomass of *Castanea sativa* Mill. coppice stands in north-west Spain. For Ecol Manage 291:417–426. https://doi.org/10.1016/j.foreco.2012.11.040
21. Paul KI, Roxburgh SH, Ritson P et al (2013) Testing allometric equations for prediction of above-ground biomass of mallee eucalypts in southern Australia. For Ecol Manage 310:1005–1015. https://doi.org/10.1016/j.foreco.2013.09.040
22. Manolis EN, Zagas TD, Poravou CA, Zagas DT (2016) Biomass assessment for sustainable bioenergy utilization in a Mediterranean forest ecosystem in northwest Greece. Ecol Eng 91:537–544. https://doi.org/10.1016/j.ecoleng.2016.02.041
23. Mosseler A, Major JE, Labrecque M, Larocque GR (2014) Allometric relationships in coppice biomass production for two North American willows (*Salix* spp.) across three different sites. For Ecol Manage 320:190–196. https://doi.org/10.1016/j.foreco.2014.02.027
24. Oliveira TS, Tomé M (2017) Improving biomass estimation for *Eucalyptus globulus* Labill at stand level in Portugal. Biomass Bioenerg 96:103–111. https://doi.org/10.1016/j.biombioe.2016.11.010
25. Brown S (1997) Estimating biomass and biomass change of tropical forests: a primer. FAO, Rome
26. Chave J, Andalo C, Brown S et al (2005) Tree allometry and improved estimation of carbon stocks and balance in tropical forests. Oecologia 145:87–99. https://doi.org/10.1007/s00442-005-0100-x
27. Cole TG, Ewel JJ (2006) Allometric equations for four valuable tropical tree species. For Ecol Manage 229:351–360. https://doi.org/10.1016/j.foreco.2006.04.017
28. Feldpausch TR, Lloyd J, Lewis SL et al (2012) Tree height integrated into pantropical forest biomass estimates. Biogeosciences 9:3381–3403. https://doi.org/10.5194/bg-9-3381-2012
29. Hairiah K, Sitompul S (2001) Methods for sampling carbon stocks above and below ground. International centre for research in agroforestry. Bogor, Indonesia
30. Mattsson E, Ostwald M, Wallin G, Nissanka SP (2016) Heterogeneity and assessment uncertainties in forest characteristics and biomass carbon stocks: important considerations for climate mitigation policies. Land Use Policy 59:84–94. https://doi.org/10.1016/j.landusepol.2016.08.026
31. Terakunpisut J, Gajaseni N, Ruankawe N (2007) Carbon sequestration potential in aboveground biomass of Thong Pha Phum national forest, Thailand. Appl Ecol Environ Res 5:93–102
32. Vieilledent G, Vaudry R, Andriamanohisoa SF et al (2012) A universal approach to estimate biomass and carbon stock in tropical forests using generic allometric models. Ecol Appl 22:572–583. https://doi.org/10.1890/11-0039.1
33. Tabacchi G, Di Cosmo L, Gasparini P (2011) Aboveground tree volume and phytomass prediction equations for forest species in Italy. Eur J Forest Res 130:911–934. https://doi.org/10.1007/s10342-011-0481-9

34. Forrester DI, Dumbrell IC, Elms SR et al (2021) Can crown variables increase the generality of individual tree biomass equations? Trees 35:15–26. https://doi.org/10.1007/s00468-020-02006-6
35. Elfving B, Ulvcrona KA, Egnell G (2017) Biomass equations for lodgepole pine in northern Sweden. Can J For Res 47:89–96. https://doi.org/10.1139/cjfr-2016-0131
36. Levine J, de Valpine P, Battles J (2021) Generalized additive models reveal among-stand variation in live tree biomass equations. Can J For Res 51:546–564. https://doi.org/10.1139/cjfr-2020-0219
37. Zhang J, Fiddler GO, Young DH et al (2021) Allometry of tree biomass and carbon partitioning in ponderosa pine plantations grown under diverse conditions. For Ecol Manage 497:119526. https://doi.org/10.1016/j.foreco.2021.119526
38. Jorge C, Tomé M, Ruiz-Peinado R et al (2023) Quercus suber allometry in the west mediterranean basin. Forests 14:649. https://doi.org/10.3390/f14030649
39. Annighöfer P, Ameztegui A, Ammer C et al (2016) Species-specific and generic biomass equations for seedlings and saplings of European tree species. Eur J Forest Res 135:313–329. https://doi.org/10.1007/s10342-016-0937-z
40. Sillett SC, Van Pelt R, Carroll AL et al (2020) Aboveground biomass dynamics and growth efficiency of Sequoia sempervirens forests. For Ecol Manage 458:117740. https://doi.org/10.1016/j.foreco.2019.117740
41. Li H, Zhao P (2013) Improving the accuracy of tree-level aboveground biomass equations with height classification at a large regional scale. For Ecol Manage 289:153–163. https://doi.org/10.1016/j.foreco.2012.10.002
42. Dillen M, Vanhellemont M, Verdonckt P et al (2016) Productivity, stand dynamics and the selection effect in a mixed willow clone short rotation coppice plantation. Biomass Bioenerg 87:46–54. https://doi.org/10.1016/j.biombioe.2016.02.013
43. Ozdemir E, Makineci E, Yilmaz E et al (2019) Biomass estimation of individual trees for coppice-originated oak forests. Eur J Forest Res 138:623–637. https://doi.org/10.1007/s10342-019-01194-2
44. Patrício MS, Monteiro ML, Tomé M (2005) Biomass equations for Castanea sativa high forest in the northwest of Portugal. Acta Horticulturae 727–732. https://doi.org/10.17660/ActaHortic.2005.693.98
45. Nord-Larsen T, Meilby H, Skovsgaard JP (2017) Simultaneous estimation of biomass models for 13 tree species: effects of compatible additivity requirements. Can J For Res 47:765–776. https://doi.org/10.1139/cjfr-2016-0430
46. Dutcă I, Mather R, Blujdea VNB et al (2018) Site-effects on biomass allometric models for early growth plantations of Norway spruce (Picea abies (L.) Karst.). Biomass Bioenerg 116:8–17. https://doi.org/10.1016/j.biombioe.2018.05.013
47. Al-Bakri JT, Taylor JC (2003) Application of NOAA AVHRR for monitoring vegetation conditions and biomass in Jordan. J Arid Environ 54:579–593. https://doi.org/10.1006/jare.2002.1081
48. Li X, Du H, Mao F et al (2018) Estimating bamboo forest aboveground biomass using EnKF-assimilated MODIS LAI spatiotemporal data and machine learning algorithms. Agric For Meteorol 256–257:445–457. https://doi.org/10.1016/j.agrformet.2018.04.002
49. Zhang Q, He HS, Liang Y et al (2018) Integrating forest inventory data and MODIS data to map species-level biomass in Chinese boreal forests. Can J For Res 48:461–479. https://doi.org/10.1139/cjfr-2017-0346
50. Blackard J, Finco M, Helmer E et al (2008) Mapping U.S. forest biomass using nationwide forest inventory data and moderate resolution information. Remote Sens Environ 112:1658–1677. https://doi.org/10.1016/j.rse.2007.08.021
51. Propastin P (2013) Large-scale mapping of aboveground biomass of tropical rainforest in Sulawesi, Indonesia, using Landsat ETM+ and MODIS data. GISci Remote Sens 50:633–651
52. Steininger MK (2000) Satellite estimation of tropical secondary forest above-ground biomass: data from Brazil and Bolivia. Int J Remote Sens 21:1139–1157. https://doi.org/10.1080/014311600210119

53. Zheng D, Rademacher J, Chen J et al (2004) Estimating aboveground biomass using Landsat 7 ETM+ data across a managed landscape in northern Wisconsin, USA. Remote Sens Environ 93:402–411. https://doi.org/10.1016/j.rse.2004.08.008
54. Gasparri NI, Parmuchi MG, Bono J et al (2010) Assessing multi-temporal Landsat 7 ETM+ images for estimating above-ground biomass in subtropical dry forests of Argentina. J Arid Environ 74:1262–1270. https://doi.org/10.1016/j.jaridenv.2010.04.007
55. Foody GM, Cutler ME, McMorrow J et al (2001) Mapping the biomass of Bornean tropical rain forest from remotely sensed data. Glob Ecol Biogeogr 10:379–387. https://doi.org/10.1046/j.1466-822X.2001.00248.x
56. Foody GM, Boyd DS, Cutler MEJ (2003) Predictive relations of tropical forest biomass from Landsat TM data and their transferability between regions. Remote Sens Environ 85:463–474. https://doi.org/10.1016/S0034-4257(03)00039-7
57. Powell SL, Cohen WB, Healey SP et al (2010) Quantification of live aboveground forest biomass dynamics with Landsat time-series and field inventory data: a comparison of empirical modeling approaches. Remote Sens Environ 114:1053–1068. https://doi.org/10.1016/j.rse.2009.12.018
58. Zhu X, Liu D (2015) Improving forest aboveground biomass estimation using seasonal Landsat NDVI time-series. ISPRS J Photogramm Remote Sens 102:222–231. https://doi.org/10.1016/j.isprsjprs.2014.08.014
59. Fernández-Manso O, Fernández-Manso A, Quintano C (2014) Estimation of aboveground biomass in Mediterranean forests by statistical modelling of ASTER fraction images. Int J Appl Earth Obs Geoinf 31:45–56. https://doi.org/10.1016/j.jag.2014.03.005
60. Heiskanen J (2006) Estimating aboveground tree biomass and leaf area index in a mountain birch forest using ASTER satellite data. Int J Remote Sens 27:1135–1158. https://doi.org/10.1080/01431160500353858
61. Poulain M, Peña M, Schmidt A et al (2010) Relationships between forest variables and remote sensing data in a *Nothofagus pumilio* forest. Geocarto Int 25:25–43. https://doi.org/10.1080/10106040902803558
62. Pham LTH, Brabyn L (2017) Monitoring mangrove biomass change in Vietnam using SPOT images and an object-based approach combined with machine learning algorithms. ISPRS J Photogramm Remote Sens 128:86–97. https://doi.org/10.1016/j.isprsjprs.2017.03.013
63. Askar NN, Phairuang W et al (2018) Estimating aboveground biomass on private forest using sentinel-2 imagery. J Sensors 2018:1–11. https://doi.org/10.1155/2018/6745629
64. Pandit S, Tsuyuki S, Dube T (2018) Estimating above-ground biomass in sub-tropical buffer zone community forests, Nepal, using sentinel 2 data. Remote Sensing 10:601. https://doi.org/10.3390/rs10040601
65. Leboeuf A, Beaudoin A, Fournier R et al (2007) A shadow fraction method for mapping biomass of northern boreal black spruce forests using QuickBird imagery. Remote Sens Environ 110:488–500. https://doi.org/10.1016/j.rse.2006.05.025
66. Sousa AMO, Gonçalves AC, Mesquita P, Marques da Silva JR (2015) Biomass estimation with high resolution satellite images: a case study of *Quercus rotundifolia*. ISPRS J Photogramm Remote Sens 101:69–79. https://doi.org/10.1016/j.isprsjprs.2014.12.004
67. Macedo FL, Sousa AMO, Gonçalves AC et al (2018) Above-ground biomass estimation for *Quercus rotundifolia* using vegetation indices derived from high spatial resolution satellite images. European J Remote Sensing 51:932–944. https://doi.org/10.1080/22797254.2018.1521250
68. Sousa AMO, Gonçalves AC, da Silva JRM (2017) Above-ground biomass estimation with high spatial resolution satellite images. In: Tumuluru JS (ed) Biomass volume estimation and valorization for energy. InTech
69. Gonçalves AC, Sousa AMO, Mesquita P (2019) Functions for aboveground biomass estimation derived from satellite images data in Mediterranean agroforestry systems. Agrofor Syst 93:1485–1500. https://doi.org/10.1007/s10457-018-0252-4
70. Gonçalves AC, Sousa AMO, Mesquita PG (2017) Estimation and dynamics of above ground biomass with very high resolution satellite images in *Pinus pinaster* stands. Biomass Bioenerg 106:146–154. https://doi.org/10.1016/j.biombioe.2017.08.026

71. Lourenço P, Godinho S, Sousa A, Gonçalves AC (2021) Estimating tree aboveground biomass using multispectral satellite-based data in Mediterranean agroforestry system using random forest algorithm. Remote Sens Appl: Soc Environ 23:100560. https://doi.org/10.1016/j.rsase.2021.100560
72. Wang M, Cao W, Guan Q et al (2018) Potential of texture metrics derived from high-resolution PLEIADES satellite data for quantifying aboveground carbon of Kandelia candel mangrove forests in Southeast China. Wetlands Ecol Manage 26:789–803. https://doi.org/10.1007/s11273-018-9610-2
73. Ploton P, Barbier N, Couteron P et al (2017) Toward a general tropical forest biomass prediction model from very high resolution optical satellite images. Remote Sens Environ 200:140–153. https://doi.org/10.1016/j.rse.2017.08.001
74. Jachowski NRA, Quak MSY, Friess DA et al (2013) Mangrove biomass estimation in Southwest Thailand using machine learning. Appl Geogr 45:311–321. https://doi.org/10.1016/j.apgeog.2013.09.024
75. Kaasalainen S, Holopainen M, Karjalainen M et al (2015) Combining lidar and synthetic aperture radar data to estimate forest biomass: status and prospects. Forests 6:252–270. https://doi.org/10.3390/f6010252
76. Santos J (2003) Airborne P-band SAR applied to the aboveground biomass studies in the Brazilian tropical rainforest. Remote Sens Environ 87:482–493. https://doi.org/10.1016/j.rse.2002.12.001
77. Solberg S, Astrup R, Gobakken T et al (2010) Estimating spruce and pine biomass with interferometric X-band SAR. Remote Sens Environ 114:2353–2360. https://doi.org/10.1016/j.rse.2010.05.011
78. Lau A, Calders K, Bartholomeus H et al (2019) Tree biomass equations from terrestrial LiDAR: a case study in Guyana. Forests 10:527. https://doi.org/10.3390/f10060527
79. Salum RB, Souza-Filho PWM, Simard M et al (2020) Improving mangrove above-ground biomass estimates using LiDAR. Estuar Coast Shelf Sci 236:106585. https://doi.org/10.1016/j.ecss.2020.106585
80. Esteban J, McRoberts R, Fernández-Landa A et al (2019) Estimating forest volume and biomass and their changes using random forests and remotely sensed data. Remote Sens 11:1944. https://doi.org/10.3390/rs11161944
81. Knapp N, Huth A, Kugler F et al (2018) Model-assisted estimation of tropical forest biomass change: a comparison of approaches. Remote Sens 10:731. https://doi.org/10.3390/rs10050731
82. Swatantran A, Dubayah R, Roberts D et al (2011) Mapping biomass and stress in the Sierra Nevada using lidar and hyperspectral data fusion. Remote Sens Environ 115:2917–2930
83. Knapp N, Fischer R, Cazcarra-Bes V, Huth A (2020) Structure metrics to generalize biomass estimation from lidar across forest types from different continents. Remote Sens Environ 237:111597. https://doi.org/10.1016/j.rse.2019.111597
84. Nelson RF, Hyde P, Johnson P et al (2007) Investigating RaDAR–LiDAR synergy in a North Carolina pine forest. Remote Sens Environ 110:98–108. https://doi.org/10.1016/j.rse.2007.02.006
85. Næsset E, Gobakken T, Solberg S et al (2011) Model-assisted regional forest biomass estimation using LiDAR and InSAR as auxiliary data: a case study from a boreal forest area. Remote Sens Environ 115:3599–3614. https://doi.org/10.1016/j.rse.2011.08.021
86. Tsui OW, Coops NC, Wulder MA et al (2012) Using multi-frequency radar and discrete-return LiDAR measurements to estimate above-ground biomass and biomass components in a coastal temperate forest. ISPRS J Photogramm Remote Sens 69:121–133. https://doi.org/10.1016/j.isprsjprs.2012.02.009
87. Montesano PM, Nelson RF, Dubayah RO et al (2014) The uncertainty of biomass estimates from LiDAR and SAR across a boreal forest structure gradient. Remote Sens Environ 154:398–407. https://doi.org/10.1016/j.rse.2014.01.027
88. Tanase MA, Panciera R, Lowell K et al (2014) Forest biomass estimation at high spatial resolution: radar versus lidar sensors. IEEE Geosci Remote Sens Lett 11:711–715. https://doi.org/10.1109/LGRS.2013.2276947

89. Omar H, Misman M, Kassim A (2017) Synergetic of PALSAR-2 and sentinel-1A SAR polarimetry for retrieving aboveground biomass in dipterocarp forest of Malaysia. Appl Sci 7:675. https://doi.org/10.3390/app7070675
90. Schlund M, Kotowska MM, Brambach F et al (2021) Spaceborne height models reveal above ground biomass changes in tropical landscapes. For Ecol Manage 497:119497. https://doi.org/10.1016/j.foreco.2021.119497
91. Shendryk I, Hellström M, Klemedtsson L, Kljun N (2014) Low-density LiDAR and optical imagery for biomass estimation over boreal forest in Sweden. Forests 5:992–1010. https://doi.org/10.3390/f5050992
92. Sun G, Ranson KJ, Guo Z et al (2011) Forest biomass mapping from lidar and radar synergies. Remote Sens Environ 115:2906–2916. https://doi.org/10.1016/j.rse.2011.03.021
93. Lucas R, Van De Kerchove R, Otero V et al (2020) Structural characterisation of mangrove forests achieved through combining multiple sources of remote sensing data. Remote Sens Environ 237:111543. https://doi.org/10.1016/j.rse.2019.111543
94. Brovkina O, Novotny J, Cienciala E et al (2017) Mapping forest aboveground biomass using airborne hyperspectral and LiDAR data in the mountainous conditions of Central Europe. Ecol Eng 100:219–230. https://doi.org/10.1016/j.ecoleng.2016.12.004
95. Persson H (2016) Estimation of boreal forest attributes from very high resolution pléiades data. Remote Sens 8:736. https://doi.org/10.3390/rs8090736
96. Basuki TM, Skidmore AK, Hussin YA, Van Duren I (2013) Estimating tropical forest biomass more accurately by integrating ALOS PALSAR and landsat-7 ETM+ data. Int J Remote Sens 34:4871–4888. https://doi.org/10.1080/01431161.2013.777486
97. Phua M-H, Johari SA, Wong OC et al (2017) Synergistic use of landsat 8 OLI image and airborne LiDAR data for above-ground biomass estimation in tropical lowland rainforests. For Ecol Manage 406:163–171. https://doi.org/10.1016/j.foreco.2017.10.007
98. Berninger A, Lohberger S, Stängel M, Siegert F (2018) SAR-based estimation of above-ground biomass and its changes in tropical forests of Kalimantan using L- and C-band. Remote Sens 10:831. https://doi.org/10.3390/rs10060831
99. Cortés L, Hernández J, Valencia D, Corvalán P (2014) Estimation of above-ground forest biomass using landsat ETM+ Aster GDEM and Lidar. Forest Res 3:1000117. https://doi.org/10.4172/2168-9776.1000117
100. Huang H, Liu C, Wang X et al (2019) Integration of multi-resource remotely sensed data and allometric models for forest aboveground biomass estimation in China. Remote Sens Environ 221:225–234. https://doi.org/10.1016/j.rse.2018.11.017
101. Chi H, Sun G, Huang J et al (2017) Estimation of forest aboveground biomass in Changbai mountain region using ICESat/GLAS and Landsat/TM data. Remote Sens 9:707. https://doi.org/10.3390/rs9070707
102. Ghosh SM, Behera MD (2018) Aboveground biomass estimation using multi-sensor data synergy and machine learning algorithms in a dense tropical forest. Appl Geogr 96:29–40. https://doi.org/10.1016/j.apgeog.2018.05.011
103. Matasci G, Hermosilla T, Wulder MA et al (2018) Large-area mapping of Canadian boreal forest cover, height, biomass and other structural attributes using Landsat composites and lidar plots. Remote Sens Environ 209:90–106. https://doi.org/10.1016/j.rse.2017.12.020
104. Kashongwe HB, Roy DP, Bwangoy JRB (2020) Democratic republic of the congo tropical forest canopy height and aboveground biomass estimation with landsat-8 operational land imager (OLI) and airborne LiDAR data: the effect of seasonal landsat image selection. Remote Sens 12:1360. https://doi.org/10.3390/rs12091360
105. Guerra-Hernández J, Narine LL, Pascual A et al (2022) Aboveground biomass mapping by integrating ICESat-2, SENTINEL-1, SENTINEL-2, ALOS2/PALSAR2, and topographic information in Mediterranean forests. GIScі Remote Sens 59:1509–1533. https://doi.org/10.1080/15481603.2022.2115599
106. Kattenborn T, Maack J, Faßnacht F et al (2015) Mapping forest biomass from space—fusion of hyperspectral EO1-hyperion data and Tandem-X and WorldView-2 canopy height models. Int J Appl Earth Obs Geoinf 35:359–367. https://doi.org/10.1016/j.jag.2014.10.008

107. Andersen H-E, Strunk J, Temesgen H et al (2012) Using multilevel remote sensing and ground data to estimate forest biomass resources in remote regions: a case study in the boreal forests of interior Alaska. Can J Remote Sens 37:596–611. https://doi.org/10.5589/m12-003
108. Tian S, Tanase MA, Panciera R et al (2013) Forest biomass estimation using radar and lidar synergies. 2013 IEEE International geoscience and remote sensing symposium—IGARSS. IEEE, Melbourne, Australia, pp 2145–2148
109. Carreiras J, Melo J, Vasconcelos M (2013) Estimating the above-ground biomass in miombo savanna woodlands (Mozambique, East Africa) using L-band synthetic aperture radar data. Remote Sens 5:1524–1548. https://doi.org/10.3390/rs5041524
110. Saatchi SS, Harris NL, Brown S et al (2011) Benchmark map of forest carbon stocks in tropical regions across three continents. Proc Natl Acad Sci 108:9899–9904
111. Chen L, Ren C, Zhang B et al (2018) Estimation of forest above-ground biomass by geographically weighted regression and machine learning with sentinel imagery. Forests 9:582. https://doi.org/10.3390/f9100582
112. Chen L, Wang Y, Ren C et al (2019) Optimal combination of predictors and algorithms for forest above-ground biomass mapping from sentinel and SRTM data. Remote Sens 11:414. https://doi.org/10.3390/rs11040414
113. Castillo JAA, Apan AA, Maraseni TN, Salmo SG (2017) Estimation and mapping of above-ground biomass of mangrove forests and their replacement land uses in the Philippines using Sentinel imagery. ISPRS J Photogramm Remote Sens 134:70–85. https://doi.org/10.1016/j.isprsjprs.2017.10.016
114. Zimbres B, Rodríguez-Veiga P, Shimbo JZ et al (2021) Mapping the stock and spatial distribution of aboveground woody biomass in the native vegetation of the Brazilian Cerrado biome. For Ecol Manage 499:119615. https://doi.org/10.1016/j.foreco.2021.119615
115. Su H, Shen W, Wang J et al (2020) Machine learning and geostatistical approaches for estimating aboveground biomass in Chinese subtropical forests. Forest Ecosystems 7:64. https://doi.org/10.1186/s40663-020-00276-7
116. Vanclay JK (1994) Modelling forest growth and yield: applications to mixed tropical forests. CAB International, Wallingford, U.K.
117. Chen C (2019) Above-ground carbon stock in merchantable trees not reduced between cycles of spruce budworm outbreaks due to changing species composition in spruce-fir forests of Maine, USA. For Ecol Manage 453:117590. https://doi.org/10.1016/j.foreco.2019.117590
118. Kangas A, Astrup R, Breidenbach J et al (2018) Remote sensing and forest inventories in Nordic countries—roadmap for the future. Scand J For Res 33:397–412. https://doi.org/10.1080/02827581.2017.1416666
119. Lu D, Chen Q, Wang G et al (2016) A survey of remote sensing-based aboveground biomass estimation methods in forest ecosystems. Int J Digital Earth 9:63–105. https://doi.org/10.1080/17538947.2014.990526

Forest Biomass as an Energy Resource

Isabel Malico

Abstract Biomass is a highly versatile and reliable source of firm, renewable energy, capable of generating heat, power and various biofuels. The technologies used to convert biomass into fuels or energy can be broadly divided into two categories: biochemical and thermochemical. Biochemical pathways for forest biomass conversion into fuels still face techno-economic challenges, requiring further research to make them economically attractive. In contrast, thermochemical conversion processes, including gasification, pyrolysis and combustion, are well suited for forest biomass conversion, with several technologies having reached a fully commercial stage. Combustion, the most common and mature thermochemical pathway, converts forest biomass into heat, power, or combined heat and power. While traditional, inefficient and polluting methods are still used for burning forest biomass, modern, cleaner, and more efficient combustion technologies are available and in use. Some pathways based on gasification and pyrolysis are also commercially viable, providing solid, liquid and gaseous biofuels. These options offer versatility across combustion systems, heat engines, fuel cells and synthesis applications. This chapter provides a comprehensive overview of forest biomass as an energy source, covering processing technologies, technology readiness levels, fuel characteristics and pre-treatment methods. It emphasizes the potential and challenges associated with using forest biomass for sustainable energy production.

Keywords Combustion · Gasification · Pyrolysis · Anaerobic digestion · Fermentation · Pre-treatments

---

I. Malico (✉)
IDMEC, Escola de Ciências e Tecnologia, Departamento de Engenharia Mecatrónica, Universidade de Évora, Rua Romão Ramalho 59, 7000-671 Évora, Portugal
e-mail: imbm@uevora.pt

A. C. Gonçalves and I. Malico (eds.), *Forest Bioenergy*, Green Energy and Technology,
https://doi.org/10.1007/978-3-031-48224-3_7

1 Introduction

Bioenergy was already used by early hominins who foraged fire for resources through the landscape around two million years ago and, later on, both early modern humans and Neanderthals resorted to sophisticated fire technologies [1]. By the end of the first industrial revolution, biomass was still the main source of energy used to support human activities–In 1850, it provided more than 80% of the global primary energy supply [2]. However, the industrial revolution, driven by access to coal, changed this paradigm [3]. Coal was available in large amounts and provided a denser fuel than fuelwood, allowing a positive feedback between its use and extraction, the steam engine, iron and steel production and the transportation system [3]. By the turn of the twentieth century, coal was already the most widely used energy source [2].

During the last century, energy end-use technologies and energy sources have diversified. First, oil emerged as an energy source, steadily increasing its share in the global energy supply and replacing coal as the main energy source around the 1970s; then, natural gas gained relevance [2]. Figure 1 shows the world total energy supply by source in the last 50 years and the today's relatively even distribution among crude and oil products, coal, natural gas and the rest of the energy sources [4, 5].

At the beginning of the twenty-first century, fossil fuels still dominated the energy landscape and were the most commonly used energy sources; however, global environmental concerns have led to an increased use of renewable energies and the re-emergence of biomass as an energy source in advanced economies. Despite this resurgence, the share of biomass is far from the values it had less than two centuries ago. The contribution of bioenergy to the world total primary energy supply is currently around 10% [6]. If the traditional bioenergy is not considered a renewable energy source, modern biomass represents only about 6% of the world energy supply, which

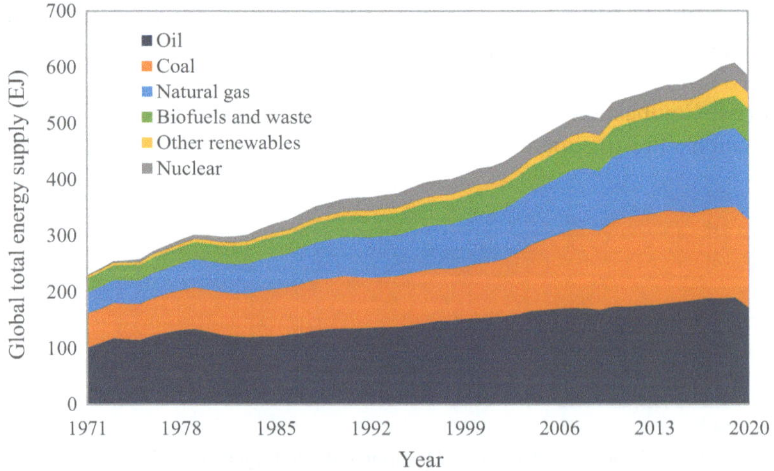

Fig. 1 World total energy supply by energy source from 1971 to 2020. (*Data source* [4, 5])

corresponds to approximately half of the global renewable energy supply (*cf.* chapter "Introduction to Forest Bioenergy"). Of all the possible sources (e.g., wood, sewage sludge, municipal solid wastes, industrial and agricultural residues), forest biomass is, by far, the most used nowadays and, therefore, has a very important contribution to the worldwide renewable energy share [7].

Some of the advantages of using biomass instead of fossil fuels may be: (*i*) its renewability; (*ii*) the fact that it is considered carbon neutral; (*iii*) its local nature, which, among others, contributes to energy security; (*iv*) the reduction of pollutant emissions and waste; (*v*) possible energy cost savings and sustainable local economic development (e.g., [8–12]). However, at this point, it is important to say that the overall sustainability of bioenergy depends on several factors, including the feedstock and conversion technology used and the location and scale of production [13, 14].

Biomass is a highly versatile and flexible source of renewable energy that can produce heat, power and fuels in a more reliable manner compared to other renewables, such as wind or solar energy, since it is not weather-dependent [15–17]. Moreover, biomass can be stored and utilised in similar ways to fossil fuels [18], which further enhances its flexibility and convenience as an energy source and partly explains its dominance as a renewable energy source.

Because of its characteristics, bioenergy plays an important role in the heat, electricity and transport sectors and is particularly relevant for hard-to-decarbonise sectors. The largest and most direct use of forest biomass is for the production of heat (for which bioenergy is very suitable), but the share of forest biomass for the production of electricity (electrical power) is also relevant [19]. On the other hand, the penetration of forest biomass as a source for the production of biofuels used in the transport sector is still not a reality, although it is a promising pathway. In 2020, 90% of the total energy consumed for transport came from biofuels [7] that were not produced from forest biomass. They are mainly first-generation biofuels, produced from food crops or vegetable oils; the share of second-generation biofuels is still very small and mainly produced from wastes, such as municipal solid wastes or animal and oil wastes [20].

This chapter presents an overview of forest biomass as an energy source. Section 2 covers the most relevant processing technologies for the conversion of forest biomass into energy and fuels, their applications and their readiness levels. The most relevant characteristics of forest biomass as a fuel are then described, followed by a description of the most relevant pre-treatment methods (upstream conversion technologies). The chapter ends with some final considerations.

2 Forest Biomass Conversion Technologies

As already mentioned, one of the advantages of forest biomass over other renewable energy sources is its versatility, since it can satisfy various needs by being converted into different products (electricity/power, heat and biofuels that may be stored for later use). The available processes involved in converting forest biomass to energy

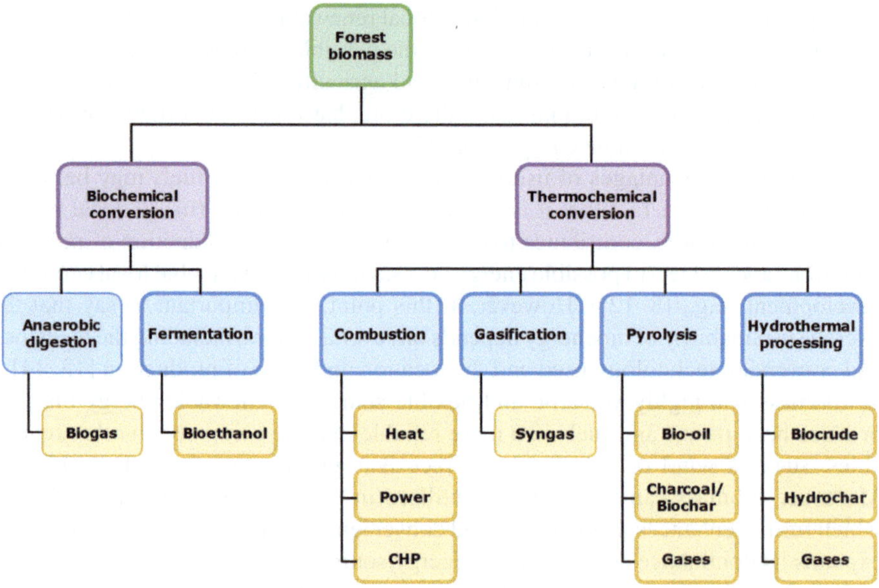

Fig. 2 Conversion paths for the production of fuels, heat and power from forest biomass

or fuels can be broadly categorised as biochemical and thermochemical (Fig. 2); however, the technological readiness level of each process is very variable.

Technology maturity is an important factor to take into consideration when choosing the most suitable technology to convert forest biomass into fuels or a useful form of energy in a specific application. Yet it is not the only one, and others, such as the properties and quantity of the resource, end-use requirements, environmental standards, local economic conditions, and support schemes, can also play an important role in determining the optimal technology choice. In this context, taking the best advantage of the forest biomass available requires in-depth knowledge of all these factors and the available conversion technologies.

Biochemical conversion processes involve the use of microorganisms to break down biomass into alternative fuel forms that are appropriate for use in different applications, including transportation. The most commonly used biochemical biomass conversion processes are anaerobic digestion and fermentation [21], which result in the production of, respectively, gaseous and liquid biofuels. Anaerobic digestion technologies are, per se, mature pathways used to convert organic matter to biogas, a gaseous fuel mixture mainly composed of methane and carbon dioxide. The process occurs in an oxygen-depleted environment and is most commonly used for the combined production of heat and power [22]. Historically, the most used feedstocks were animal manure and sewage sludge from wastewater treatment plants [23], but the last decades have seen a rise in the market for biogas from solid substrates [24], and, more recently, lignocellulosic biomass, including forest residues and wood from energy crops, has drawn much attention as a possible feedstock for anaerobic digestion [25]. However, the digestion of lignocellulosic biomass still faces several challenges related to the complexity of its structure [25], and more research is essential to make this conversion pathway attractive [26, 27].

Fermentation technologies are also mature when used to convert organic matter rich in directly fermentable sugars or starch into ethanol, a liquid biofuel [28]; however, the techno-economic challenges for the conversion of lignocellulosic biomass by fermentation are complex, and further research is required [28]. Major issues involved in the fermentation of lignocellulosic biomass are: (*i*) the need for extensive pre-treatment to yield fermentable sugars [29]; (*ii*) the low concentrations of ethanol, which require large amounts of energy for downstream separations and ethanol purification [30] (*iii*) the presence of both pentose and hexose sugars in the fermentation broth [31]; and (*iv*) the presence of toxic compounds that may act as inhibitors [31]. The technology has reached a demonstration stage at an industrial scale, but the few existing projects struggle to thrive [32]. It is foreseen that policies aimed at decarbonising the transport sector and moving away from refined oil products will push oil refineries to potentially adopt entirely new business models, among them those based on the biochemical conversion of biomass to biofuels, typically via fermentation systems [33]. Additionally, minimum blending rates for advanced biofuels imposed in some countries will directly benefit the expansion of cellulose ethanol [34].

Combustion, by far the most common biomass conversion route [35, 36], is one of the available thermochemical pathways for the conversion of biomass into energy. It is a mature technology widely used to convert biomass into heat, power or heat and power simultaneously (combined heat and power, CHP) [37]. The conversion of forest biomass into energy by combustion involves the reactions between biomass and an oxidant, typically air, to produce heat (and, subsequently, power, if desired). The process can be carried out by means of a wide range of, often very different, technologies that cover a broad spectrum of capacities. Chaps. Biomass for Domestic Heat and Biomass for Industrial and District Heating describe some of these technologies.

Essentially, two distinct ways of burning forest biomass exist: one so-called traditional and the other modern [38]. In 2021, the traditional use of biomass (e.g., firewood, charcoal), generally burned in basic and inefficient processes and a source of environmental concern, was still significant worldwide [6] (*cf.* chapter "Biomass for Domestic Heat"). Nevertheless, technology and market developments have led to commercially available, highly improved biomass combustion systems [39], which are mainly employed in countries with advanced economies and technological capabilities that are actively pursuing the development of modern combustion technologies (e.g., in Northern Europe and North America) [40].

Together with combustion, gasification and pyrolysis constitute the other commercially available forest biomass thermochemical conversion technologies [41]. Gasification is an endothermic process used to convert biomass into a low molecular weight combustible gaseous mixture that varies in composition and name depending on several factors, such as the gasifying agent and gasifying conditions used (typical temperatures are 800–1500 °C [42]). When the gasifying agent is air, the typical gasification product is producer gas, which mainly contains carbon monoxide, hydrogen, nitrogen, carbon dioxide, methane and water vapour [43]. On the other hand, syngas, a gas mixture of predominantly hydrogen and carbon monoxide, is the product of the gasification process if oxygen and steam are used as gasifying agents and the gasification is followed by gas separation [44]. Carbon dioxide can also be a gasifying agent, but the process currently has some limitations and has not been established at an industrial scale [45]. Usually, the term syngas is widely used and understood in the context of gasification when referring to the gaseous products from any kind of gasification process, irrespective of the proportions of the various product components. This is the term used throughout this book, even though it is not strictly correct.

The heat required for gasification can be obtained by the partial oxidation of the feedstock (in this case, the process is called autothermal or direct) or indirectly provided by the gasifying agent or heat exchangers (the gasification process is named allothermal or indirect gasification). The latter typically results in syngas with a larger heating value (12–20 $MJ{\cdot}Nm^{-3}$), whereas when partial oxidation exists, the syngas has a significantly inferior heating value (4–7 $MJ{\cdot}Nm^{-3}$ when air is the gasifying agent and 10–12 $MJ{\cdot}Nm^{-3}$ when oxygen is used) [46].

Syngas is a more versatile fuel than the original solid forest biomass [41]. The combustible mixture produced through forest biomass gasification can be used as a fuel in a combustion system [43, 47–49], fed into a fuel cell [50, 51] or into a Stirling engine [52–54] to produce energy, or, if its nitrogen content is low [49], for the synthesis of substitute natural gas [55], high-quality liquid fuels [12, 31, 56, 57], chemicals [31, 58] or purified to produce hydrogen [59]. Of all these possible routes, forest biomass gasification used in conjunction with combustion is the only one that is commercially available [44]. Heat, power and CHP systems are available on the market [60–63], but nowadays small-scale CHP gasification units dominate [64]. Even though forest biomass gasification has the potential to produce a variety of value-added fuels and chemicals, commercial deployment of the existing technologies has been hindered by a range of intractable issues [65] and the technologies have only reached the demonstration phase [63, 66–68].

If forest biomass is heated and thermally decomposed in an inert atmosphere, the thermochemical process is termed pyrolysis [69]. However, partial oxidation may still occur, for example, because biomass naturally contains oxygen or because old kilns lack tightness [70]. As biomass is heated, its temperature will increase. In a first stage, up to 150–200 °C, the water contained in the fuel will vaporise and around 200 °C pyrolytic decomposition starts, ending at temperatures of 500–700 °C [71]. Most of the reactions involved in this process are endothermic, so heat must be externally supplied [71].

In pyrolysis, three types of products may be obtained: bio-oil, non-condensable gases and charcoal/biochar (the distinction between charcoal and biochar mainly lies in terms of intended uses [70]). Depending on the process conditions, the formation of a particular phase can be favoured. If the heating rate is high and the residence time is short, the liquid phase (bio-oil) is the main product obtained, and the process is named fast pyrolysis [72, 73]. On the other hand, if the process occurs at a moderate temperature with a low heating rate and a longer residence time, it is called slow pyrolysis and the solid phase (charcoal or biochar) is sought after [72–74]. The process can occur at intermediate or more extreme conditions (intermediate and flash pyrolysis), but these are not so common [73]. The range of reaction temperatures in pyrolysis is wide but typically lower than those in gasification (300–1000 °C, [73]).

The slow thermal degradation of forest biomass is the conventional pyrolytic process. Indeed, carbonisation, which is a slow pyrolysis with maximum temperatures between 350–700 °C [71], has been used for the production of charcoal since prehistory [75] and is a commercial technology still used nowadays in many countries, especially in Africa, Asia and the Americas [76]. Charcoal is mainly used as a fuel for cooking, heating and steel production [70]. The traditional process consists of slowly heating a pile of wood covered with plant material and earth to make charcoal [77]. In the traditional process, the yields of the various gaseous, liquid, and solid products are not controlled; the yield of charcoal is low; internal heating based on the partial combustion of the wood is usually employed; and the gases produced during pyrolysis are not used [76, 77]. The quality and yield of charcoal/biochar can be enhanced through the use of improved technology and operating conditions [78, 79]. While the majority of earlier studies on slow pyrolysis used an inert gas as the pyrolysis medium, often nitrogen, using carbon dioxide to substitute the inert gas in pyrolysis has attracted a lot of attention in recent years because of its apparent benefits under certain operating conditions, among which are an increase in biochar yield and an enhancement of important biochar characteristics for environmental and agricultural applications [80].

Other types of reactors that operate at moderate temperatures (~500 °C) and short residence times (<2 s) have been developed to maximise bio-oil yields [72, 78, 81]. Bio-oil can be a substitute for heavy fuel oil or coal, light fuel oil, gas oil, and vacuum gas oil when directly combusted in boilers, furnaces, gas turbines or compression-ignition engines [81, 82]. This is a relatively simple process, but it has low added value [81], and the low quality of bio-oil as a fuel makes its commercialisation difficult [31]. Bio-oil is composed of highly reactive oxygen-containing species, which result in its immiscibility with hydrocarbon fuels, a low energy density and instability during storage [82]. Upgrading bio-oil to replace gasoline and diesel seems more attractive since these are high-quality fuels; however, further process development is required [31, 81]. The production of higher-value chemicals is also an interesting option to improve the competitiveness of fast pyrolysis, but further research is also needed [81]. Another strategy for the utilisation of bio-oil could be its co-processing in conventional refineries [31]. This would reduce the costs of introducing second-generation biofuels into the market since existing refineries would be used.

The common strategies based on fast pyrolysis concentrate on the valorisation of the bio-oil, while the other by-products are usually used to generate process heat, despite the availability of more valuable uses [12]. For instance, biochar can be used for soil amendment [83–85], long-term carbon sequestration [73, 86] or the production of activated carbon [87, 88] among other valuable products. On the other hand, the non-condensable gases are hydrogen-rich and can be purified to hydrogen [12]. The integration of all these processes in a single plant may be a way of improving the profitability of pyrolysis plants [12]. Only a few forest biomass fast pyrolysis plants have so far reached a commercial stage [89]. The technology has been proven and commercialised for natural gas and heating oil substitution, but great care must be given to the logistics of feedstock acquisition and product applications [81] and economic feasibility is the key factor in the development of commercial pyrolysis [90]. On the other hand, technologies for upgrading fast pyrolysis bio-oil to drop-in fuels and coproducts have not reached the commercial stage [81, 91].

A last note on gasification and pyrolysis of forest biomass: While they are thermochemical processes, they can be used in conjunction with biochemical processes, such as fermentation or enzymatic hydrolysis, to convert syngas and bio-oil, respectively, into biofuels, biochemicals or other value-added products. This presents some advantages since, for example, fermentation is more flexible, requires lower temperatures and pressures and requires less gas cleaning than thermochemical processes [12].

Another promising thermochemical conversion technology is hydrothermal processing, which involves processing biomass in water slurries at moderate or high temperatures (160–750 °C) and high pressures (2–28 MPa) to convert organic materials into useful fuels or chemicals [92–95]. Depending on the process temperatures and pressures, high-value plant chemicals (e.g., resins, phenolics), carbohydrate, hydrochar (a solid carbonised product), liquid hydrocarbons (biocrude) or gaseous products are obtained [94, 96]. Hydrothermal processing is suitable for treating wet biomass materials without drying [92]. Three main process groups are hydrothermal carbonisation, hydrothermal liquefaction and hydrothermal gasification [93]. At this point, however, more research is needed before large-scale hydrothermal processing plants are developed [94, 95, 97].

Table 1 summarises relevant available pathways for the production of bioenergy or biofuels, some of their typical advantages and disadvantages compared to the other bioenergy routes and their commercial application, when appropriate. Only modern combustion systems were considered when producing this table, but it should be mentioned that the advantages and disadvantages mentioned in the table should be critically interpreted. Within each pathway, a variety of available technologies and energy system configurations exist. Therefore, what is generally an advantage or disadvantage for a certain conversion pathway may not be so for a specific system configuration within this pathway. Other important factors to take into consideration are different feedstock characteristics (even when focusing the analysis on forest biomass alone), the specific context of the application (e.g., location, scale) or that, sometimes, a problem can be turned into an advantage (e.g., even though the ash produced in biomass power plants raises environmental concerns and handling issues, opportunities also arise and ash might be utilised not only in conventional applications such as agriculture, construction and the cement industry but also in novel applications such as nanotechnology in industrial catalysis and environmental applications [98]).

Table 1 Summary of relevant available conversion processes for bioenergy or biofuel production from forest biomass

Process	Brief description	Advantages	Disadvantages	Commercially available?
Combustion for heat, power or combined heat and power production	Forest biomass is oxidised, usually with air, to produce heat that can be used directly and/or converted to power	Simple technology Mature technology Relative low capital costs Fuel flexibility Good scalability Reduced risk Allows the use of existent fossil-fuel based technologies	Relatively low electrical efficiency Flue gas treatment systems are required Ash production, which requires proper handling and disposal	Yes. The process is well-established to produce heat, electricity or combined heat and power
Gasification associated with combustion for heat, power or combined heat and power production	Forest biomass is converted to a combustible gas that can be combusted to produce heat that can be used directly and/or converted to power	Higher electrical efficiencies compared to direct combustion systems Fuel flexibility Syngas is more versatile than forest biomass and, therefore, can potentially be used in more efficient and cleaner energy systems Allows the use of existent fossil-fuel based technologies	High capital cost Technical complexity Gas cleaning required when used for power or combined heat and power production Stirling engines may be sensible to impurities in the syngas	Yes. The process is already commercially used to produce heat, electricity, or combined heat and power (the last two associated with internal combustion engines)
Gasification associated with fuel cells for power or combined heat and power production	Forest biomass is converted to a combustible gas that, after clean up, can be fed into a fuel cell to produce power or combined heat and power	Increased electrical efficiency when compared to gasification-combustion systems	High capital cost Technical complexity Requires intensive gas clean up upstream of a fuel cell Degradation of fuel cells	No

(continued)

Table 1 (continued)

Process	Brief description	Advantages	Disadvantages	Commercially available?
Gasification for fuel production	Forest biomass is converted into a combustible gas, syngas, that can be further processed to other advanced biofuels	Syngas can be used directly or upgraded to multiple fuels that can be used in a broad variety of applications, including transportation Syngas and upgraded fuels can be stored and transported Growing demand for advanced renewable transportation fuels	High capital cost Technical complexity Stringent syngas quality requirements Syngas transportation and storage involves high capital costs Syngas needs further upgrading in many added value applications	No
Slow pyrolysis for fuel production	Forest biomass is slowly heated in an inert atmosphere to mainly produce charcoal/bio-char	Charcoal has high energy density Charcoal can be more efficiently stored and transported than forest biomass Bio-char can be used for a variety of applications	In the traditional process small yields of charcoal are produced	Yes. Slow pyrolysis is commercially used for charcoal production
Fast pyrolysis for fuel production	Forest biomass is rapidly heated in an inert atmosphere to mainly produce bio-oil that can be further processed to other advanced biofuels	Bio-oil can be used directly or upgraded to multiple fuels that can be used in a broad variety of applications, including transportation Liquid biofuels can be more efficiently stored and transported than forest biomass Technology well developed Growing demand for advanced renewable transportation fuels	High capital costs Technical complexity Bio-oil needs further upgrading in many added value applications	Yes, but only a few forest biomass fast pyrolysis plants have so far reached a commercial stage

(continued)

Table 1 (continued)

Process	Brief description	Advantages	Disadvantages	Commercially available?
Anaerobic digestion for heat, power or combined heat and power production	Forest biomass is digested by microorganisms in an oxygen depleted environment to produce biogas that can be directly used to produce heat, power or combined heat and power	Biogas is a versatile fuel that can be used in a broad variety of conversion processes Produces a nutrient-rich digestate that can be used as fertiliser	Lignocellulosic biomass is not easily digested Relatively slow process Sensitive to inhibitors (e.g., heavy metals present in wood waste and those resulting from pre-treatments)	No
Anaerobic digestion for fuel production	Forest biomass is digested by microorganisms in an oxygen depleted environment to produce biogas that can be further processed to other biofuels	Biogas can be upgraded to biomethane, a valuable and versatile renewable gas, substitute of natural gas Simpler process design than other thermochemical or biochemical process to convert biomass to biofuels Produces a nutrient-rich digestate that can be used as fertiliser	Lignocellulosic biomass is not easily digested Relatively slow process Sensitive to inhibitors (e.g., to heavy metals present in wood waste and those resulting from pre-treatments) High capital costs	No
Fermentation for fuel production	Forest biomass is converted into biofuels by microorganisms such as yeast and bacteria	Produces versatile fuels that can be used in a broad variety of applications, including transportation Growing demand for advanced renewable transportation fuels	Lignocellulosic biomass is not easily fermented Extensive pre-treatment is required Energy and water intensive Technical complexity High capital and operating costs Relatively slow process	No

3 Properties of Forest Biomass Fuels

The choice of the most suitable biomass energy conversion pathways is highly dependent on the properties of the resource and, therefore, knowledge of these properties is fundamental to taking the best advantage of forest biomass as an energy source. Commonly, forest biomass is used for energy purposes with little or no processing (i.e., raw biomass, essentially in the form it is produced). However, there is a growing interest in upgraded solid biofuels derived from forest woody biomass, including pellets and torrefied biomass. These are densified solid fuels that have higher energy content and bulk density than raw biomass and meet specific fuel quality standards, allowing for efficient conversion into energy. Additionally, because of their higher density, they lead to reduced transportation costs and are easier to handle and store. Densified solid biomass products are seeing increasing demand and offer a sizable untapped market in North America, Europe and some parts of Asia [99].

Dry woody biomass originated from forests, industrial or consumer activities is mainly composed of cellulose, hemicellulose and lignin [100]. The combination of those three polymers is collectively called lignocellulose. Significant variability in the relative share of these three main constituents exists within different species and between anatomical fractions of a specific species, but as a general rule, the variability within woody biomass is lower than between different biomass types [101]. For example, Williams et al. [101] report that for 241 samples of woody biomass (23 species), the average cellulose, hemicellulose and lignin contents are 51.2% ± 8.7%, 21.0% ± 8.7% and 26.1% ± 5.3%, respectively, while for 251 samples of corn stover and 488 samples of *Sorghum*, herbaceous feedstocks, they are 34.3% ± 2.5%, 20.7% ± 2.0% and 15.2% ± 1.6%, and 28.6% ± 2.6%, 15.4% ± 1.6% and 12.2% ± 1.9%, respectively.

Cellulose is the main wood constituent, followed by hemicellulose and lignin. Both hardwoods and softwoods contain a relatively high percentage of lignin [100], a polymer that is extremely resistant to chemical and enzymatic degradation [102]. Additionally, the complex crystalline structure of cellulose also contributes to the resistance of wood to biological degradation [27]. The recalcitrant nature of lignocellulosic materials favours thermochemical conversion processes, which is why the most mature biomass conversion technologies rely on combustion (*cf.* Sect. 2).

The following sub-sections describe the composition of forest biomass fuels and their properties that affect their conversion into energy or fuels.

Ultimate Analysis

Knowledge of the elemental composition of forest biomass, determined through ultimate analysis, is very important for its effective conversion into energy and fuels. Carbon, C, is the most abundant organic element in dry raw wood (about 50 wt%), followed by oxygen, O, (about 42 wt%) and hydrogen, H, (about 6 wt%) [100, 101,

103] (wt% means weight percent). The carbon content of softwood species tends to be slightly higher than that of hardwood species because of the different lignin and extractive content [104]. Compared to other conventional fuels, namely coal, raw biomass presents high O/C and H/C ratios [103]. The higher raw forest biomass O/C ratio leads to a lower heating value (*cf.* section "Energy Content"). However, the chemical exergy increases with the O/C ratio [105]. Additionally, a higher O/C ratio may potentially result in higher greenhouse gas emissions [106], with the consequent environmental impacts. When raw woody biomass is subjected to a thermal treatment, such as pyrolysis, its O/C and H/C ratios decrease and, consequently, its higher heating value increases.

Typically, forest biomass has very low concentrations of sulphur (usually in quantities less than 1 wt% on a dry basis) [103], generally below 0.05 wt% [107]; however, black liquor, a residue of the pulp industry (*cf.* chapter "Sources and Distribution of Forest Biomass for Energy"), presents a relatively high sulphur content (3.0–5.7 wt% on a dry basis [108]). This element plays a crucial role in the formation of pollutants, namely sulphur oxide (SO_x) emissions. However, although some of the sulphur present in the biomass contributes to the formation of SO_x, ash constituents in the flue gas, especially calcium, potassium and sodium, which are frequently found in quite high concentrations in biomass, largely retain SO_x [109]. Chemical reactions of sulphur with ash lead to fouling and slagging [103] and, at low temperatures, the presence of sulfuric acid in the flue gases may lead to corrosion, even though the concentrations of sulfuric acid in low-sulphur biomass are usually very small and it is the presence of hygroscopic salts in deposits that are responsible for low-temperature corrosion [109].

The nitrogen content of forest biomass is also important for environmental reasons. Like sulphur, nitrogen is important in the formation of pollutant emissions (in this case, nitrogen oxides, NO_x). The main mechanism for the formation of NO_x is the fuel nitrogen mechanism, since the majority of industrial biomass combustors operate at relatively low temperatures and thermal NO_x represents only a small contribution to the overall NO_x emissions [103]. However, the nitrogen content of woody biomass is relatively low. It is typically less than 1 wt% on a dry basis [103] and, for example, Williams et al. [101] report, for 192 samples of woody biomass, comprising several varieties of hardwoods, softwoods and other wood varieties, an average nitrogen content of 0.32%.

Another element that is important for environmental reasons is chlorine, Cl. Like sulphur and nitrogen, it is also present in forest biomass in small quantities (usually in quantities less than 1 wt% on a dry basis) [103]. Chloride content largely depends on management and soil rather than plant type or variety [107]. It contributes to fouling, slagging, corrosion and emissions of important air pollutants, including hydrogen chloride, HCl, which plays a role in the formation of dioxins and furans [35, 103].

Forest biomass also contains other inorganic elements in relatively high concentrations. For the thermochemical conversion processes, for example, the ash and salt-forming elements of biomass, such as silicon, potassium, sodium, calcium or magnesium, are crucial [110]. These elements contribute to slag formation and ash fouling and, therefore, have important consequences for the design and operation of conversion equipment. Calcium and magnesium typically increase the ash-melting temperature, while potassium and sodium decrease it [35]. These two alkali metals, in combination with sulphur and chlorine, are also important because of corrosion [110].

Table 2. presents the ultimate analyses for a few selected biomass fuels. The data was obtained from an online database maintained by TNO Biobased and Circular Technologies [111]. The table presents the characterisation of raw and processed forest biomass fuel and black liquor. Since biomass has a significant potential for co-firing with coal and/or replacing it (*cf.* chapter "Biomass for Industrial and District Heating"), the properties of a medium-rank coal are also presented.

Proximate Analysis

In the proximate analysis, the fuel is broken down into moisture, volatiles, fixed carbon and ash. Water and ash are incombustible substances, while volatiles and fixed carbon are combustible (even though they also include incombustible components). Volatile content is determined by standardised methods and refers to the fraction of the solid biomass, excluding moisture, that is converted to gas when biomass is heated. Ash is the name given to the solid residue that is formed during the combustion of biomass with air and is also determined by standardised methods. It is not exactly equal to the inorganic content of the original biomass because of the oxidation process involved in its determination, but small corrections can be made for a more exact calculation of the inorganic content of the biomass [44]. The fixed carbon content, the mass remaining after the release of volatiles excluding the ash and moisture contents, is determined by material balance.

Table 2. lists the results of proximate analyses (on a dry basis) for several selected solid biomass fuels, together with the corresponding higher heating values. The volatile matter and fixed carbon give an indication of how easily biomass can be ignited and then gasified or oxidised [112]. Raw wood has relatively high concentrations of volatile matter and a relatively low content of fixed carbon. These are considered very relevant advantages for thermochemical energy conversion [113]. On the other hand, when the content of volatile matter is high, the velocity of combustion is higher, the combustion is more difficult to control, and a larger and better-designed combustion chamber is needed for complete combustion and low pollutant emissions typical of incomplete combustion [113]. In small appliances where the combustion process is poor (e.g., fireplaces, stoves), the emission of pollutants from forest biomass is many times higher than that of coal power plants [114]. Advanced combustion technologies based on air staging enable more efficient and cleaner combustion (*cf.* chapter "Biomass for Domestic Heat").

Upgraded wood fuels (pellets, charcoal and biochar) have a higher fixed carbon content than the original biomass. Milder thermal treatments result in a smaller loss of volatiles, while more intense treatments lead to fuels that are similar to coal in terms of the ratio between volatiles and fixed carbon. For example, the raw pine wood and

Table 2 Properties of selected biomass fuels and bituminous coal. (*Data source* [111])

	Pine wood	*Eucalyptus* wood	Bark (spruce)	Wood pellets	Wood chips (willow)
Database ID	126	699	3566	2808	1091
Proximate analysis (wt% db)					
Fixed carbon	14.07	12.86	11.21	18.01	17.62
Volatile matter	85.70	86.61	85.03	81.49	80.77
Ash	0.23	0.53	3.76	0.50	1.61
Ultimate analysis (wt% db)					
Carbon	51.58	49.51	52.68	50.15	48.70
Hydrogen	5.78	5.75	6.73	6.07	5.84
Oxygen	42.32	43.98	32.25	43.18	43.40
Nitrogen	0.06	0.14	0.41	0.09	0.41
Sulphur	0.01	0.03	0.17	0.01	0.04
Higher heating value (MJ·kg $^{-1}$ *db)*	20.56	19.22	20.57	19.93	19.60
	Black liquor	Torrefied pellets (poplar)	Biochar (oak, 400 °C)	Charcoal	Bituminous coal
Database ID	1394	3529	3534	1954	1145
Proximate analysis (wt% db)					
Fixed carbon	13.00	26.10	65.20	89.60	61.30
Volatile matter	46.80	72.20	24.00	9.38	32.50
Ash	40.20	1.70	10.80	1.02	6.20
Ultimate analysis (wt% db)					
Carbon	29.20	53.81	71.40	92.04	76.73
Hydrogen	4.40	6.21	3.30	2.45	4.69
Oxygen	31.10	39.76	14, 10	2.96	10.52
Nitrogen	0.14	0.10	0.34	0.53	1.41
Sulphur	4.90	0.02	0.04	1.00	0.40
Higher heating value (MJ·kg $^{-1}$ *db)*	11.15	22.20	24.25	34.39	31.60

db dry basis

Eucalyptus samples presented in Table 2. have volatile matter to fixed carbon ratios above 8, while torrefied pellets have 2.8, biochar has 0.4 (slightly smaller than that of the bituminous coal sample) and charcoal has 0.1 (very close to the volatile matter to fixed carbon ratio of an anthracite coal (the highest rank of coal) sample—sample #1144 of the Phylis2 database and not presented in the table).

Compared to coal, raw biomass behaves very differently when it comes to the release of volatiles and, in particular, the oxidation of char [109]. Biomass chars are more reactive to oxidation by O_2, CO_2, and H_2O than coal chars because of their high porosity, large internal specific surface area and the presence of catalytically important ash-forming elements, often potassium [109].

The ash content of raw woody biomass is typically low when compared to coal (Table 2.), but different tree species and tree components have different ash contents and, for example, those of bark [104], stumps [107] or tropical woods [115] are significantly higher. The age of the trees at harvest is also an important factor because of the varying proportions of wood and bark with age and the number of small branches [35]. Residual woody biomass from industrial processes that contain contaminants such as adhesives and coating particles (*cf.* chapter "Sources and Distribution of Forest Biomass for Energy") typically presents higher ash contents [107].

Table 3. presents the ash composition for five forest biomass samples (two for pine to show the variability within the same wood type). The values are merely indicative because the minerals contained in the biomass vary widely between and within species, depending on the soil, growth rate and age of the plants [104].

The minerals contained in forest biomass can be (*i*) natural components of the feedstock (such as calcium, Ca, potassium, K, or silicon, Si) or (*ii*) artificially introduced during the biomass supply chain (mostly silica, SiO_2, during harvest [101]). The silica content is usually very small in wood from temperate species but more important in tropical woods and if it is present in wood in more than 0.5%, it can damage cutting tools [115].

Knowledge of the quantity and type of inorganic components present in the feedstock is important because of fouling, slag formation, corrosion, the formation of aerosols emitted with the flue gases and the possibilities for using the ashes produced during combustion [35]. Because of all the ash-related problems, fuels with low ash content are preferable: they (*i*) usually lead to lower dust emissions, (*ii*) require

Table 3. Ash composition of selected forest biomass fuels. (*Data source* [111])

	Eucalyptus wood	Pine wood 1	Pine wood 2	Bark (spruce)	Wood pellets
Database ID	699	122	1786	3158	2808
Ash composition (wt% ash)					
Al_2O_3	7.87	2.52	10.91	1.08	2.83
CaO	26.52	32.90	29.05	39.18	25.18
MgO	7.25	1.55	4.73	5.14	5.47
Na_2O	4.98	0.94	0.63	0.33	1.08
K_2O	7.20	6.75	13.06	7.59	10.48
P_2O_5	29.11	–	5.27	4.12	2.98
Fe_2O_3	–	2.00	4.48	0.14	2.29
SiO_2	17.83	–	46.06	1.50	16.04

simpler conversion equipment maintenance and design and simpler pollutant abatement equipment, and (*iii*) simplify ash utilisation and disposal, making ash transport and storage easier [35]. However, high levels of alkaline earth elements (Ca, K) in biomass are also beneficial for a number of processes, including: (*i*) soil amendment and fertilisation; (*ii*) production of building materials, adsorbents and ceramics; (*iii*) mineral synthesis or (*iv*) recovery of valuable components [113]. Phosphorous, P, and magnesium, Mg, are also relevant soil fertilisers [35]. Another advantage of the alkaline and alkaline earth metals present in biomass is that their catalytic reactivity allows lower gasification temperatures, which leads to higher gasification efficiencies [71].

Ash may also contain significant amounts of heavy metals, such as cadmium, Cd, lead, Pb, and zinc, Zn, which may be particularly important when ash is used. The amounts of heavy metals are particularly important for waste wood materials [35]. Also, the longer the rotation time, the larger the accumulation of heavy metals, so forest residues, which are typically associated with long rotation times, show accumulations of heavy metals [35]. Cadmium, Cd, is the most environmentally relevant element in raw biomass [35] and Pb and Zn may be responsible for corrosion [109].

Moisture Content

The moisture content of biomass quantifies the amount of water contained in a given biomass material. It is usually expressed in percentage and can be defined in terms of wet basis (wb), dry basis (db) or dry and ash-free basis (dab). The moisture content on a wet basis (M, Eq. 1) is given by the ratio between the mass of water (m_{H2O}) and the total mass of the moist material (m_{wb}) and is the measure used in the marketing of wood fuels [116].

$$\mathrm{M} = \frac{m_{\mathrm{H_2O}}}{m_{\mathrm{wb}}} = \frac{m_{\mathrm{H_2O}}}{m_{\mathrm{H_2O}} + m_{\mathrm{odb}}} 100 \tag{1}$$

where m_{odb} is the mass of the oven-dry biomass. On the other hand, the moisture content on a dry basis is given by the ratio between the mass of water and the mass of oven-dry material and is a term commonly used in the wood industry [116].

The moisture content of forest biomass is highly variable but typically higher than that of peat and coal, which show less variability and moisture content values up to around 20% [113]. Freshly harvested green forest biomass generally contains 50% to 60% moisture on a wet basis [116]. However, the amount of water varies between different tree components and with the seasons. Typically, leaves and branches present a higher moisture content than stumps and stems [117, 118]. Furthermore, the moisture content of freshly harvested forest biomass is reduced if stored and allowed to dry naturally in the sun and wind. Wood stored for a summer or for several years has, respectively, 25–35% and 15–25% moisture content on a wet basis [116].

The industrial residues of the wood-based industries also vary widely. For example, dry-wood processing residues may have moisture contents below 10% [119], strong black liquor before entering the recovery boilers in modern mills has a moisture content of about 20% wb and weak black liquor before concentration has a moisture content of 80–85% wb [120].

Upgraded solid biofuels produced from forest woody biomass present lower moisture contents than the original feedstocks. For a start, in the production of pellets and briquettes, the raw material should have a low moisture content of around 8–12% wb for pellets and 12–14% wb for briquettes [35]. Furthermore, pelletising reduces the moisture content of the initial feedstock by around 1.5–3%. [121]. This results in a fuel with a homogeneous and low moisture content. As a disadvantage, like raw wood, this type of fuel has a hydrophilic nature and, therefore, cannot be stored outside for long periods [122]. Forest biomass subjected to torrefaction and slow pyrolysis also presents low moisture contents [123–125]. First, when the temperature is increased, the free water present in biomass evaporates and at temperatures above 200 °C the physically bound water is released [125]. An advantage of torrefied biomass is that it is hydrophobic, which allows for long-term storage [122].

Moisture content is a critical parameter for the choice of the conversion technology. Typically, forest biomass is not the biomass resource with the highest moisture content and, therefore, it is not the first choice for biochemical conversion technologies such as fermentation and biodigestion, which favour high-moisture content biomass. Its most obvious use is in thermal conversion technologies, namely combustion, gasification and pyrolysis. Generally, combustion and gasification technologies are capable of dealing with forest biomass with a high moisture content (up to 60% wb). For example, fluidised bed and moving grate furnaces and updraft gasifiers allow for up to 60% moisture, while understoker furnaces allow for up to 50% [126, 127]. However, some technologies require dried feedstocks and may even need an additional gas fuel ignition burner for start-up. This is the case with pulverised fuel firing units that are limited to fuels with up to 20% moisture content [126]. On the other hand, downdraft gasifiers require biomass with moisture contents below 25% [126]. Also, low moisture contents (maximum 10%) should be specified for fast pyrolysis processes to improve bio-oil yield and quality [128], while slow pyrolysis allows feedstocks with a moisture up to 40–50% [129].

The moisture content of biomass has a very significant impact on its energy content and density, as will be shown in the next sections. For small residential heating systems, the moisture content of the wood should be limited to 25–30% wb, but for larger capacities (e.g., industrial heating, district heating), the use of fuels with a higher water content is common for economic reasons [116]. In this case, efficiency and low pollutant emissions are achieved by the technical characteristics of the energy systems [116]. However, higher moisture contents reduce the combustion temperatures, increase the residence time needed for complete combustion, increase the volume of flue gas produced per energy unit and thus require larger equipment dimensions [35].

Storage durability is also influenced by the moisture content of biomass. For example, fresh wood chips with a moisture content above 50% wb are not suitable for long-term storage and should not be stored in closed rooms [116]. Also, firewood with a high moisture content should not be stored in closed rooms because, in these conditions, the water contained in the biomass cannot evaporate and fungi and bacteria degrade the organic material, which results in dry-matter loss and constitutes a health hazard [116].

Another typical problem related to high-moisture-content forest biomass may be the necessity of using more energy to reduce the size of the raw feedstock [130]. For example, hammer mill grinding of wood and wood pellets with high water content requires substantially more energy than if a low-moisture feedstock is used [131].

Energy Content

The heating value, also referred to as calorific value, is a measure of the energy content of a fuel. It is defined as the energy released per unit mass of the fuel during complete combustion with air and can be expressed in several ways, depending on whether it is determined at constant pressure or at constant volume and on whether the water contained in the products of combustion is condensed or not. If the water formed during combustion is in the liquid phase, the heating value is called higher heating value. This is the maximum amount of energy that the fuel can potentially deliver. If the water contained in the flue gases is in the gaseous phase, the heating value is called lower heating value and, as the name indicates, it is lower than when the water in the flue gas is liquid because there is no heat release associated with the condensation of water.

The elemental composition of biomass is related to its heating value–the higher the carbon, hydrogen and sulphur contents, the higher the heat released during combustion per unit mass of fuel. Oxygen, nitrogen and inorganic elements have the opposite effect (Eq. 2). Usually, forest biomass has a low ash content compared to other solid fuels [113]. However, some particular feedstocks have a high content of inorganic materials. For example, a typical value for the ash content of black liquor is 45%, which results in low heating values (13–15 $MJ{\cdot}kg^{-1}$ of dry fuel) [120].

Several empirical formulas were developed to express the higher heating value as a function of the elemental composition of the fuel [132]. For example, Channiwala and Parikh [132] proposed the following correlation for the higher heating value in MJ per kg of oven-dry fuel (HHV_{dry}):

$$\begin{aligned} HHV_{dry} = {} & 0.3491 w_C + 1.1783 w_H + 0.1005 w_S - 0.1034 w_O - 0.0151 w_N \\ & - 0.0211 w_{Ash} \end{aligned} \tag{2}$$

where w_i is the content of carbon (C), hydrogen (H), sulphur (S), oxygen (O), nitrogen (N) and ash in wt% of oven-dry fuel. This formula was validated with measured data for biomass fuels, showing an average absolute error of 1.94% and a bias error of –0.17% [132].

The lower heating value of an oven-dry biomass (LHV_{dry}) can be calculated from the HHV_{dry} if one knows the hydrogen content of the fuel (Eq. 3).

$$LHV_{dry} = HHV_{dry} - 2.443 \cdot 8.936 \frac{w_H}{100} \quad (3)$$

Because raw forest biomass has a lower carbon concentration and higher volatile matter and oxygen contents than coal (*cf.* sections "Ultimate Analysis" and "Proximate Analysis"), it is characterised by smaller heating values. This fact, combined with a low density (*cf.* section "Density"), leads to low values of energy density.

Different types of oven-dry wood have relatively little variance in their higher heating values (e.g., Table 2. and Fig. 3). As a reference, the higher heating value of raw dry softwoods is typically 20–22 $MJ\cdot kg^{-1}$ and of hardwoods, 19–21 $MJ\cdot kg^{-1}$ [104]. However, in reality, raw wood is frequently not oven-dried and still contains some water, which has a significant impact on the value of the energy content per unit mass of fuel.

Figure 3 presents the higher and lower heating values defined per unit mass of wet biomass for two wood samples with various moisture contents (the ultimate and proximate analyses for the two samples are presented in Table 2.). For their determination, the moisture content of the biomass needs to be known and, additionally, for the calculation of the lower heating value, the energy associated with the vaporisation of the water present in the flue gas has to be considered (Eqs. 4 and 5).

$$HHV_{wb} = HHV_{dry}(1 - M/100) \quad (4)$$

$$LHV_{wb} = LHV_{dry}(1 - M/100) - 2.443M/100 \quad (5)$$

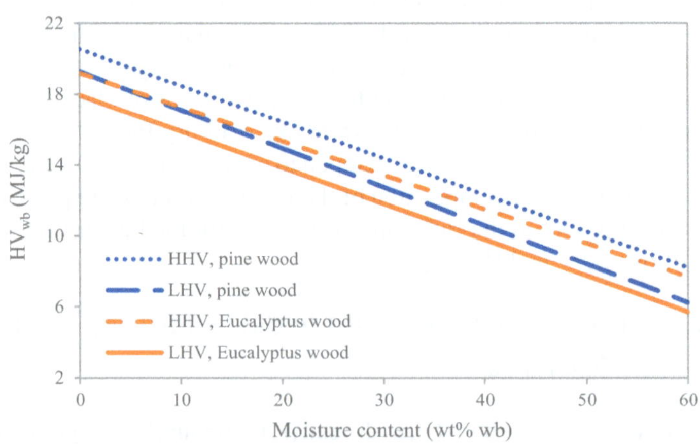

Fig. 3 Higher and lower heating values for wood (HV_{wb}) as a function of the moisture content

where HHV_{wb} and LHV_{wb} are, respectively, the higher and lower heating values of a biomass fuel with a moisture content M (wt% wb) defined per unit mass of wet biomass.

The moisture content of the biomass results in a substantial decrease in the HHV_{wb} and LHV_{wb} (Fig. 3). The reduction in HHV_{wb} with the moisture content is explained by the fact that the energy released during complete combustion with liquid water in the flue gases is divided by a larger mass of fuel, which contains water. In terms of energy release, the heat lost to evaporate the water present in the fuel is recovered when the water in the flue gases condensates.

It is important, when reporting heating values, to clearly indicate whether they are higher or lower heating values and if the biomass is oven-dry or contains water (and in this circumstance also the moisture content of the biomass). This is not always the case. Also, when reporting efficiencies of conversion systems, it is important to state the basis for their calculation. The efficiency of a conversion process (η) can generically be expressed as:

$$\eta = \frac{E_{\text{useful}}}{E_{\text{biomass}}} \tag{6}$$

In Eq. 6, E_{useful} is the useful energy delivered by the conversion system and E_{biomass} refers to the energy content of the biomass. Both the lower heating value or the higher heating value of the biomass are used; the efficiency calculated with the latter being smaller.

Density

When referring to the density (i.e., the ratio between mass and volume) of solid biomass, several definitions may be applied. Wood density was historically measured at ambient air moisture after air drying, but nowadays it is usually measured at a fixed moisture content (for example, the international standard is that the mass and volume of the samples are both measured at 12% moisture db) [133]. This metric is relevant for the traditional wood-based industries, but for bioenergy studies, the most useful measure of density is the basic density, which refers to the ratio between the mass of oven-dry biomass and the volume of green biomass. Since both the mass and the volume of wood increase with its water content up to around 30% moisture [104], density values measured at different moisture contents should be converted into basic densities (above 30% moisture only mass increases).

Another important density measure for energy conversion systems is the bulk density, which is defined as the mass of a portion of material constituted by several pieces/particles divided by the volume they occupy (consisting of solid volume, inter-particle void volume and internal pore volume). However, this is a challenging property to measure consistently, which is affected by several factors such as moisture content, material size or methodology used [134]. In contrast to bulk density, which is not an intrinsic property of the material, particle density, defined as the density of a single particle, is an intrinsic property of the solid.

The density of wood has a wide variation, depending on the tree species, growth conditions and part of the tree. Fast-growing, short-lived species have generally lower wood densities than slow-growing, long-lived species [133]. Basic wood densities can be as low as about 200 kg·m^{-3} and as high as about 1100 kg·m^{-3} [133], tropical woods showing wider variations [104].

The bulk density of forest biomass also has a large variation from around 200 kg·m^{-3} for loose materials (e.g., sawdust from sawmills) to more than 800 kg·m^{-3} for densified materials (e.g., pellets) (Table 4).

Raw forest biomass has a much lower bulk density than coal (typically one fifth that of coal [135]). This, combined with also lower heating value, results in an overall lower energy density than coal. As a consequence, larger material handling systems and more space for transport and storage are needed, which results in more complex supply chains and higher costs. To overcome this disadvantage, in some applications, the energy density of biomass is increased through compaction and/or thermal pre-treatments.

Table 4 Typical moisture and bulk density of selected forest biomass fuels

Feedstock	Moisture (wt% wb)	Bulk density (kg·m^{-3} wb)	Source
Stemwood from final fellings (hardwoods)	48.3	360 (320–420)	[136]
Stemwood from final fellings (softwoods)	53.9 (30–55)	330 (310–350)	[136]
Logging residues (hardwoods)[a]	48.3 (25–50)	250 (200–400)	[136]
Logging residues (softwoods)[a]	53.6 (25–55)	250 (200–350)	[136]
Stumps from final fellings[b]	30 (35–40)	250 (200–400)	[136]
Sawdust from sawmills[b]	50 (6–55)	150–300	[136]
Sawmill residues (excluding sawdust)[b]	50 (6–55)	236 (150–300)	[136]
Residues from industries producing semi-finished wood-based panels and further wood processing industries	10 (6–15)	236 (150–300)	[136]
Bark	50 (40–60)	236 (150–300)	[136]
Wood pellets	7.7	591 (520–640)	[137]
Wood pellets	5.0	765 (728–808)	[138]

[a] Logging residues are composed of tops and branches both from final fellings or thinnings and stemwood from thinnings

[b] From hardwoods and softwoods

Other Properties

The size, size distribution and shape of the biomass material are also important physical properties to consider in the conversion of forest biomass into energy and fuels, especially when it comes to the fuel-feeding system and conversion technology used. Forest biomass is available in many sizes and shapes: as bulk material, constituted by several small pieces (e.g., sawdust, wood chips, pellets) or as larger fuel materials (e.g., log wood). Table 5 lists the main woody biomass fuels traded and their typical sizes.

Feedstock size affects its collection, transport, storage, handling and feeding to the conversion system and the conversion system itself. Some technologies are designed to handle large pieces (e.g., wood log stoves for residential heat, *cf.* chapter "Biomass for Domestic Heat"), others very small pieces (e.g., pulverised fuel combustion for power production, *cf.* chapters "Biomass for Industrial and District Heating and Biomass for Power Production and Cogeneration"). Table 6 presents the particle size requirements of thermochemical conversion technologies. The residence time of the biomass material in the conversion equipment depends on the size of the fuel. Typically, the smaller the biomass feedstock, the shorter the residence time because the increase in the ratio between the reactive surface and the volume of the feedstock enhances heat and mass transfer [71].

Other than the size itself, biomass material size distribution is also an important characteristic of the fuel. Some feedstocks are quite homogeneous in size (e.g., pellets, briquettes), while others are not (e.g., hog fuel). Some types of equipment, such as fixed bed combustors, are well-suitable for non-uniform feedstock sizes [110], while other, like updraft gasifiers, require a tight control of the feedstock particle size [71]. Additionally, the variability in the size of the biomass material affects its flowability and can pose problems to the feeding system, which can reduce the performance and increase maintenance costs.

Table 5 Main traded woody biofuels (based on [139])

Biofuel	Description	Typical size
Log wood	Cut fuelwood produced directly from the forest	Length $\geq$ 500 mm
Firewood	Cut and split oven-ready fuelwood usually used in household wood burning appliances	15 cm < length < 100 cm
Wood briquettes	Densified woody biomass made with or without additives in a pre-determined geometric form	At least two dimensions > 25 mm
Wood pellets	Densified woody biomass made with or without additives usually with a cylindrical form	5 mm < length < 40 mm Diameter $\leq$ 25 mm
Wood chips	Chipped woody biomass with a subrectangular shape, typically in the form of pieces with a defined particle size produced by mechanical treatment with sharp tools	5 mm < length < 50 mm

Table 6 Particle size requirements of several thermochemical processes (based on [71])

Particle size	Process
Large	Fixed-bed combustion
Medium to large	Moving bed gasification
Medium	Fluidised bed combustion
Small to large	Slow pyrolysis
Small to medium	Fluidised bed gasification, hydrothermal carbonisation
Small	Suspension combustion, entrained flow gasification, intermediate pyrolysis, fast pyrolysis (except ablative pyrolysis), hydrothermal liquefaction, hydrothermal gasification

The size of forest biomass may need to be reduced to meet the energy conversion equipment requirements [131] or to upgrade it to densified biomass products [140]. Because of its fibrous nature, raw biomass presents poor grindability [141], resulting in significant energy requirements for size reduction [123]. Thermal pre-treatment may improve this property, as it will be presented below.

4 Biomass Pre-treatment

Before conversion processes, either to heat and/or power or to biofuels, forest biomass is almost always subjected to pre-treatment, i.e., operations that are performed prior to the main conversion in order to improve the properties of the feedstock. The extent of the pre-treatment and the technologies or set of technologies chosen depend on the subsequent conversion technology.

Pre-treatment technologies may be categorised in physical, chemical, physio-chemical or biological methods (Fig. 4).

Separation of forest biomass into different components is sometimes desirable and/or required. Delimbing and debarking of trees is a common practice in the forest industry [142], with primary forest industries often processing wood that has had its branches and bark removed. Branches and bark might be left on the stands to mitigate the nutrient exports [143] (*cf.* chapter "Stand Structure and Biomass"). Another example of separation is the one that occurs at recycling facilities to identify and separate wood treated with preservatives from wood waste. This separation allows the untreated fraction to be used as fuel or mulch, for example, without the added burden from wood preservative chemicals [144].

Some forest biomass by-products are already small in size (e.g., sawdust, cork powder). However, in most cases, forest biomass undergoes mechanical size reduction to align with the requirements of the conversion technology used (*cf.*

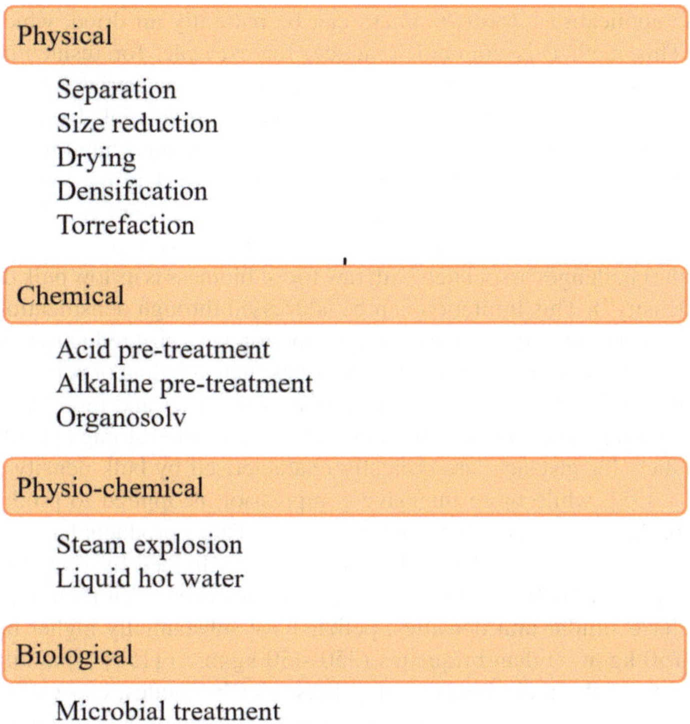

Fig. 4 Forest biomass pre-treatment technologies

section "Other Properties"). Additionally, smaller-sized biomass is easier to handle, transport and requires less storage space. Drawbacks of size reduction are that it can consumes considerable amounts of energy [145] and may have an impact on the durability and longevity of forest biomass during storage [146]. Therefore, it is essential to carefully assess the trade-offs between size reduction and conversion efficiency. In general, the smaller the forest biomass particles, the higher the accessible surface area and the reaction rates of either biochemical or thermochemical conversion processes [27, 102, 147]. Several methods exist to reduce the size of forest biomass, including grinding, chipping or crushing.

Another commonly used pre-treatment step is drying, which involves the removal of moisture from biomass as vapour. While drying may be a requirement for some thermochemical conversion processes, it is generally not required for biochemical processes (*cf.* section "Moisture Content"). Moisture reduction offers several advantages in forest biomass handling, transport and storage increases and enhances the efficiency of thermochemical conversion processes. For instance, in direct combustion systems, every 10% increase in biomass moisture content decreases efficiency by approximately 2% [148]. However, drying often requires capital investment and energy. Therefore, finding the optimal balance between the cost of drying and its benefits is crucial.

In some applications, forest biomass can be naturally air-dried, which is cost-effective. This method is common practice, for example, for residential heating using firewood. However, when biomass is processed industrially for energy or fuels, specialised equipment, such as kiln dryers or industrial dryers, is usually required to achieve faster moisture reduction, often following a natural drying stage. In these situations, feedstock drying is energy intensive. Drying does not require high-temperature heat, and methods like recovering waste heat from flue gases [38] or heat engines [149] can be used to improve the energy balance.

One of the challenges associated with raw forest biomass is its low bulk density (*cf.* section "Density"). This limitation can be addressed through densification, a physical pre-treatment method often used in conjunction with size reduction and drying. Densification can be achieved through techniques such as baling or pelletising, which involve different levels of compression. It results in easier and more cost-effective feedstock handling and transportation, as well as simplified storage [150]. Bales of forest residues, for instance, are typically characterised by bulk density of around 450 kg·m^{-3} [35], while more intensive compression is applied to produce densified solid biofuels such as pellets and briquettes. These products have higher unit densities (the mass of an individual pellet/briquette divided by its volume), typically ranging from 1000 to 1200 kg·m^{-3} [140]. However, even though pellets and briquettes have similar unit densities, pellets have substantially higher bulk densities (700–750 kg·m^{-3}) than briquettes (350–450 kg·m^{-3}) [151]. This difference is related to one of the main benefits of pellets over briquettes, especially in large-scale operations: Pellets combine the convenient handling of bulk materials with the high-density processing advantages offered by densified products [99].

Wood pellets are dense, cylindrical solid fuel with a diameter and a length not exceeding, respectively, 25 mm and 40 mm (Table 5). They have uniform characteristics in terms of size, shape and density. Pellets are produced by compressing dried and comminuted small wood particles in pellet mills. This compression leads to an increase in temperatures caused by friction, which softens the lignin and resins, allowing them to bind the wood fibres together. Most often, primary and secondary residues from the forest industries are used as feedstock for pellet production [152]. In some cases, additives may be used to enhance pellet quality [116]. Premium pellets have low ash content (<0.5%) and low moisture content (around 8%) [153], while industrial pellets (utility-grade pellets) have lower quality, with ash content that can exceed 3% [116].

Briquettes are another form of densified forest biomass fuel. The process of producing briquettes is similar to that of pellets, with the main difference being their shape and size. Typically, briquettes have a cylindrical or six-sided shape and are larger in size (Table 5). The feedstock used for briquette production also usually comes from the forest industries. Additionally, densified forest biomass is also produced in the form of logs.

Typical lower heating values of wood pellets and briquettes range from 17 to 18 MJ·kg^{-3} [138, 140]. The difference from raw biomass is mostly justified by the lower moisture and ash content of the densified biomass. If some carbonisation occurs in the densification process, the heating value is slightly higher (20–22 MJ·kg $^{-3}$) [140].

Wood pellets, briquettes and densified logs are standardised, which ensures they are a consistent feedstock. Consistent fuel quality is an advantage that make densified solid biofuels suitable for small-, medium- and large-scale applications [154] and that results in cleaner [153] and more efficient [137] combustion. Also handling, transport and storage of these densified fuels are easier, more efficient and cheaper [155]. The market for pellets and briquettes is well-established and developed in some regions of the world (e.g., Europe [156]). However, despite these advantages, the cost of densified fuels is substantially higher than that of other wood products [157].

Another process to increase the density of forest biomass is torrefaction, which can be classified as a thermal pre-treatment. At this point it is relevant to mention that the distinction between pre-treatments and conversion technologies is not without discussion. Torrefaction is a form of pyrolysis and as such it could be, and it is often, listed as a conversion technology (but so could pelletisation or briquetting, which result in marketable solid biofuels).

Torrefaction involves heating biomass to temperatures generally between 200 °C and 300 °C [158]. During this process, biomass will lose water and a part of the compounds that are volatile at this temperature range [159]. Torrefied biomass presents several advantages over raw biomass: (*i*) higher heating value and higher mass and energy densities, improving its handling, transport and storage; (*ii*) improved grindability, offering advantages in conversion processes requiring small feedstock sizes; (*iii*) hydrophobicity, which improves storability and results in a lower risk of biological degradation; (*iv*) improved and closer to coal proximate and ultimate compositions, which makes it a better substitute for coal; (*v*) reduction of Cl concentrations, leading to improved corrosion behaviour; (*vi*) improved quality of bio-oils obtained from pyrolysis of torrefied biomass, albeit with lower yield [31, 122, 123, 125]. Torrefaction can be combined with densification to produce, for example, torrefied pellets. Some commercial demonstration industrial scale plants are in operation, but the technology has not reached full commercialisation [160]. To achieve economically viability at a commercial-scale, several technical challenges need to be addressed, such as improving control over critical process parameters, particularly temperature and addressing issues related to fuel flexibility and scaling up the torrefaction system [161].

Steam explosion is a suitable physio-chemical pre-treatment step for both thermochemical and biochemical conversion technologies [31]. In this process, forest biomass is subjected to high pressure steam (7–50 bar) and temperatures ranging from 160–260 °C for short periods, followed by rapid decompression [31]. Steam explosion results in particle size reduction, improves enzyme accessibility, and enhances biomass densification [31, 99]. This pre-treatment also increases the heating value of biomass, reduces its water and volatile matter content, increases bulk density and hydrophobicity, and improves grindability [162]. However, steam explosion can lead to the formation of inhibitors and toxic compounds when used as a pre-treatment for biochemical conversion technologies; therefore, microorganisms more tolerant to inhibitors are required [31]. It is cost-effective and has a low environmental impact [31]. Steam explosion has been a long-standing technique in the wood industries, employed for decades in the production of hardboard [163].

Chemical and biological pre-treatments are valuable techniques for enhancing the conversion of forest biomass into biofuels, but their application is primarily focused on biochemical conversion technologies, which will not be covered in the subsequent chapters. These pre-treatments play a crucial role in breaking down complex biomass structures for more efficient enzymatic or microbial conversion processes.

5 Final Considerations

In the early twenty-first century, fossil fuels continue to dominate the energy landscape and are the most commonly used energy sources. Nevertheless, growing environmental concerns prompted a shift towards renewable energies and the re-emergence of biomass as an energy source in advanced economies. Currently, bioenergy contributes around 10% to the world primary energy supply and even when traditional bioenergy uses are excluded, biomass accounts, by far, for the largest share of the global renewable energy supply.

Forest biomass is a highly versatile and flexible source of renewable energy that can produce heat, power and fuels in a more reliable manner than other intermittent renewables, such as wind or solar energy, since it is not weather-dependent. Because of its characteristics, bioenergy plays an important role in the heat, electricity and transport sectors and is particularly relevant for hard-to-decarbonise sectors.

Forest biomass can be converted into energy or fuels through two distinct types of processes: biochemical and thermochemical. Biochemical conversion processes, such as anaerobic digestion and fermentation, yield gaseous and liquid biofuels. However, commercial-scale conversion of lignocellulosic biomass using these technologies has not yet been accomplished. Thermochemical processes, including combustion, gasification and pyrolysis, are currently available on the market. Combustion is widely used for heat and power production, with modern, efficient systems replacing traditional ones (despite these technological developments, the traditional use of biomass is still a reality, especially in low- and middle-income countries). Gasification generates syngas suitable for various applications, offering versatility in utilising the original solid forest biomass. However, among the various gasification pathways, only heat, power and predominantly combined heat and power production have reached commercialisation. Pyrolysis, on the other hand, produces bio-oil, gases and biochar, holding the potential for the commercial production of advanced biofuels. Nonetheless, except for traditional carbonisation, these technologies still face technical and economic challenges. Additionally, a fourth thermochemical conversion process, hydrothermal processing, is an emerging technology that requires further research before large-scale deployment can be realised.

The choice of the most suitable forest biomass energy conversion pathway for a specific application depends on several factors, including the properties of the biomass feedstock used. This requires a thorough understanding of biomass properties for an effective utilisation of forest biomass as an energy source. While forest biomass is frequently used for energy with minimal processing (raw biomass), there is a growing interest in upgraded solid biofuels derived from woody biomass, like pellets, charcoal and torrefied biomass. These upgraded solid fuels offer higher energy content, increased bulk density, and adherence to fuel quality standards, resulting in efficient energy conversion. They also reduce transportation costs and simplify fuel handling and storage.

Key properties influencing the thermochemical conversion of biomass into energy or fuels include elemental composition, ash content, volatile matter content, moisture content, heating value and bulk density. Generally, forest biomass contains a significant amount of carbon, followed by oxygen and hydrogen, with smaller quantities of other elements, like nitrogen, sulphur and chlorine. Woody biomass typically presents low ash content. The composition and ash content impact combustion systems, affecting factors such as fouling, slagging, corrosion, and pollution. Raw biomass typically has a high volatile matter content, making it suitable for various thermochemical processes, but its high oxygen content results in lower heating values and energy densities. Additionally, high moisture content can decrease combustion efficiency and heating value. The typically low bulk density of raw wood fuels has an impact on handling, transport and storage of biomass. Raw forest biomass often undergoes pre-treatment with the objective of improving its properties.

Acknowledgements The work was supported by *Fundação para a Ciência e a Tecnologia*, through IDMEC, under LAETA [project UIDB/50022/2020].

References

1. Gowlett JAJ (2016) The discovery of fire by humans: a long and convoluted process. Philos Trans R Soc B Biol Sci 371:20150164. https://doi.org/10.1098/rstb.2015.0164
2. Grübler A, Nakićenović N (1996) Decarbonizing the global energy system. Technol Forecast Soc Change 53:97–110
3. Fischer-Kowalski M, Schaffartzik A (2015) Energy availability and energy sources as determinants of societal development in a long-term perspective. MRS Energy Sustain 2:1. https://doi.org/10.1557/mre.2015.2
4. IEA (2022) Energy statistics data browser. https://www.iea.org/data-and-statistics/data-tools/energy-statistics-data-browser. Accessed 7 Apr 2023
5. IEA (2023) Interactive Sankey diagram. https://www.iea.org/sankey. Accessed 7 Apr 2023
6. IEA (2022) World energy outlook 2022. International Energy Agency, Paris
7. WBA (2022) Global bioenergy statistics 2022. World Bioenergy Association, Stockholm
8. Hoogwijk M, Faaij A, van den Broek R et al (2003) Exploration of the ranges of the global potential of biomass for energy. Biomass Bioenergy 25:119–133. https://doi.org/10.1016/S0961-9534(02)00191-5
9. Demirbas A (2005) Potential applications of renewable energy sources, biomass combustion problems in boiler power systems and combustion related environmental issues. Prog Energy Combust Sci 31:171–192. https://doi.org/10.1016/j.pecs.2005.02.002

10. Abbasi T, Abbasi SA (2010) Biomass energy and the environmental impacts associated with its production and utilization. Renew Sustain Energy Rev 14:919–937. https://doi.org/10.1016/j.rser.2009.11.006
11. Singh J (2015) Overview of electric power potential of surplus agricultural biomass from economic, social, environmental and technical perspective–a case study of Punjab. Renew Sustain Energy Rev 42:286–297. https://doi.org/10.1016/j.rser.2014.10.015
12. Giwa T, Akbari M, Kumar A (2023) Techno-economic assessment of an integrated biorefinery producing bio-oil, ethanol, and hydrogen. Fuel 332:126022. https://doi.org/10.1016/j.fuel.2022.126022
13. Röder M, Thiffault E, Martínez-Alonso C et al (2019) Understanding the timing and variation of greenhouse gas emissions of forest bioenergy systems. Biomass Bioenergy 121:99–114. https://doi.org/10.1016/j.biombioe.2018.12.019
14. Welfle A, Röder M (2022) Mapping the sustainability of bioenergy to maximise benefits, mitigate risks and drive progress toward the sustainable development goals. Renew Energy 191:493–509. https://doi.org/10.1016/j.renene.2022.03.150
15. Keoleian GA, Volk TA (2005) Renewable energy from willow biomass crops: life cycle energy, environmental and economic performance. Crit Rev Plant Sci 24:385–406. https://doi.org/10.1080/07352680500316334
16. Spiecker S, Weber C (2014) The future of the European electricity system and the impact of fluctuating renewable energy–a scenario analysis. Energy Policy 65:185–197. https://doi.org/10.1016/j.enpol.2013.10.032
17. Luderer G, Krey V, Calvin K et al (2014) The role of renewable energy in climate stabilization: results from the EMF27 scenarios. Clim Change 123:427–441. https://doi.org/10.1007/s10584-013-0924-z
18. Demirbaş A (2006) Global renewable energy resources. Energy Sources Part Recovery Util Environ Eff 28:779–792. https://doi.org/10.1080/00908310600718742
19. Gonçalves AC, Malico I, Sousa AMO (2018) Solid biomass from forest trees to energy: a review. In: Jacob-Lopes E, Queiroz Zepka L (eds) Renewable resources and biorefineries. Intech, London, UK, pp 23–46
20. IEA (2022) Renewables 2022. Analysis and forecast to 2027. International Energy Agency, Paris
21. McKendry P (2002) Energy production from biomass (part 2): conversion technologies. Bioresour Technol 83:47–54.https://doi.org/10.1016/S0960-8524(01)00119-5
22. Vasco-Correa J, Khanal S, Manandhar A, Shah A (2018) Anaerobic digestion for bioenergy production: global status, environmental and techno-economic implications, and government policies. Bioresour Technol 247:1015–1026.https://doi.org/10.1016/j.biortech.2017.09.004
23. Steffen R, Szolar O, Braun R (1998) Feedstocks for anaerobic digestion. University of Agricultural Sciences, Vienna
24. Rajendran K, Murthy GS (2019) Techno-economic and life cycle assessments of anaerobic digestion–a review. Biocatal Agric Biotechnol 20:101207. https://doi.org/10.1016/j.bcab.2019.101207
25. Sawatdeenarunat C, Surendra KC, Takara D et al (2015) Anaerobic digestion of lignocellulosic biomass: challenges and opportunities. Bioresour Technol 178:178–186. https://doi.org/10.1016/j.biortech.2014.09.103
26. Paul S, Dutta A (2018) Challenges and opportunities of lignocellulosic biomass for anaerobic digestion. Resour Conserv Recycl 130:164–174.https://doi.org/10.1016/j.resconrec.2017.12.005
27. Abraham A, Mathew AK, Park H et al (2020) Pretreatment strategies for enhanced biogas production from lignocellulosic biomass. Bioresour Technol 301:122725. https://doi.org/10.1016/j.biortech.2019.122725
28. Hahn-Hägerdal B, Galbe M, Gorwa-Grauslund MF et al (2006) Bio-ethanol–the fuel of tomorrow from the residues of today. Trends Biotechnol 24:549–556. https://doi.org/10.1016/j.tibtech.2006.10.004

29. von Schenck A, Berglin N, Uusitalo J (2013) Ethanol from Nordic wood raw material by simplified alkaline soda cooking pre-treatment. Appl Energy 102:229–240. https://doi.org/10.1016/j.apenergy.2012.10.003
30. Kumakiri I, Yokota M, Tanaka R et al (2021) Process intensification in bio-ethanol production–recent developments in membrane separation. Processes 9:1028. https://doi.org/10.3390/pr9061028
31. Papadokonstantakis S, Johnsson F (2018) Biomass conversion technologies–Definitions. D3.1 Report on definition of parameters for defining biomass conversion technologies. Chalmers University of Technology, Gothenburg
32. Brown A, Ebadian M, Saddler J et al (2020) The role of renewable transport fuels in decarbonizing road transport. Part 2–production technologies and costs. In: Bacovsky D (ed). International energy agency, p 122
33. Griffiths S, Sovacool BK, Kim J et al (2022) Decarbonizing the oil refining industry: a systematic review of sociotechnical systems, technological innovations, and policy options. Energy Res Soc Sci 89:102542. https://doi.org/10.1016/j.erss.2022.102542
34. Traverso L, Colangeli M, Morese M et al (2020) Opportunities and constraints for implementation of cellulosic ethanol value chains in Europe. Biomass Bioenergy 141:105692. https://doi.org/10.1016/j.biombioe.2020.105692
35. van Loo S, Koppejan J (2008) The handbook of biomass combustion and co-firing. Earthscan, London; Sterling, VA
36. Obernberger I (2009) Reached developments of biomass combustion technologies and future outlook. In: Proceedings of the 17th european biomass conference. Hamburg, Germany, pp 20–37
37. Eisentraut A, Brown A (2012) Technology roadmap: bioenergy for heat and power. International Energy Agency, Paris
38. Bauen AW (1999) Gasification-based biomass fuel cycles: an economic and environmental analysis at the regional level. PhD Thesis, King's College London
39. Faaij A (2006) Modern biomass conversion technologies. Mitig Adapt Strateg Glob Change 11:343–375. https://doi.org/10.1007/s11027-005-9004-7
40. Kar T, Keles S (2016) Environmental impacts of biomass combustion for heating and electricity generation. J Eng Res Appl Sci 5:458–465
41. Demirbaş A (2001) Biomass resource facilities and biomass conversion processing for fuels and chemicals. Energy Convers Manag 42:1357–1378. https://doi.org/10.1016/S0196-8904(00)00137-0
42. Landälv I, Maniatis K, Waldheim L et al (2018) Building up the future. Technology status and reliability of the value chains. European Commission, Luxembourg
43. Tinaut FV, Melgar A, Horrillo A, Díez de la Rosa A (2006) Method for predicting the performance of an internal combustion engine fuelled by producer gas and other low heating value gases. Fuel Process Technol 87:135–142. https://doi.org/10.1016/j.fuproc.2005.08.009
44. Bain RL, Broer K (2011) Gasification. In: Brown RC (ed) Thermochemical processing of biomass. Conversion into fuels, chemicals and power. John Wiley & Son, Chichester, pp 47–77
45. Parvez AM, Afzal MT, Victor Hebb TG, Schmid M (2020) Utilization of CO_2 in thermochemical conversion of biomass for enhanced product properties: a review. J CO2 Util 40:101217. https://doi.org/10.1016/j.jcou.2020.101217
46. Göransson K, Söderlind U, He J, Zhang W (2011) Review of syngas production via biomass DFBGs. Renew Sustain Energy Rev 15:482–492. https://doi.org/10.1016/j.rser.2010.09.032
47. Pröll T, Siefert IG, Friedl A, Hofbauer H (2005) Removal of NH_3 from biomass gasification producer gas by water condensing in an organic solvent scrubber. Ind Eng Chem Res 44:1576–1584. https://doi.org/10.1021/ie049669v
48. Al-attab KA, Zainal ZA (2014) Performance of a biomass fueled two-stage micro gas turbine (MGT) system with hot air production heat recovery unit. Appl Therm Eng 70:61–70. https://doi.org/10.1016/j.applthermaleng.2014.04.030

49. Schulzke T (2019) Biomass gasification: conversion of forest residues into heat, electricity and base chemicals. Chem Pap 73:1833–1852. https://doi.org/10.1007/s11696-019-00801-1
50. Archer SA, Steinberger-Wilckens R (2018) Systematic analysis of biomass derived fuels for fuel cells. Int J Hydrog Energy 43:23178–23192. https://doi.org/10.1016/j.ijhydene.2018.10.161
51. Subotić V, Baldinelli A, Barelli L et al (2019) Applicability of the SOFC technology for coupling with biomass-gasifier systems: short- and long-term experimental study on SOFC performance and degradation behaviour. Appl Energy 256:113904. https://doi.org/10.1016/j.apenergy.2019.113904
52. Najafi G, Hoseini SS, De Goey LPH, Yusaf T (2020) Optimization of combustion in micro combined heat and power (mCHP) system with the biomass-Stirling engine using SiO_2 and Al_2O_3 nanofluids. Appl Therm Eng 169:114936. https://doi.org/10.1016/j.applthermaleng.2020.114936
53. Schneider T, Ruf F, Müller D, Karl J (2021) Performance of a fluidized bed-fired Stirling engine as micro-scale combined heat and power system on wood pellets. Appl Therm Eng 189:116712. https://doi.org/10.1016/j.applthermaleng.2021.116712
54. Borisov I, Khalatov A, Paschenko D (2022) The biomass fueled micro-scale CHP unit with Stirling engine and two-stage vortex combustion chamber. Heat Mass Transf 58:1091–1103. https://doi.org/10.1007/s00231-021-03165-z
55. van der Meijden CM, Veringa HJ, Rabou LPLM (2010) The production of synthetic natural gas (SNG): a comparison of three wood gasification systems for energy balance and overall efficiency. Biomass Bioenergy 34:302–311. https://doi.org/10.1016/j.biombioe.2009.11.001
56. Yan Q, Yu F, Liu J et al (2013) Catalytic conversion wood syngas to synthetic aviation turbine fuels over a multifunctional catalyst. Bioresour Technol 127:281–290. https://doi.org/10.1016/j.biortech.2012.09.069
57. Dupuis DP, Grim RG, Nelson E et al (2019) High-octane gasoline from biomass: experimental, economic, and environmental assessment. Appl Energy 241:25–33. https://doi.org/10.1016/j.apenergy.2019.02.064
58. Bengelsdorf FR, Dürre P (2017) Gas fermentation for commodity chemicals and fuels. Microb Biotechnol 10:1167–1170. https://doi.org/10.1111/1751-7915.12763
59. García CA, Betancourt R, Cardona CA (2017) Stand-alone and biorefinery pathways to produce hydrogen through gasification and dark fermentation using *Pinus Patula*. J Environ Manag 203:689–703. https://doi.org/10.1016/j.jenvman.2016.04.001
60. Peterson D, Haase S (2009) Market assessment of biomass gasification and combustion technology for small- and medium-scale applications. National Renewable Energy Laboratory, Golden, CO
61. Quirion-Blais O, Malladi KT, Sowlati T et al (2019) Analysis of feedstock requirement for the expansion of a biomass-fed district heating system considering daily variations in heat demand and biomass quality. Energy Convers Manag 187:554–564. https://doi.org/10.1016/j.enconman.2019.03.036
62. Anca-Couce A, Hochenauer C, Scharler R (2021) Bioenergy technologies, uses, market and future trends with Austria as a case study. Renew Sustain Energy Rev 135:110237. https://doi.org/10.1016/j.rser.2020.110237
63. Hrbek J (2022) Status report on thermal gasification of biomass and waste 2021. International Energy Agency
64. Hrbek J (2020) Past, present and future of thermal gasification of biomass and waste. Acta Innov 35:5–20. https://doi.org/10.32933/ActaInnovations.35.1
65. Thomson R, Kwong P, Ahmad E, Nigam KDP (2020) Clean syngas from small commercial biomass gasifiers; a review of gasifier development, recent advances and performance evaluation. Int J Hydrog Energy 45:21087–21111. https://doi.org/10.1016/j.ijhydene.2020.05.160
66. Miedema JH, van der Windt HJ, Moll HC (2018) Opportunities and barriers for biomass gasification for green gas in the Dutch residential sector. Energies 11:2969. https://doi.org/10.3390/en11112969

67. di Gruttola F, Borello D (2021) Analysis of the EU secondary biomass availability and conversion processes to produce advanced biofuels: use of existing databases for assessing a metric evaluation for the 2025 perspective. Sustainability 13:7882. https://doi.org/10.3390/su13147882
68. AECOM, Fichtner Consulting Engineers (2021) Advanced gasification technologies–Review and benchmarking, Department for Business, Energy & Industrial Strategy
69. Bridgwater AV (2003) Renewable fuels and chemicals by thermal processing of biomass. Chem Eng J 91:87–102. https://doi.org/10.1016/S1385-8947(02)00142-0
70. Hagemann N, Spokas K, Schmidt H-P et al (2018) Activated carbon, biochar and charcoal: linkages and synergies across pyrogenic carbon's ABCs. Water 10:182. https://doi.org/10.3390/w10020182
71. Pollex A, Ortwein A, Kaltschmitt M (2012) Thermo-chemical conversion of solid biofuels: conversion technologies and their classification. Biomass Convers Biorefinery 2:21–39. https://doi.org/10.1007/s13399-011-0025-z
72. Bridgwater AV (2012) Review of fast pyrolysis of biomass and product upgrading. Biomass Bioenergy 38:68–94. https://doi.org/10.1016/j.biombioe.2011.01.048
73. Roy P, Dias G (2017) Prospects for pyrolysis technologies in the bioenergy sector: a review. Renew Sustain Energy Rev 77:59–69. https://doi.org/10.1016/j.rser.2017.03.136
74. Homagain K, Shahi C, Luckai N, Sharma M (2014) Biochar-based bioenergy and its environmental impact in Northwestern Ontario Canada: a review. J For Res 25:737–748. https://doi.org/10.1007/s11676-014-0522-6
75. Antal MJ, Grønli M (2003) The art, science, and technology of charcoal production. Ind Eng Chem Res 42:1619–1640. https://doi.org/10.1021/ie0207919
76. Rodrigues T, Braghini Junior A (2019) Charcoal: a discussion on carbonization kilns. J Anal Appl Pyrolysis 143:104670. https://doi.org/10.1016/j.jaap.2019.104670
77. Schenkel Y, Bertaux P, Vanwijnbserghe S, Carre J (1998) An evaluation of the mound kiln carbonization technique. Biomass Bioenergy 14:505–516. https://doi.org/10.1016/S0961-9534(97)10033-2
78. Manyà JJ (2012) Pyrolysis for biochar purposes: a review to establish current knowledge gaps and research needs. Environ Sci Technol 46:7939–7954. https://doi.org/10.1021/es301029g
79. Garcia-Nunez JA, Pelaez-Samaniego MR, Garcia-Perez ME et al (2017) Historical developments of pyrolysis reactors: a review. Energy Fuels 31:5751–5775. https://doi.org/10.1021/acs.energyfuels.7b00641
80. Premchand P, Demichelis F, Chiaramonti D et al (2023) Biochar production from slow pyrolysis of biomass under CO_2 atmosphere: a review on the effect of CO_2 medium on biochar production, characterisation, and environmental applications. J Environ Chem Eng 11:110009. https://doi.org/10.1016/j.jece.2023.110009
81. Oasmaa A, Lehto J, Solantausta Y, Kallio S (2021) Historical review on VTT fast pyrolysis bio-oil production and upgrading. Energy Fuels 35:5683–5695. https://doi.org/10.1021/acs.energyfuels.1c00177
82. Lehto J, Oasmaa A, Solantausta Y et al (2014) Review of fuel oil quality and combustion of fast pyrolysis bio-oils from lignocellulosic biomass. Appl Energy 116:178–190. https://doi.org/10.1016/j.apenergy.2013.11.040
83. Hunt J, DuPonte M, Sato D, Kawabata A (2010) The basics of biochar: a natural soil amendment. 30:1–6
84. McElligott K, Page-Dumroese D, Coleman M (2011) Bioenergy production systems and biochar application in forests: potential for renewable energy, soil enhancement, and carbon sequestration. Department of Agriculture, Forest Service, Rocky Mountain Research Station, Fort Collins, CO
85. Grutzmacher P, Puga AP, Bibar MPS et al (2018) Carbon stability and mitigation of fertilizer induced N_2O emissions in soil amended with biochar. Sci Total Environ 625:1459–1466. https://doi.org/10.1016/j.scitotenv.2017.12.196
86. Rockwood DL, Ellis M, Liu R, et al (2020) Forest trees for biochar and carbon sequestration: production and benefits. In: Abdelhafez AA, Abbas MHH (eds) Applications of biochar for environmental safety. Intech, London, UK, pp 27–46

87. Zhang T, Walawender W, Fan L et al (2004) Preparation of activated carbon from forest and agricultural residues through CO activation. Chem Eng J 105:53–59. https://doi.org/10.1016/j.cej.2004.06.011
88. Azargohar R, Dalai AK (2005) Biochar as a precursor of activated carbon. In: McMillan JD, Adney WS, Mielenz JR, Klasson KT (eds) Proceedings of the twenty-seventh symposium on biotechnology for fuels and chemical. Humana Press, Denver, Colorado, pp 762–773
89. Giudicianni P, Gargiulo V, Grottola CM et al (2021) Inherent metal elements in biomass pyrolysis: a review. Energy Fuels 35:5407–5478. https://doi.org/10.1021/acs.energyfuels.0c04046
90. Yogalakshmi KN, Poornima DT, Sivashanmugam P et al (2022) Lignocellulosic biomass-based pyrolysis: a comprehensive review. Chemosphere 286:131824. https://doi.org/10.1016/j.chemosphere.2021.131824
91. Sorunmu Y, Billen P, Spatari S (2020) A review of thermochemical upgrading of pyrolysis bio-oil: techno-economic analysis, life cycle assessment, and technology readiness. GCB Bioenergy 12:4–18. https://doi.org/10.1111/gcbb.12658
92. Elliot DC (2011) Hydrothermal processing. In: Brown RC (ed) Thermochemical processing of biomass. Conversion into fuels, chemicals and power. John Wiley & Son, Chichester, UK, pp 200–231
93. Tekin K, Karagöz S, Bektaş S (2014) A review of hydrothermal biomass processing. Renew Sustain Energy Rev 40:673–687. https://doi.org/10.1016/j.rser.2014.07.216
94. Khan N, Mohan S, Dinesha P (2021) Regimes of hydrochar yield from hydrothermal degradation of various lignocellulosic biomass: a review. J Clean Prod 288:125629. https://doi.org/10.1016/j.jclepro.2020.125629
95. Cao Y, He M, Dutta S et al (2021) Hydrothermal carbonization and liquefaction for sustainable production of hydrochar and aromatics. Renew Sustain Energy Rev 152:111722. https://doi.org/10.1016/j.rser.2021.111722
96. Brown RC (2011) Introduction to thermochemical processing of biomass into fuels, chemicals, and power. In: Brown RC (ed) Thermochemical processing of biomass. Conversion into fuels, chemicals and power. John Wiley & Son, Chichester, UK, pp 1–12
97. Kumar M, Olajire Oyedun A, Kumar A (2018) A review on the current status of various hydrothermal technologies on biomass feedstock. Renew Sustain Energy Rev 81:1742–1770. https://doi.org/10.1016/j.rser.2017.05.270
98. Munawar MA, Khoja AH, Naqvi SR et al (2021) Challenges and opportunities in biomass ash management and its utilization in novel applications. Renew Sustain Energy Rev 150:111451. https://doi.org/10.1016/j.rser.2021.111451
99. Bajwa DS, Peterson T, Sharma N et al (2018) A review of densified solid biomass for energy production. Renew Sustain Energy Rev 96:296–305. https://doi.org/10.1016/j.rser.2018.07.040
100. Rowell R, Pettersen R, Tshabalala M (2012) Cell wall chemistry. In: Handbook of wood chemistry and wood composites, 2nd edn. CRC Press, pp 33–72
101. Williams CL, Emerson RM, Tumuluru JS (2017) Biomass compositional analysis for conversion to renewable fuels and chemicals. In: Tumuluru JS (ed) Biomass volume estimation and valorization for energy. InTech, Rijeka, Croatia, pp 251–270
102. Lee J (1997) Biological conversion of lignocellulosic biomass to ethanol. J Biotechnol 56:1–24. https://doi.org/10.1016/S0168-1656(97)00073-4
103. Jenkins BM, Baxter LL, Miles TR Jr, Miles TR (1998) Combustion properties of biomass. Fuel Process Technol 54:17–46
104. Ragland KW, Aerts DJ, Baker AJ (1991) Properties of wood for combustion analysis. Bioresour Technol v:161–168. https://doi.org/10.1016/0960-8524(91)90205-X
105. Prins M, Ptasinski K, Janssen F (2007) From coal to biomass gasification: comparison of thermodynamic efficiency. Energy 32:1248–1259. https://doi.org/10.1016/j.energy.2006.07.017
106. Li W, Dang Q, Brown RC et al (2017) The impacts of biomass properties on pyrolysis yields, economic and environmental performance of the pyrolysis-bioenergy-biochar platform to

carbon negative energy. Bioresour Technol 241:959–968. https://doi.org/10.1016/j.biortech.2017.06.049
107. Elbersen W, Alakangas E, Elbersen B et al (2016) Explanatory note accompanying the database for standardized biomass characterization (and minimal biomass quality requirement for each biomass conversion technology). S2Biom project
108. Cardoso M, de Oliveira ÉD, Passos ML (2009) Chemical composition and physical properties of black liquors and their effects on liquor recovery operation in Brazilian pulp mills. Fuel 88:756–763. https://doi.org/10.1016/j.fuel.2008.10.016
109. Hupa M, Karlström O, Vainio E (2017) Biomass combustion technology development–it is all about chemical details. Proc Combust Inst 36:113–134. https://doi.org/10.1016/j.proci.2016.06.152
110. Obernberger I (1998) Decentralized biomass combustion: state of the art and future development. Biomass Bioenergy 14:33–56. https://doi.org/10.1016/S0961-9534(97)00034-2
111. TNO (2023) Phyllis2, database for (treated) biomass, algae, feedstocks for biogas production and biochar. https://phyllis.nl/. Accessed 23 Jun 2023
112. McKendry P (2002) Energy production from biomass (part 1): overview of biomass. Bioresour Technol 83:37–46. https://doi.org/10.1016/S0960-8524(01)00118-3
113. Vassilev SV, Vassileva CG, Vassilev VS (2015) Advantages and disadvantages of composition and properties of biomass in comparison with coal: an overview. Fuel 158:330–350. https://doi.org/10.1016/j.fuel.2015.05.050
114. Kubica K, Paradiz B, Dilara P (2007) Small combustion installations: techniques, emissions and measures for emission reduction. Publications Office of the European Union, Luxembourg
115. Pettersen RC (1984) The chemical composition of wood. In: Rowell R (ed) The chemistry of solid wood. American Chemical Society, Washington, DC, pp 57–126
116. Krajnc N (2015) Wood fuels handbook. Food and Agriculture Organization of the United Nations, Pristina, Kosovo
117. Hytönen J, Kaakkurivaara N, Kaakkurivaara T, Nurmi J (2018) Biomass equations for rubber tree (*Hevea brasiliensis*) components in Southern Thailand. J Trop For Sci 30:588–596. https://doi.org/10.26525/jtfs2018.30.4.588596
118. Soleymani M, Shokrpoor S, Jaafarzadeh N (2023) A comprehensive study of essential properties of *Conocarpus erectus* as a potential bioenergy crop. Int J Environ Sci Technol 20:6147–6160. https://doi.org/10.1007/s13762-023-04878-w
119. Brasil ACM, Brasil A, Malico I (2020) Evaluation of the electrical energy potential of woody biomass in the near region of the hydropower plant Tucuruí-Brazil. Waste Biomass Valorization 11:2297–2307. https://doi.org/10.1007/s12649-018-0407-6
120. Magdeldin M, Järvinen M (2020) Supercritical water gasification of Kraft black liquor: process design, analysis, pulp mill integration and economic evaluation. Appl Energy 262:114558. https://doi.org/10.1016/j.apenergy.2020.114558
121. Stegelmeier M, Schmitt VEM, Kaltschmitt M (2011) Pelletizing of autumn leaves—possibilities and limits. Biomass Convers Biorefinery 1:173–187. https://doi.org/10.1007/s13399-011-0016-0
122. Niu Y, Lv Y, Lei Y et al (2019) Biomass torrefaction: properties, applications, challenges, and economy. Renew Sustain Energy Rev 115:109395. https://doi.org/10.1016/j.rser.2019.109395
123. Phanphanich M, Mani S (2011) Impact of torrefaction on the grindability and fuel characteristics of forest biomass. Bioresour Technol 102:1246–1253. https://doi.org/10.1016/j.biortech.2010.08.028
124. Fagernäs L, Kuoppala E, Tiilikkala K, Oasmaa A (2012) Chemical composition of birch wood slow pyrolysis products. Energy Fuels 26:1275–1283. https://doi.org/10.1021/ef2018836
125. Keipi T, Tolvanen H, Kokko L, Raiko R (2014) The effect of torrefaction on the chlorine content and heating value of eight woody biomass samples. Biomass Bioenergy 66:232–239. https://doi.org/10.1016/j.biombioe.2014.02.015
126. Quaak P, Knoef H, Stassen H (1999) Energy from biomass: a review of combustion and gasification technologies. The World Bank, Washington, DC

127. Nussbaumer T (2003) Combustion and co-combustion of biomass: fundamentals, technologies, and primary measures for emission reduction. Energy Fuels 17:1510–1521. https://doi.org/10.1021/ef030031q
128. Bridgwater AV, Toft AJ, Brammer JG (2002) A techno-economic comparison of power production by biomass fast pyrolysis with gasification and combustion. Renew Sustain Energy Rev 6:181–246. https://doi.org/10.1016/S1364-0321(01)00010-7
129. Hornung A, Stenzel F, Grunwald J (2021) Biochar—just a black matter is not enough. Biomass Convers Biorefinery. https://doi.org/10.1007/s13399-021-01284-5
130. Kenney KL, Smith WA, Gresham GL, Westover TL (2013) Understanding biomass feedstock variability. Biofuels 4:111–127. https://doi.org/10.4155/bfs.12.83
131. Temmerman M, Jensen PD, Hébert J (2013) Von Rittinger theory adapted to wood chip and pellet milling, in a laboratory scale hammermill. Biomass Bioenergy 56:70–81. https://doi.org/10.1016/j.biombioe.2013.04.020
132. Channiwala SA, Parikh PP (2002) A unified correlation for estimating HHV of solid, liquid and gaseous fuels. Fuel 81:1051–1063. https://doi.org/10.1016/S0016-2361(01)00131-4
133. Vieilledent G, Fischer FJ, Chave J et al (2018) New formula and conversion factor to compute basic wood density of tree species using a global wood technology database. Am J Bot 105:1653–1661. https://doi.org/10.1002/ajb2.1175
134. Eisenbies MH, Volk TA, Therasme O, Hallen K (2019) Three bulk density measurement methods provide different results for commercial scale harvests of willow biomass chips. Biomass Bioenergy 124:64–73. https://doi.org/10.1016/j.biombioe.2019.03.015
135. Baxter L (2005) Biomass-coal co-combustion: opportunity for affordable renewable energy. Fuel 84:1295–1302. https://doi.org/10.1016/j.fuel.2004.09.023
136. S2Biom (2016) Database for standardized biomass characterization. https://s2biom.wenr.wur.nl/web/guest/biomass-characteristics. Accessed 14 Apr 2023
137. Obernberger I, Thek G (2004) Physical characterisation and chemical composition of densified biomass fuels with regard to their combustion behaviour. Biomass Bioenergy 27:653–669. https://doi.org/10.1016/j.biombioe.2003.07.006
138. Tumuluru JS, Sokhansanj S, Lim CJ et al (2010) Quality of wood pellets produced in British Columbia for export. Appl Eng Agric 26:1013–1020. https://doi.org/10.13031/2013.35902
139. ISO (2022) ISO 16559:2022. Solid biofuels—vocabulary. https://www.iso.org/standard/75261.html. Accessed 31 Aug 2023
140. Tumuluru JS, Wright CT, Hess JR, Kenney KL (2011) A review of biomass densification systems to develop uniform feedstock commodities for bioenergy application. Biofuels Bioprod Biorefining 5:683–707. https://doi.org/10.1002/bbb.324
141. Abdullah H, Wu H (2009) Biochar as a fuel: 1. Properties and grindability of biochars produced from the pyrolysis of mallee wood under slow-heating conditions. Energy Fuels 23:4174–4181. https://doi.org/10.1021/ef900494t
142. Frombo F, Minciardi R, Robba M et al (2009) Planning woody biomass logistics for energy production: a strategic decision model. Biomass Bioenergy 33:372–383. https://doi.org/10.1016/j.biombioe.2008.09.008
143. Berndes G, Abt B, Asikainen A et al (2016) Forest biomass, carbon neutrality and climate change mitigation. European Forest Institute
144. Hasan AR, Solo-Gabriele H, Townsend T (2011) Online sorting of recovered wood waste by automated XRF-technology: part II. Sorting efficiencies. Waste Manag 31:695–704. https://doi.org/10.1016/j.wasman.2010.10.024
145. Naimi LJ, Sokhansanj S, Mani S et al (2006) Cost and performance of woody biomass size reduction for energy production. In: 2006 CSBE/SCGAB. American society of agricultural and biological engineers, Edmonton, AB Canada
146. Mirkouei A, Haapala KR, Sessions J, Murthy GS (2017) A review and future directions in techno-economic modeling and optimization of upstream forest biomass to bio-oil supply chains. Renew Sustain Energy Rev 67:15–35. https://doi.org/10.1016/j.rser.2016.08.053
147. Kumar A, Jones D, Hanna M (2009) Thermochemical biomass gasification: a review of the current status of the technology. Energies 2:556–581. https://doi.org/10.3390/en20300556

148. Energy and Environmental Analysis, Eastern Research Group (2007) Biomass combined heat and power catalog of technologies. Environmental Protection Agency, U. S
149. Bridgwater AV (1995) The technical and economic feasibility of biomass gasification for power generation. Fuel 74:631–653
150. Schipfer F, Mäki E, Schmieder U et al (2022) Status of and expectations for flexible bioenergy to support resource efficiency and to accelerate the energy transition. Renew Sustain Energy Rev 158:112094. https://doi.org/10.1016/j.rser.2022.112094
151. Tumuluru JS (2018) Effect of pellet die diameter on density and durability of pellets made from high moisture woody and herbaceous biomass. Carbon Resour Convers 1:44–54. https://doi.org/10.1016/j.crcon.2018.06.002
152. Camia A, Giuntoli J, Jonsson R et al (2021) The use of woody biomass for energy production in the EU. Publications Office of the European Union, Luxembourg
153. Beauchemin PA, Tampier M (2008) Emissions from wood-fired combustion equipment. Envirochem Services Inc., North Vancouver, B. C.
154. Obernberger I, Thek G (2010) The pellet handbook: the production and thermal utilisation of pellets. Earthscan, London, UK
155. Sikkema R, Steiner M, Junginger M et al (2011) The European wood pellet markets: current status and prospects for 2020. Biofuels Bioprod Biorefining 5:250–278. https://doi.org/10.1002/bbb.277
156. Nunes LJR, Matias JCO, Catalão JPS (2016) Wood pellets as a sustainable energy alternative in Portugal. Renew Energy 85:1011–1016. https://doi.org/10.1016/j.renene.2015.07.065
157. Kristöfel C, Strasser C, Schmid E, Morawetz UB (2016) The wood pellet market in Austria: a structural market model analysis. Energy Policy 88:402–412. https://doi.org/10.1016/j.enpol.2015.10.039
158. Kota KB, Shenbagaraj S, Sharma PK et al (2022) Biomass torrefaction: an overview of process and technology assessment based on global readiness level. Fuel 324:124663. https://doi.org/10.1016/j.fuel.2022.124663
159. Koppejan J, Cremers M (2019) Biomass pre-treatment for bioenergy. Policy Report. IEA Bioenergy
160. Wild M, Calderón C (2021) Torrefied biomass and where is the sector currently standing in terms of research, technology development, and implementation. Front Energy Res 9:678492. https://doi.org/10.3389/fenrg.2021.678492
161. Brachi P, Tumuluru JS, Nhuchhen DR, Chen W-H (2021) Editorial: "Torrefaction pretreatment for biomass upgrading: fundamentals and technologies." Front Energy Res 9:769625. https://doi.org/10.3389/fenrg.2021.769625
162. Livingston WR, Middelkamp J, Willeboer W et al (2016) The status of large scale biomass firing. IEA Bioenergy
163. Biermann CJ, Schultz TP, McGinnis GD (1984) Rapid steam hydrolysis/extraction of mixed hardwoods as a biomass pretreatment. J Wood Chem Technol 4:111–128

Biomass for Domestic Heat

Isabel Malico

Abstract Biomass is an important source of energy in the residential sector and meets a significant proportion of the energy needs of one-third of the world's population. In many low- and middle-income countries, particularly in rural areas, it is still used in a traditional way and provides the basic energy needs of the population, such as cooking and water and space heating. In many poor regions where forests are abundant, wood, of all the possible biomass resources, is the dominant fuel. It can be obtained at low or no monetary cost and burns in simple and inexpensive equipment. However, the consequences of this traditional use of biomass are several: indoor and outdoor pollution, impacts on health, pressures on forest resources and increased burden on women and children. Developments in residential wood fuel energy technologies are driven by the need for higher efficiency and fewer environmental impacts. As a consequence of such developments, today, highly efficient and cleaner equipment is used to provide energy from forest resources, but still mainly in high-income countries. This chapter reviews the use of energy from forest biomass in the residential sector and the different available conversion technologies, from the traditional to the most advanced ones.

Keywords Residential sector · Households · Space heating · Cooking · Bioenergy · Conversion technologies

1 Introduction

The residential sector is one of the most important sectors in terms of energy consumption, having accounted for about 89 EJ of final energy consumption in the world in 2020 [1]. This corresponds to around 25% of the total global final energy consumption in that year (Fig. 1) and contributed to approximately 12% of total direct greenhouse gas emissions from end-use sectors; direct means that indirect emissions

I. Malico (✉)
IDMEC, Escola de Ciências e Tecnologia, Departamento de Engenharia Mecatrónica, Universidade de Évora, Rua Romão Ramalho 59, 7000-671 Évora, Portugal
e-mail: imbm@uevora.pt

A. C. Gonçalves and I. Malico (eds.), *Forest Bioenergy*, Green Energy and Technology,
https://doi.org/10.1007/978-3-031-48224-3_8

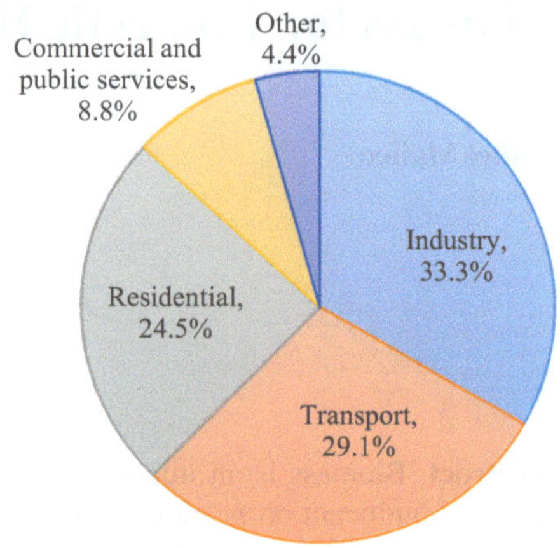

Fig. 1 Share of the different end-use sectors in the total final energy consumption in the world in 2020. (*Data source* [1])

from the electricity and heat generation consumed in households are not considered [1].

A sector that has much in common with the residential sector, as far as energy consumption and mitigation options are concerned, is the commercial and public service sector [2]. Combining the two, their share in the global final energy consumption in 2020 was 33%, and the share in the total emissions from end-use sectors was 16% [1].

In the residential and in the commercial and public service sectors, the consumption of energy for lighting, heating, cooling and powering electronic devices and other equipment typically plays a relevant role, even though the classes of buildings, the type of equipment and the dynamics of both sectors differ. Moreover, in both, there is a significant potential for energy savings and efficiency improvements and for a higher incorporation of renewable energies [3]. To promote these goals, in many countries, both sectors are subjected to energy-related regulations and policies, such as building energy codes, energy standards for appliances and equipment and incentives for the adoption of energy efficiency measures and renewable energies [2]. This chapter will focus on the residential sector, even if, from what has been said, much is also relevant for the commercial and public service sector.

The relative importance of households (excluding transport) in the total final energy consumption varies among the different world regions. The highest share occurs in Africa (Fig. 2), a region where a substantial part of the population does not have access to modern energy services and relies on traditional uses of biomass (*cf.* chapter "Sources and Distribution of Forest Biomass for Energy"). The lowest share occurs in the group of Asian countries belonging to the Organisation for Economic Cooperation and Development (OECD).

Energy consumption in the residential sector is dependent on a combination of different climatic, technological and socio-economic aspects that vary regionally [4–10]. Globally, most of the energy in the residential and service sectors is consumed in the form of heat, especially for space heating [3], but the importance of the latter is

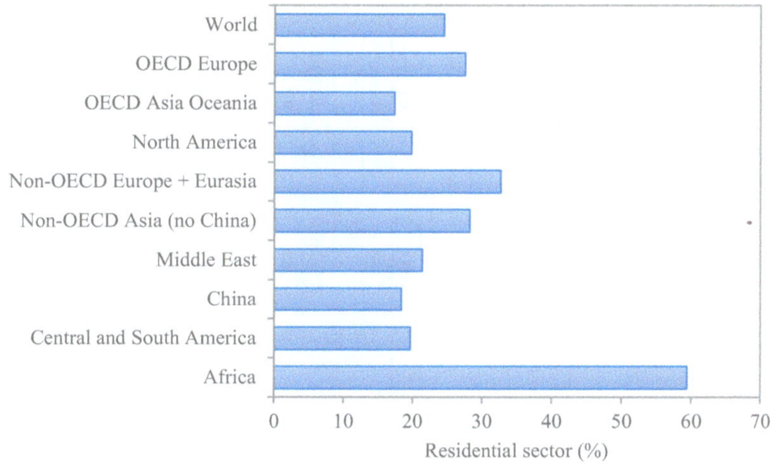

Fig. 2 Share of the residential sector in the total final energy consumption for different regions in 2020. (*Data source* [1])

obviously highly dependent on climate. For instance, according to the latest data from the Residential Energy Consumption Survey made in the United States of America, space heating accounted for more than half of the energy consumed in households located in the colder northern zones. In contrast, in hotter regions of the United States, its share dropped to around one-fifth [10]. Despite this, even in the hotter climate zones, on average, space and water heating combined represented the largest share of energy consumed by American households [10].

In 2020, globally, biofuels and waste were the most used energy sources in households, representing 32% of the energy consumed in the residential sector [1] (Table 1). Nevertheless, the relative importance of biofuels has been decreasing in the last decades, while that of electricity and natural gas has been increasing (in 2020, they accounted for, respectively, 26% and 23% of the energy consumed in households, whereas in 1990, only 14% and 18%, respectively) [1]. An even larger relative growth came from renewable energy sources (RES) other than bioenergy, such as solar, which had an almost 19-fold increase in the last 30 years [1]. On the other side, coal is the only energy source that saw a decrease in consumption in the residential sector (the household coal consumption in 2020 was 36% of that of 1990 [1]).

Despite the decreasing share of biofuels and wastes in residential energy consumption over the last three decades, the global consumption of these fuels in the residential sector increased from 26 to 28 EJ from 1990 to 2020 [1]. With the exception of China, where biogas represents a noteworthy share of the biofuels and wastes consumed in households (9% in 2020), in the other world regions, primary solid biofuels constitute almost all (>99%) of the biofuels and wastes used up in the residential sector [1]. They are particularly important in Africa, where in 2020, 85% of the energy in households came from primary solid biofuels, such as firewood or charcoal [1]. Woody materials also constitute the most important source of energy in households

Table 1 Share (in %) of the different energy sources in the world residential energy consumption in 1990 and 2020. (*Data source* [1])

	Coal	Oil products	Natural gas	Biofuels and waste	Other RES	Electricity	Heat
1990							
World	10.1	13.0	18.4	41.0	0.2	14.3	3.2
Africa	1.0	5.9	0.7	89.2	0.0	3.2	0.0
Central and South America	0.4	23.1	7.5	51.6	0.0	17.5	0.0
China	28.1	1.0	0.6	68.8	0.0	0.8	0.7
Middle East	0.0	59.0	11.5	0.8	0.2	28.4	0.0
Non-OECD Asia excluding China	1.4	10.0	1.0	83.8	0.0	3.6	0.2
Non-OECD Europe and Eurasia	13.8	9.6	42.9	6.9	0.0	10.4	16.4
North America	0.6	14.6	44.6	4.2	0.0	35.9	0.0
OECD Asia and Oceania	14.6	27.3	16.2	3.9	2.7	35.4	0.1
OECD Europe	12.4	23.2	27.3	10.2	0.2	19.3	7.4
2020							
World	2.6	10.2	22.5	31.7	2.1	25.6	5.3
Africa	0.8	4.3	3.4	85.4	0.0	6.0	0.0
Central and South America	0.1	17.5	14.0	34.6	1.0	32.8	0.0
China	8.9	10.2	12.9	22.0	9.8	27.0	9.2
Middle East	0.0	13.3	47.7	0.5	0.2	38.2	0.0
Non-OECD Asia excluding China	1.5	14.7	3.8	58.6	0.3	20.9	0.1
Non-OECD Europe and Eurasia	3.3	8.8	42.9	5.9	0.1	13.4	25.8
North America	0.0	6.6	39.7	6.1	0.5	47.1	0.0
OECD Asia and Oceania	0.3	19.0	28.5	1.8	1.4	46.4	2.6
OECD Europe	2.8	11.5	37.8	13.6	1.4	26.2	6.8

in Central and South America and non-OECD Asia excluding China (respectively, 35% and 59% of the energy consumed in the residential sectors of these two world regions in 2020 [1]).

Africa was also the region that consumed the largest share of the primary solid biofuels used in the residential sector worldwide (63%), followed by non-OECD Asia including China (41%). Despite the large importance that biomass and waste has for the households in Central and South America, their share of the consumption of primary solid biofuels used in the world is small (4%).

In 2020, around one third of the world population (*circa* 2.4 billion people), mostly in Asia and Sub-Saharan Africa, lacked access to clean fuels and technologies for cooking [11], the most universal residential energy service [12]. It is in these two regions that the traditional use of biomass for space heating and cooking is mostly concentrated (in 2020, traditional use of biomass accounted for 24 EJ, a value that has remained practically stable since 1990) [3, 11].

A different scenario occurs in OECD countries, which mainly consume electricity or natural gas in households. In these countries, the importance of primary solid biomass in the residential sector is relatively small; OECD Europe being the region with the largest share of biofuel consumption (14% in 2020 [1]). Several OECD European countries have a strong penetration of biomass in the residential sector, but others do not. Estonia (40%), Slovenia (39%), Latvia (38%), Czechia (29%), Lithuania (28%) and Portugal (25%), all have shares of primary solid biofuels in the final energy consumption in households in 2021 above or equal to 25%, while Ireland (1%), Luxembourg (3%), The Netherlands (4%), Turkey (6%) and Belgium (7%) are below 10% [13].

The remaining part of this chapter presents the most relevant forest biomass heat production technologies used in the residential sector. It starts with a description of the traditional uses of biomass and their environmental implications, followed by the presentation of more modern technologies.

2 Traditional Uses of Biomass

Wood has been used for cooking, heating and lighting for millennia and plays a vital role as a fuel for human beings since primordial times. In some parts of the world it is still used in traditional ways; the term "traditional use of biomass" referring to the direct combustion of solid biomass in inefficient and polluting equipment [3, 14]. This includes using wood products, charcoal, agricultural residues and animal dung, often produced in unsustainable ways and not commercialised [14, 15].

The traditional use of biomass is still preponderant in households in rural areas of low- and middle-income countries, but its share in the total primary energy consumption tends to decrease with the increase in gross domestic product per capita and degree of urbanisation and industrialisation [15]. In countries with a high percentage of land covered by forests (e.g., Cambodia and Laos), traditional biomass is mainly composed of wood fuel and charcoal, whereas in countries where forest area is proportionally lower (e.g., China and India), wood fuel is not as important as animal dung or agricultural residues [15].

Biomass traditionally used is local in nature, low-cost and does not require processing before use [16]. Globally, it is mostly utilised for cooking and water and space heating [17], with the relative importance of the different end-uses varying among regions because of climatic conditions: Space heating is less important in lowland areas than in highland areas [16].

Open fires and inefficient traditional stoves (International Organisation for Standardization, ISO tiers 0 and 1), often with no or inefficient chimneys, are the basic technologies associated with the traditional use of biomass [3]. Examples of popular traditional wood-fired stoves are three-stone fires and mud stoves, whose typical efficiencies range from 9 to 23% (Table 2, based on [18–29]). They are typically home-made, created or assembled for free with local materials, which results in different types and designs worldwide [30].

Whereas the biomass used in rural households in areas of low- and middle-income countries is essentially unprocessed, charcoal is the main cooking fuel for millions of urban households in sub-Saharan Africa and is also relevant, although relatively less important, in South Asia and Latin America [31]. Traditionally, charcoal is produced from wood in simple earth and earth mound kilns, with typical efficiencies between 10 and 34%. The efficiencies are often below 20%, depending on the tree species, moisture content of the wood and skill of the charcoal producer [22]. As with unprocessed wood, traditionally, charcoal is burned in simple stoves with low efficiencies, granted that the typical efficiencies of traditional charcoal cooking stoves are generally slightly higher than those of basic wood stoves [22]. The low efficiencies of basic charcoal stoves associated with the low efficiencies of charcoal processing result in an even higher overall impact on wood resources than the use of wood fuel. However, charcoal is increasingly used in urban areas (worldwide, in 1990, 3% of the population in urban areas relied primarily on charcoal for cooking, while in 2020, this proportion increased to 5% [32]).

It is in Africa that charcoal is a far more important part of the cooking fuel mix. In 2020, 15% of the African population relied primarily on charcoal for cooking (3%

Table 2 Types and efficiencies of traditional and improved biomass cooking stoves and charcoal kilns. For comparison typical efficiencies of stoves fuelled with biogas and liquefied petroleum gas (LPG) are given

Type of stove	*Fuel*	*Efficiency (%)*
Three-stone fires	Wood	9–18
Mud stoves	Wood	10–23
Improved stoves	Wood	13–52
Charcoal stoves	Charcoal	15–47
Gas stove	Biogas	57
Gas stove	LPG	54–57
Type of kiln	*Feedstock*	*Efficiency (%)*
Traditional charcoal kilns	Wood	10–34
Improved charcoal kilns	Wood	12–45

in the Eastern Mediterranean, 1.5% in America and less than 1% in the other world regions) [32]. In urban areas in Africa, this share steadily increased from 16% in 1990 to 27% in 2020, while the relevance of unprocessed biomass decreased, which resulted in charcoal surpassing wood fuel in terms of the number of users (Fig. 3). In rural areas in Africa, a vast majority of the population continues to rely primarily on unprocessed biomass (86% in 2020), despite a slight increase in the proportion of the population that primarily relies on charcoal (7% in 2020).

Charcoal is more convenient for urban households because of its higher density compared to wood (*cf.* chapter "Forest Biomass as an Energy Resource"), which makes it easier to store, handle and transport. Additionally, high-quality charcoal generally emits less fine particulate matter [26] and, therefore, helps to improve indoor air quality and reduce health risks associated with prolonged exposure to smoke.

Traditional cooking with biomass is often intertwined with a wide range of other end-uses, such as water preparation, hygiene, lighting or space heating [33]. The latter is particularly important in cold regions, where the same equipment that is used for cooking is frequently used for space heating [34]. In low- and middle-income countries, open and semi-open fires and simple stoves with a primary air inlet are simultaneously used for heating and cooking with no or inefficient chimneys [35]. They are associated with indoor air pollution, have low thermal efficiencies, only provide localised heating, and offer low comfort levels [36].

Dedicated (i.e., not used for cooking) traditional heating stoves with chimneys are used in low- and middle-income countries as well, but also present low efficiencies (for example, Li et al. [37] reported efficiencies of 24% for one of the most popular household stoves used in China fuelled with biomass briquets made of sawdust).

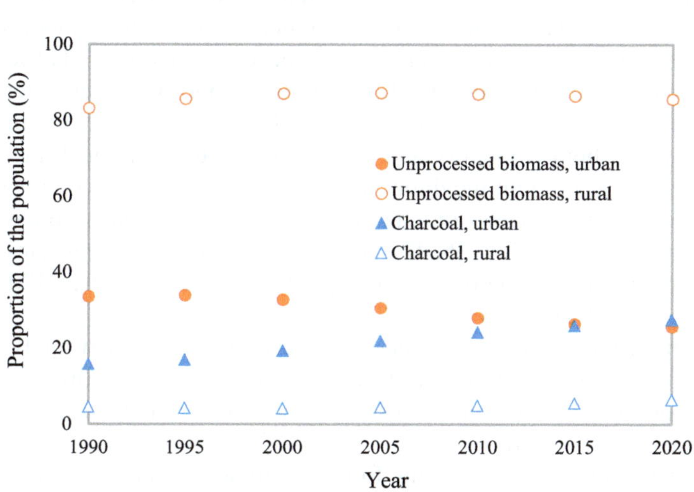

Fig. 3 Proportion of the African population with primary reliance on unprocessed biomass and charcoal for cooking. (*Data source* [32])

The research landscape may vary, but in general, there tend to be more studies on traditional biomass cooking compared to traditional biomass heating.

Traditional, inefficient heating with biomass is not limited to low- or middle-income countries. Heating with firewood in open chimney hearths/fireplaces is still a popular practice in Europe, North America, Australia and New Zealand [35], often as a supplemental heat source and in some cases only used for ambience [38]. Despite the fact that the flue gases emitted by fireplaces are removed outdoors through a chimney, they provide heat locally and in an uncontrolled and inefficient manner. Their efficiency does not exceed 20% [39], but because of air infiltration induced by the draft of the fireplace chimney, the net efficiencies are much lower than gross efficiencies and in cold weather can even be negative [40].

Impacts of Traditional Use of Biomass

Incomplete combustion, characteristic of inefficient biomass burning, results in high levels of indoor pollutants, especially when biomass is also used for space heating in poorly ventilated homes [16]. The smoke produced by biomass combustion includes a large number of chemical compounds with documented adverse health effects, including, but not limited to, carbon monoxide, particulate matter, benzene and polycyclic aromatic carbons.

Pollutant emission factors, a measure of the magnitude of the emissions from a specific source, depend on various parameters, such as the energy conversion equipment used, fuel type, fuel moisture content and operating conditions [41]. Also, different emission sampling strategies may result in different emission factors [42]. These emission factors are essential as input for atmospheric models [42], for the development of emissions inventories at local, regional or global levels [19], to approximate pollutants exposures [23], and to support the development of policies and strategies aimed at reducing pollution [43]. Several studies report emission factors (based on both laboratory and field measurements, although the former is more common) for different fuel/biomass conversion technology combinations and operating conditions. Reviews on emission factors of traditional biomass use can be found, for example, in [17, 35, 42, 44]. For reference, Table 3 presents emission factors for carbon monoxide (CO) and respirable particulate matter ($PM_{2.5}$) based on the works of [23, 26, 29, 43].

Emissions from incomplete biomass combustion have been strongly linked to adverse health problems like chronic obstructive lung disease, acute lower respiratory tract infections, lung cancer, tuberculosis, asthma, cardiovascular diseases, cataracts or low birthweight [45–48].

Additionally, pollutants resulting from incomplete biomass combustion also impact the climate [17, 49]. Many of the pollutants released in the inefficient traditional use of biomass (carbon monoxide, nitrous oxide, methane and other hydrocarbon emissions) have a higher impact on the climate than carbon dioxide because of their higher global warming potentials [19]. Moreover, when referring to the

Table 3 CO and $PM_{2.5}$ emission factors for traditional and improved biomass cooking stoves and traditional open fireplaces

Type of stove	*Fuel*	*CO (g/MJ$_{delivered}$)*	*$PM_{2.5}$ (mg/MJ$_{delivered}$)*
Three-stone fires	Wood	11.2–15.5	719–1498
Improved stoves	Wood	0.1–10.9	25–914
Charcoal stoves	Charcoal	6.4–38.5	33–644
Open fireplaces	*Fuel*	*CO ($g \cdot kg^{-1}$)[a]*	*$PM_{2.5}$ ($g \cdot kg^{-1}$)[a]*
Traditional open fireplaces	Wood	74–114	5.8–14

[a] Expressed in relation to the mass of wood burned on a dry basis

greenhouse gas emissions associated with the use of traditional biomass, the non-renewable fraction of the total wood fuel harvested (fNRB) has to be mentioned. If the quantity of biomass harvested exceeds the annual increment in a specific area, then the biomass harvested is considered non-renewable biomass, and the carbon dioxide emitted in the combustion of the fNRB should be considered for the calculation of greenhouse gas emissions. The accurate estimation of fNRB is a complex, difficult and challenging task because it requires detailed local knowledge of the consumption of wood fuel and of the rates of forest biomass growth (see, for example, [17, 50]).

Particularly important to the climate impact of the traditional use of biomass is the soot emitted by the incomplete combustion of biomass. Soot, also known as black carbon, is a significant air pollutant and a strong absorber of solar radiation, with both local and global impacts [51].

The low efficiencies associated with the traditional use of biomass lead to a huge waste of resources and the need for large amounts of biomass to satisfy the energy needs of the population. Often, it is women and children, especially girls, who are responsible for fuelwood collection. This activity may take a considerable amount of time, reduce schooling hours and attendance and result in several health problems [52–54]. Another consequence of the high demand for wood fuel associated with the traditional use of biomass is the pressure on forest resources. Even though the link between the demand for wood fuel and large-scale deforestation is controversial, the traditional use of biomass has been associated with deforestation in specific places and the consequent environmental degradation [55, 56]. Particularly relevant to the latter seems to be the rapid rise in charcoal consumption, mainly by the urban population in Africa, and the fact that its production is concentrated in order to supply large urban centres [55]. Generally, there is a lack of data-driven evidence on charcoal production and its impacts on forest resources, but recently, several studies based on remote sensing have associated charcoal production as the main driver of forest degradation in sub-Saharan Africa [57]. According to Mwampamba et al. [31], even though charcoal production can be the driver of deforestation, it is most likely a cause of forest degradation (Schoene et al. [58] provide definitions for forest degradation and deforestation).

Improvements on the Traditional Use of Biomass

The traditional use of biomass holds significant potential for improvement [22]. Recognising both this potential and the negative impacts associated with traditional biomass use has spurred the development of improved biomass conversion equipment.

The research conducted in recent decades has primarily focused on reducing pollutants emitted from traditional single-pot cookstoves [59], resulting in the introduction of improved stoves that are normally more efficient, safer, and emit fewer pollutants. Numerous designs with a wide range of performances have been developed worldwide (e.g., [60, 61]). Many offer substantially better efficiencies and lower pollutant emissions than the traditional ones, but some are still relatively inefficient and pollutant (Tables 2 and 3).

Common features of improved stoves are: (*i*) inclusion of a grate under the fuel, (*ii*) enclosing the fire with isolated, low density and specific heat walls, and (*iii*) considering a short internal channel to direct the flue gases at the cooking pot [62]. Over the course of several decades, the designs of cookstoves have undergone significant changes. In the 1980s, cookstoves were primarily characterised by high-mass stoves constructed using local materials [26]. However, as time progressed, there was a shift towards rocket stoves in the 1990s and 2000s, which offered improved efficiency and performance [26]. In more recent years, the development of cookstoves has focused on the careful engineering of rocket, gasifier and forced draft stoves [26].

As cookstove designs have evolved, there has been a corresponding need to develop and update laboratory and field-testing protocols. These protocols include the water boiling test, the controlled cooking test and the kitchen performance test, among others [26]. In 2012, the ISO International Workshop Agreement, IWA 11, was published and served as a preliminary system for rating cookstove performance in laboratories, categorising it in five tiers [63]. These protocols have been revised, and ISO 19867-1:2018 now defines the laboratory measurement and evaluation methods for particulate and gaseous air pollutant emissions, energy efficiency, safety and durability of cookstoves [64]. Additionally, ISO 19869:2019 provides field testing methods to evaluate cooking system performance in real-world conditions [65].

Improved stoves have been introduced in low- and middle-income countries since the 1970s, first in Africa and then in Latin America and Asia [62]. Millions of improved cookstoves have been distributed [66]–in China alone, around 200 million [67]. Progress has been made, but the success of these programs has been limited worldwide and the transition from traditional biomass to modern energy systems has been difficult [17].

Cookstove programs aimed at mitigating the impacts of residential solid fuel combustion have often fallen short of their intended goals, despite promising outcomes observed during laboratory-based cookstove development and testing [68, 69]. Low valuation of the improved cookstoves by households seems to be part of the cause, as reported by Hanna et al. [68], who evaluated an improved stove program run in India. They followed households that received new stoves for four years and showed evidence that these households did not use the improved stoves regularly or appropriately, did not invest sufficiently on maintenance and usage rates declined over time. Differences in stove operation and fuel characteristics also explain the discrepancy between field measurements and laboratory-based results [69]. Additionally, a wide range of different technologies and fuels fall under the term improved stoves, and differences in performance are significant (Tables 2 and 3). Even though the pollutant emissions of advanced biomass stoves are still higher than those of stoves fuelled with gaseous fuels, like LPG and natural gas, advanced pellet stoves performed better than wood and charcoal stoves, approaching LPG, as far as reducing pollutant exposures is concerned [69]. However, users are required to either process their fuels or utilise pre-processed fuels.

Because of their lower pollutant emissions, gaseous fuels have been promoted in numerous countries [70]. Despite the fact that commercial fossil fuel stoves are cleaner and more efficient, rural and low-income urban households do not have easy access and cannot afford commercial fossil fuels [69, 70]. Much of the focus of research and programs is rather on replacing wood fuels with alternative energy sources and not so much on how to secure continued wood fuel supply [55] and use biomass resources more efficiently and in a less polluting way [69], helping people that depend on biomass and cannot afford alternative fuels. According to Arnold et al. [55], it became increasingly evident that there is a need for more holistic support for local tree and forest management as well as addressing local energy requirements. This need arises within the context of broader interventions aimed at promoting development and enhancing livelihoods. Cookstove programs need to reflect local needs and practices and acknowledge the existence of stacking, i.e., that households use multiple devices and fuels along with improved cooking stoves [71]. Recognising that cooking systems encompass more than the stove, important factors to consider are: adherence to local customs, reliability, ease of maintenance, employment opportunities and a comprehensive evaluation of the entire supply chain [70].

3 Modern Uses of Biomass

In high-income countries, residential biomass is almost exclusively used for heating purposes and modern biomass appliances are sometimes seen as luxurious pieces of equipment that not everyone can afford [72]. On the other hand, biomass heating is still commonly used in rural areas in high-income countries and often remains the main source of heating for low- and middle-income people [72]. The seeming contradiction surrounding biomass use arises from the diverse contexts and variations within its usage. Among high-income countries, it is in the group of the European countries where most biomass is used in the residential sector (Table 1).

The need for more efficient and cleaner energy conversion technologies that provide residential heat led to the development of technologically advanced biomass heating technologies with substantially better performance than older technology

[73]. All the traditional and modern biomass installations in high-income countries, such as the United States of America or European Union countries, are obliged to have a chimney that directs the flue gases to the exterior [72]. Nevertheless, residential biomass burning constitutes a significant contributor to atmospheric contaminants in many high-income countries [42] because of older technology and poor firing habits [73].

Residential heating can be achieved through two main methods: small-scale heating appliances and district heating systems. Small-scale heating appliances are specifically designed for individual buildings or smaller spaces, providing localised heating solutions. On the other hand, district heating systems are designed to serve larger areas, such as neighbourhoods. District heating systems have considerable larger capacities and involve a centralised approach to efficiently distribute heat to multiple buildings and consumers, employing technologies similar to those used in industrial heating; therefore, they will be discussed in chapter "Biomass for Industrial and District Heating".

Small-scale residential heating systems can be broadly categorised into two main types: local heating systems and central heating systems (Table 4). While both aim to provide warmth and comfort to homes, they differ in their approaches to heat distribution and overall system design. Local heating systems are designed to provide heat within a specific area or room, whereas central heating systems are designed to simultaneously heat multiple spaces (an entire household or building) from a central heat source located in a suitable room.

Local heating systems rely on equipment that provide heat directly in the room being heated and are suitable for small and intermittently used areas [39]. Their main advantages are the low capital cost, relatively easy installation and simple operation; however, they cannot provide heat uniformly to the room [39]. Some of them, namely stoves, store heat in high-thermal-mass systems in order to provide heat to the room over a longer period of time. Examples of local biomass heating systems are the traditional open fireplace, wood stoves, fireplace inserts and wood pellet stoves.

Table 4 Classification of common modern small-scale solid biomass residential heating appliances

Unit location	Fuel	Operation mode	Unit type
Local heating systems	Wood	Batch	Wood stove
			Fireplace inserts
			Zero-clearance inserts
	Pellets	Automatic	Pellet stoves
			Fireplace inserts
			Zero-clearance inserts
Central heating systems	Wood	Continuous	Biomass boilers
	Wood chips	Automatic	Biomass boilers
	Pellets	Automatic	Biomass boilers
			Forced-air furnaces

Central heating systems, on the other hand, generate heat in a central unit and distribute it to multiple rooms through a piping system. They can be categorised into two main types: hydronic and forced-air systems. In hydronic systems, usually water is used as the heat transfer medium, which is then circulated to radiators (the most common heat emitters in hydronic systems), convectors or underfloor heating pipes [39]. On the other hand, forced-air heating systems use air as the heat transfer medium and deliver it to the rooms through diffusers. Additionally, biomass central heating systems can also provide hot water for sanitary use. Examples of central biomass heating systems comprise wood pellets or wood log boilers.

Small-scale solid biomass heating appliances can also be classified based on their feeding systems, which determine how the fuel is fed into the combustion chamber. Essentially, two types exist: batch feeding systems and automatic feeding systems. Batch feeding systems or manual feeding systems, require users to load the biomass fuel into the combustion chamber manually, while automatic feeding systems supply biomass to the combustion chamber automatically. Typically, in modern wood log boilers, fuel is manually fed into the combustion chamber, but the fuel is continuously fed to the combustion zone by gravity [74].

Fuel type can also be a criterion for the classification of solid biomass residential heating appliances. Small-scale biomass heating appliances use almost exclusively woody biomass [75]. Wood logs, pellets and briquettes are the most common fuels used in modern systems [42]. Some residential biomass heating appliances are designed to operate with a specific fuel, while others allow the use of multiple fuels.

The diversity of small biomass heating systems installed in high-income countries is high, from very simple, older, inefficient installations to modern ones with significantly better performance. Fixed-bed combustion is the most common technology in residential heating systems [76], with staged combustion being typically used in advanced, small-scale residential heating appliances [77]. The next sections describe popular modern biomass heating systems available on the market.

Wood Stoves

Wood stoves are heating appliances designed to burn firewood or briquettes as a fuel source, with the main purpose of generating heat. They are the most popular small-scale biomass heating appliances in the European Union countries and the United States of America [78, 79]. Wood stoves are free-standing appliances that heat the room locally by radiation and convection to their surroundings. The key components of wood stoves are the firebox, combustion systems, flue pipe and chimney connection and air inlet controls.

The firebox is the main chamber, where the wood is burned. It is typically made of durable materials such as cast iron, steel or soapstone [78]. It usually contains a glass window that provides a view of the burning fire. Wood stoves operate in batch mode and typically have a door at the front that can be opened to add wood and remove ashes. However, some models are fed through the top [78].

Modern wood stoves incorporate staged combustion, which allows for efficient and controlled combustion of the wood [75]. Staged-combustion ensures proper oxygen supply for combustion: primary air is directed to the fuel bed to achieve rich combustion, and afterwards, secondary air is supplied to support further lean combustion of the gases and particulate matter that are released during the initial burning process [75]. The burn rate of wood stoves is regulated by controlling the amount of primary combustion air [80]. Typically, secondary air is preheated, and air may be supplied close to the door to protect the glass door from soot and tar [79]. The control or air supply may be manual or automatic with the use of integrated or retrofitted automatic combustion air supply regulation devices [79]. The use of automatic combustion air regulation devices promotes the optimisation of efficiency and emission performance [79]. Numerous designs for wood stoves exist, depending on the flow paths of the air [80]. As an illustrative example of a possible design, Fig. 4 presents a rough sketch of a down-draught wood stove, i.e., a stove where the flow of the primary air into the combustion chamber comes from above the fire.

Many wood stoves are equipped with an ash box located beneath the firebox. This removable tray collects ashes and makes cleaning the stove easier. However, some stoves have no ash grates, and ashes need to be collected directly from the hearth [80].

Wood stoves may include a catalytic combustor in the path of the flue gases as part of their design. The catalytic combustor, which is a device made of a ceramic foam or honeycomb structure coated with a catalyst, enables the combustion of the incomplete combustion products at a lower temperature through a catalytic chemical reaction [79, 81]. Typical advantages are higher efficiencies and lower emissions at

Fig. 4 Rough sketch of a down-draught wood stove (P.A.—primary air; S.A.—secondary air; F.G.—flue gases)

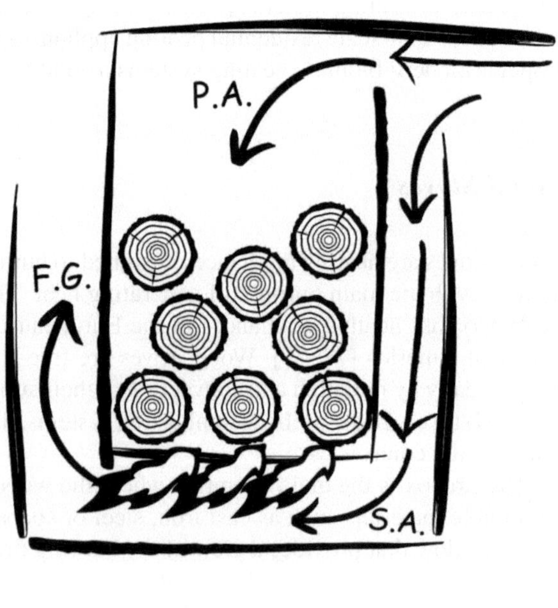

Table 5 Average CO, NO_x and particulate matter emission factors (PM EF) and efficiencies for room heaters from European manufacturers (based on data from [83])

Technology	Efficiency[a] (%)	CO EF ($mg \cdot MJ^{-1}$)	NO_x EF ($mg \cdot MJ^{-1}$)	PM EF ($mg \cdot MJ^{-1}$)
Wood stoves	80	828	91	30
Advanced wood stoves	86	865	87	32
Fireplace inserts	79	842	124	31
Advanced fireplace inserts	86	785	124	27
Pellet stoves	88	128	93	17
Advanced pellet stoves	94	106	94	15

[a] Based on the lower heating value

lower burn rates [81]. However, operating and optimising the efficiency of catalytic wood stoves can be more challenging compared to non-catalytic wood stoves [81].

Modern wood stoves have typical overall thermal efficiencies (based on the high heating value, HHV) between 60 and 80% [82]. Efficiencies and emission factors from a comprehensive study that covers biomass heating appliances from 76 manufacturers from 13 European countries are presented in Table 5.

Other Stoves

Alternatives to traditional masonry fireplaces were developed in order to improve their efficiency and emissions. Fireplace inserts are designed to fit into an existing masonry fireplace opening, while zero-clearance fireplaces are stand-alone units designed to be enclosed. They typically consist of a metal firebox with an insulated glass front door and include the supply of both primary and secondary combustion air to promote better combustion. As in free-standing stoves, the wood is batch-fed, and some models are equipped with catalytic combustors to increase efficiency and reduce emissions caused by incomplete combustion.

Both fireplace inserts and zero-clearance fireplaces resemble stoves, as far as design and combustion principles are concerned, producing comparable emissions as wood stoves [77]. Modern closed fireplaces have typical overall thermal efficiencies (based on the HHV) between 60 and 80% [82]. Table 5 presents average efficiencies and emission factors for fireplace inserts from manufacturers in 13 European countries.

Pellet stoves are a specific type of advanced stove that burns pellets and is suited for continuous operation [84]. They feature a hopper that stores the pellets and a fuel feeder that automatically feeds the pellets into the combustion chamber. Under-feeding and top-feeding are the most common strategies [84]. Additionally, pellet stoves are designed with automatic control of the combustion air and fuel, optimising combustion [85]. As a consequence, and also because of the characteristics of the fuel used (e.g., low moisture content), these type of stoves generally offer lower emissions of incomplete combustion products, but are more expensive than wood log stoves [75]. Multi-fuel units, that combine the use of wood logs with pellets, are also available on the market [75]. Some pellet stoves integrate heat exchangers to produce hot water [84].

A comprehensive study that covers biomass heating appliances from manufacturers from 13 European countries reports an average efficiency of 94% (based on the lower heating value, LHV) for advanced pellet stoves, compared to 88% for other (not so advanced) pellet stoves (Table 5, [83]). Compared to wood stoves and fireplace inserts, pellet stoves exhibit substantially lower average CO and particulate matter emissions (Table 5). Average NO_x emissions from wood and pellet stoves are similar but lower than NO_x from fireplace inserts (Table 5).

Boilers

While biomass stoves are primarily used for direct space heating, generating heat locally within a specific area, biomass boilers are appliances that burn biomass to heat water, which is then distributed throughout multiple rooms for space heating. As a consequence, the capacities of biomass boilers are larger than those of stoves, typically ranging from around 10 kW_{th} to 50 kW_{th}, although smaller and larger boilers exist on the market [86]. Some systems incorporate the burner and the boiler in a single integrated unit, while others are composed of two separate units [86].

Biomass boilers can burn wood logs, briquettes, pellets or woodchips and can broadly be classified as over-fire or under-fire boilers. Over-fire boilers that burn wood logs or briquettes are the simplest and cheapest of the two and are therefore commonly used in households [74]. However, they are no longer the state-of-the-art technology for wood log boilers [84]. In over-fire boilers, the primary air enters the combustion chamber at the bottom, while the secondary air enters at the top, above the fuel bed (Fig. 5). The flue gases then flow through a heat exchanger and transfer energy to heat water. Over-fire boilers normally do not require ventilation and are characterised by relatively low temperatures and the fact that combustion is incomplete, especially when the boilers are operated at partial load [77]. However, some are connected to water tanks, which enables them to operate at nominal power and, consequently, reduce emissions [80].

The efficiency of over-fire wood log boilers is similar to that of traditional stoves (between 50 and 65%) [74].

The state-of-the-art combustion principle for wood log boilers is under-fire burning [84]. As the name indicates, in under-fire boilers, the first stage of wood combustion occurs at the bottom of the fuel bed in the first part of the combustion

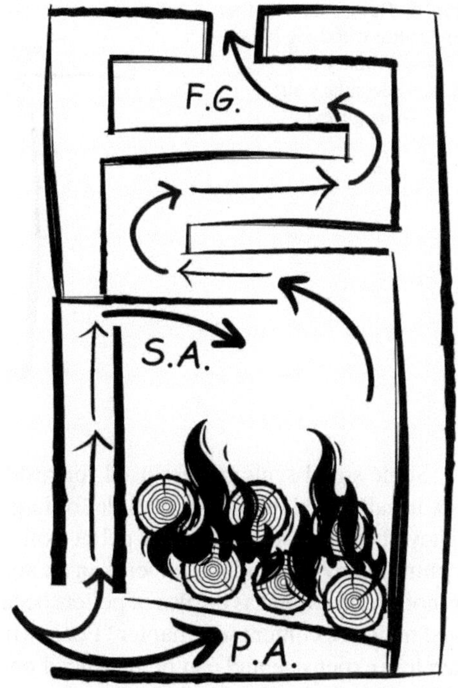

Fig. 5 Rough sketch of an over-fire wood log boiler (P.A.—primary air; S.A.—secondary air; F.G.—flue gases)

chamber (Fig. 6). This compartment is also where the fuel is stored [77]. After partial combustion, the incomplete combustion products are drawn downward or sideways, and a second combustion stage takes place in a different compartment [84]. In modern appliances, primary and secondary air are introduced in the combustion chambers by a fan [84] and oxygen sensors and an advanced control system are used to promote efficient and clean combustion [77]. The use of storage tanks reduces cycling and, therefore, increases the efficiency and reduces emissions associated with wood-log boilers [38]. Modern wood boilers have typical overall thermal efficiencies (based on the high heating value, HHV) between 70 and 90% [82].

Pellet boilers are characterised by the automatic supply of pellets to the primary combustion zone. The pellets are stored in a fuel tank and are usually fed from the top, even though other feeding systems (e.g., horizontally-fed) are also used [86]. Contrary to wood-log boilers, the burning rate is controlled by the supply of fuel and not of primary air [80]. Modern pellet boilers also have an automatic supply of primary and secondary combustion air, which is done by fans. The fully automatic and continuous combustion of an upgraded fuel enables very high efficiencies and lower emissions, similar to those of liquid fuel boilers [74]. Modern pellet boilers have typical overall thermal efficiencies (based on the high heating value, HHV) between 60 and 90% [82]. Typical current emissions are $EF_{CO} = 30$ mg·MJ^{-1}, $EF_{NOx} = 95$ mg·MJ^{-1}, $EF_{PM} = 20$ mg·MJ^{-1} [75].

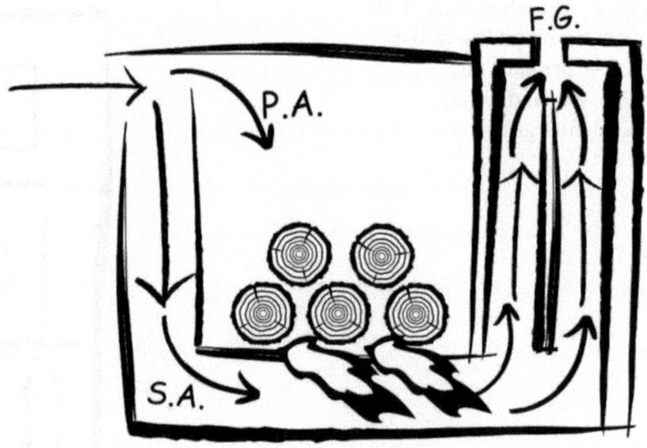

Fig. 6 Rough sketch of an under-fire wood log boiler (P.A.—primary air; S.A.—secondary air; F.G.—flue gases)

Some small-scale boilers used for residential heating burn wood chips, but they are usually used in the countryside for larger houses [80] and higher heat flows [84]. They share the advantages of pellet boilers: automatic operation with combustion controlled by fuel supply, rather than air supply [80]. However, the use of wood chips is not as convenient as the use of pellets because the fuel presents a higher bulk density and moisture content (*cf.* chapter "Forest Biomass as an Energy Resource"). Yet, they are less expensive and can be produced onsite [84]. The technology used in modern wood chip boilers is similar to that of modern pellet boilers [80], but horizontal feed burners and underfeed burners are more common than top feed burners [86].

Other Technologies

In regions where natural gas is an important energy source in the residential sector, one of the strategies to increase the integration of renewable energies in households may be the substitution of natural gas by green gas [87]. This would take advantage of the existing infrastructure for distribution and consumption of natural gas and would require no technological adjustments on the demand side. As Miedema et al. [87] argue, these adjustments are limited by social and economic barriers: (i) tenants may show little interest in the energy performance of their house; (ii) owners face little financial means for investments in another technology; and (iii) may feel no economic incentive to switch from natural gas to another heating technology. In this context, a technological change on the supply side would be easier.

Green gas can potentially be obtained by anaerobic digestion or gasification of forest biomass, but none of the technologies is yet commercially available (*cf.* chapter "Forest Biomass as an Energy Resource").

4 Final Considerations

In terms of energy consumption, the residential sector is among the most significant, accounting for 25% of the total global final energy consumption and contributing to approximately 12% of total direct greenhouse gas emissions from end-use sectors. The relative importance of households in total final energy consumption varies among the different world regions, with the highest share occurring in Africa and the lowest share in Asian and Oceanian countries belonging to the OECD. Energy consumption in the residential sector is dependent on a combination of different climatic, technological and socioeconomic aspects, but globally, heating is the end-use that consumes the most energy.

Primary solid biofuels are currently the most commonly used energy sources in households, representing 32% of the energy consumed in the residential sector worldwide in 2020. They comprise different fuels such as unprocessed wood, charcoal, pellets, briquettes or agriculture residues, but globally, woody biomass, in its various forms, is the most used solid biofuel in the residential sector (in some regions, with smaller shares of forest area, forest biomass tends to have less relevance than agricultural biomass and animal dung).

The use of biomass in the residential sector is very diverse. In low- and middle-income countries, it is mostly used in a traditional, inefficient way by households, while in high-income countries, highly-efficient and low-emission appliances are readily available on the market and are often used with high-quality, upgraded wood products.

Open fires and inefficient conventional stoves are the basic technologies associated with the traditional use of biomass. In rural areas in low- and middle-income countries, biomass is primarily used with little or no processing, while charcoal, produced from wood in simple earth and earth mound kilns, is the main cooking fuel for millions of urban households in sub-Saharan Africa, South Asia and Latin America. Charcoal is more convenient for urban households due to its higher density and lower fine particulate matter emissions.

Traditional cooking with biomass is often intertwined with other end-uses, such as water preparation, hygiene, lighting or space heating. In low- and middle-income countries, open and semi-open fires and simple stoves are used for heating and cooking with no or inefficient chimneys. Dedicated (i.e., not used for cooking) traditional heating stoves with chimneys are used in low- and middle-income countries as well but also present low efficiencies. Traditional, inefficient heating with biomass is not limited to low- or mid-income countries; in high-income countries, it is associated with poor communities or used as a complement to other heating appliances.

The traditional use of biomass is characterised by incomplete combustion, which results in high levels of indoor pollutants, especially when used in poorly ventilated homes. Pollutant emissions depend on parameters such as energy conversion equipment, fuel type, fuel moisture content and operating conditions. The traditional use of biomass has been linked to adverse health problems, negative impacts on the climate, forest degradation and social and gender inequalities. Improved biomass conversion equipment has been developed to reduce fuel consumption and pollutants emitted from traditional technologies. Progress has been made, but the transition from traditional biomass to modern energy systems has been difficult because of limited access to modern energy infrastructures, affordability and cost considerations, a lack of awareness and knowledge, cultural and behavioural factors and/or various technical and operational challenges.

In high-income countries, biomass is almost exclusively used for residential heating. Both biomass district heating systems (addressed in the next chapter) and small-scale heating appliances are used. The diversity of residential heating systems available on the market is high, but modern appliances that burn solid biomass typically have in common the use of staged combustion, which enables a more efficient and cleaner conversion of biomass. Staged combustion is characterised by sequential combustion in multiple zones. In the first zone (primary combustion zone), the solid biomass is burned with a limited supply of air, which promotes the release of volatile gases and the partial combustion of the fuel. The products of the primary combustion zone are then further combusted in a different zone of the combustion chamber, where additional air is introduced. Advanced appliances incorporate forced draft of combustion air and advanced control systems. Common fuels used in modern heating appliances are wood logs, briquettes and pellets.

Small-scale biomass heating appliances broadly fall into two categories: central heating and local heating systems. Local heating appliances are designed to provide heat in a specific area or room. They are suitable for small and intermittently used areas with relatively low capital cost, relatively easy installation and simple operation. Examples of free-standing appliances are wood stoves and pellet stoves, and examples of appliances designed to be enclosed are fireplace inserts and zero clearance fireplaces. On the other hand, central heating systems can heat multiple rooms at once. They include a central heat source located in a suitable location and a distribution system that transfers heat to the various rooms. The heat transfer medium is usually water, but air is also used. Pellet boilers are among the cleanest and most efficient small-scale biomass heating appliances.

Acknowledgements The work was supported by *Fundação para a Ciência e a Tecnologia*, through IDMEC, under LAETA [project UIDB/50022/2020]. Isabel Malico expresses her gratitude to Lucas Saffian for drawing the rough sketches included in this chapter.

References

1. IEA (2022) Energy statistics data browser. https://www.iea.org/data-and-statistics/data-tools/energy-statistics-data-browser. Accessed 8 May 2023
2. Sathaye JA, Meyers S (1995) Residential and commercial sectors. In: Greenhouse gas mitigation assessment: a guidebook. Springer, Dordrecht
3. IEA (2022) World energy outlook 2022. International Energy Agency, Paris
4. IEA (2004) Oil crises and climate challenges: 30 years of energy use in IEA countries. International Energy Agency, Paris

5. Zhang Q (2004) Residential energy consumption in China and its comparison with Japan, Canada, and USA. Energy Build 36:1217–1225. https://doi.org/10.1016/j.enbuild.2003.08.002
6. Howden-Chapman P, Viggers H, Chapman R et al (2009) Warm homes: drivers of the demand for heating in the residential sector in New Zealand. Energy Policy 37:3387–3399. https://doi.org/10.1016/j.enpol.2008.12.023
7. Ürge-Vorsatz D, Cabeza LF, Serrano S et al (2015) Heating and cooling energy trends and drivers in buildings. Renew Sustain Energy Rev 41:85–98. https://doi.org/10.1016/j.rser.2014.08.039
8. Reyes R, Nelson H, Navarro F, Retes C (2015) The firewood dilemma: Human health in a broader context of well-being in Chile. Energy Sustain Dev 28:75–87. https://doi.org/10.1016/j.esd.2015.07.005
9. Reames TG (2016) Targeting energy justice: exploring spatial, racial/ethnic and socioeconomic disparities in urban residential heating energy efficiency. Energy Policy 97:549–558. https://doi.org/10.1016/j.enpol.2016.07.048
10. S2Biom (2016) Database for standardized biomass characterization. https://s2biom.wenr.wur.nl/web/guest/biomass-characteristics. Accessed 14 Apr 2023
11. IEA, IRENA, UNSD, et al (2022) Tracking SDG 7: the energy progress report. World Bank, Washington DC
12. Malico I, Mujeebu MA (2015) Potential of porous media combustion technology for household applications. Int J Adv Thermofluid Res 1:45–62. https://doi.org/10.51141//IJATR.2015.1.4
13. Eurostat (2023) Eurostat database. https://ec.europa.eu/eurostat/web/main/data/database. Accessed 13 May 2023
14. Goldemberg J, Teixeira Coelho S (2004) Renewable energy—traditional biomass versus modern biomass. Energy Policy 32:711–714. https://doi.org/10.1016/S0301-4215(02)00340-3
15. Victor NM, Victor DG (2002) Macro patterns in the use of traditional biomass fuels. Stanford University, Stanford
16. Karekezi S, Lata K, Coelho ST (2004) Traditional biomass energy: improving its use and moving to modern energy use. In: Thematic background paper for the international conference for renewable energies. Bonn, pp 1–57
17. Masera OR, Bailis R, Drigo R et al (2015) Environmental burden of traditional bioenergy use. Annu Rev Environ Resour 40:121–150. https://doi.org/10.1146/annurev-environ-102014-021318
18. Schenkel Y, Bertaux P, Vanwijnbserghe S, Carre J (1998) An evaluation of the mound kiln carbonization technique. Biomass Bioenergy 14:505–516. https://doi.org/10.1016/S0961-9534(97)10033-2
19. Smith KR, Uma R, Kishore VVN et al (2000) Greenhouse implications of household stoves: an analysis for India. Annu Rev Energy Environ 25:741–763. https://doi.org/10.1146/annurev.energy.25.1.741
20. Pennise DM, Smith KR, Kithinji JP et al (2001) Emissions of greenhouse gases and other airborne pollutants from charcoal making in Kenya and Brazil. J Geophys Res Atmospheres 106:24143–24155. https://doi.org/10.1029/2000JD000041
21. Bhattacharya SC, Albina DO, Salam PA (2002) Emission factors of wood and charcoal-fired cookstoves. Biomass Bioenergy 23:453–469
22. Smeets E, Johnson FX, Ballard-Tremeer G (2012) Keynote introduction: traditional and improved use of biomass for energy in Africa. In: Janssen R, Rutz D (eds) Bioenergy for sustainable development in Africa. Springer, Dordrecht, pp 3–12
23. Jetter J, Zhao Y, Smith KR et al (2012) Pollutant emissions and energy efficiency under controlled conditions for household biomass cookstoves and implications for metrics useful in setting international test standards. Environ Sci Technol 46:10827–10834. https://doi.org/10.1021/es301693f
24. Singh P, Gundimeda H (2014) Life cycle energy analysis (LCEA) of cooking fuel sources used in India households. Energy Environ Eng 2:20–30. https://doi.org/10.13189/eee.2014.020103
25. Sweeney D (2015) Field evaluation of alternative and traditional cooking fuels in Haiti. Massachussetts Institute of Technology, Cambridge

26. Still D, Bentson S, Li H (2015) Results of laboratory testing of 15 cookstove designs in accordance with the ISO/IWA tiers of performance. EcoHealth 12:12–24. https://doi.org/10.1007/s10393-014-0955-6
27. Jagger P, Pedit J, Bittner A et al (2018) Fuel efficiency and air pollutant concentrations of wood-burning improved cookstoves in Malawi: implications for scaling-up cookstove programs. Energy Sustain Dev 41:112–120. https://doi.org/10.1016/j.esd.2017.08.007
28. Rodrigues T, Braghini Junior A (2019) Charcoal: a discussion on carbonization kilns. J Anal Appl Pyrolysis 143:104670. https://doi.org/10.1016/j.jaap.2019.104670
29. Champion WM, Hays MD, Williams C et al (2021) Cookstove emissions and performance evaluation using a new ISO protocol and comparison of results with previous test protocols. Environ Sci Technol 55:15333–15342. https://doi.org/10.1021/acs.est.1c03390
30. Adria O, Bethge J (2013) What users can save with energy-efficient cooking stoves and ovens. Wuppertal Institute for Climate, Environment and Energy, Wuppertal
31. Mwampamba TH, Ghilardi A, Sander K, Chaix KJ (2013) Dispelling common misconceptions to improve attitudes and policy outlook on charcoal in developing countries. Energy Sustain Dev 17:75–85. https://doi.org/10.1016/j.esd.2013.01.001
32. WHO (2023) Household energy database. https://www.who.int/data/gho/data/themes/air-pollution/who-household-energy-db. Accessed 25 May 2023
33. Grabher HF, Rau H, Ledermann ST, Haberl H (2023) Beyond cooking: an energy services perspective on household energy use in low and middle income countries. Energy Res Soc Sci 97:102946. https://doi.org/10.1016/j.erss.2023.102946
34. WHO (2016) Burning opportunity: clean household energy for health, sustainable development, and wellbeing of women and children. World Health Organization, Geneva
35. Carvalho RL, Jensen OM, Tarelho LAC (2016) Mapping the performance of wood-burning stoves by installations worldwide. Energy Build 127:658–679. https://doi.org/10.1016/j.enbuild.2016.06.010
36. World Bank (2020) Fueling Kyrgyzstan's transition to clean household heating solutions. World Bank, Washington, DC
37. Li Q, Jiang J, Qi J et al (2016) Improving the energy efficiency of stoves to reduce pollutant emissions from household solid fuel combustion in China. Environ Sci Technol Lett 3:369–374. https://doi.org/10.1021/acs.estlett.6b00324
38. Trojanowski R, Fthenakis V (2019) Nanoparticle emissions from residential wood combustion: a critical literature review, characterization, and recommendations. Renew Sustain Energy Rev 103:515–528. https://doi.org/10.1016/j.rser.2019.01.007
39. Martinopoulos G, Papakostas KT, Papadopoulos AM (2018) A comparative review of heating systems in EU countries, based on efficiency and fuel cost. Renew Sustain Energy Rev 90:687–699. https://doi.org/10.1016/j.rser.2018.03.060
40. Sonderegger RC, Condon PE, Modera MP (1980) In-situ measurements of residential energy performance using electric co-heating. In: Proceedings of the ASHRAE semiannual meeting. Los Angeles, CA
41. Bhattacharya SC, Albina DO, Khaing AM (2002) Effects of selected parameters on performance and emission of biomass-fired cookstoves. Biomass Bioenergy 23:387–395
42. Vicente ED, Alves CA (2018) An overview of particulate emissions from residential biomass combustion. Atmospheric Res 199:159–185. https://doi.org/10.1016/j.atmosres.2017.08.027
43. Calvo AI, Martins V, Nunes T et al (2015) Residential wood combustion in two domestic devices: relationship of different parameters throughout the combustion cycle. Atmos Environ 116:72–82. https://doi.org/10.1016/j.atmosenv.2015.06.012
44. Shen H, Luo Z, Xiong R et al (2021) A critical review of pollutant emission factors from fuel combustion in home stoves. Environ Int 157:106841. https://doi.org/10.1016/j.envint.2021.106841
45. Naeher LP, Brauer M, Lipsett M et al (2007) Woodsmoke health effects: a review. Inhal Toxicol 19:67–106. https://doi.org/10.1080/08958370600985875
46. Fullerton DG, Bruce N, Gordon SB (2008) Indoor air pollution from biomass fuel smoke is a major health concern in the developing world. Trans R Soc Trop Med Hyg 102:843–851. https://doi.org/10.1016/j.trstmh.2008.05.028

47. Noubiap JJN, Essouma M, Bigna JJR (2015) Targeting household air pollution for curbing the cardiovascular disease burden: a health priority in sub-Saharan Africa. J Clin Hypertens 17:825–829. https://doi.org/10.1111/jch.12610
48. Bede-Ojimadu O, Orisakwe OE (2020) Exposure to wood smoke and associated health effects in sub-Saharan Africa: a systematic review. Ann Glob Health 86:32. https://doi.org/10.5334/aogh.2725
49. Lelieveld J, Klingmüller K, Pozzer A et al (2019) Effects of fossil fuel and total anthropogenic emission removal on public health and climate. Proc Natl Acad Sci 116:7192–7197. https://doi.org/10.1073/pnas.1819989116
50. Bailis R, Pemberton-Pigott C (2020) Fraction of non-renewable biomass in emission crediting in clean and efficient cooking projects. A review of concepts, rules, and challenges. The World Bank, Washington DC
51. Malico I, Nepomuceno Pereira S, Costa MJ (2017) Black carbon trends in southwestern Iberia in the context of the financial and economic crisis. The role of bioenergy. Environ Sci Pollut Res 24:476–488. https://doi.org/10.1007/s11356-016-7805-8
52. Oparaocha S, Dutta S (2011) Gender and energy for sustainable development. Curr Opin Environ Sustain 3:265–271. https://doi.org/10.1016/j.cosust.2011.07.003
53. Oluwole O, Otaniyi OO, Ana GA, Olopade CO (2012) Indoor air pollution from biomass fuels: a major health hazard in developing countries. J Public Health 20:565–575. https://doi.org/10.1007/s10389-012-0511-1
54. Kammila S, Kappen JF, Rysankova D et al (2014) Clean and improved cooking in Sub-Sahran Africa, 2nd edn. The World Bank, Washington, DC
55. Arnold M, Köhlin G, Persson R, Shepherd G (2003) Fuelwood revisited: what has changed in the last decade? Center for International Forestry Research, Bogor
56. Bailis R, Wang Y, Drigo R et al (2017) Getting the numbers right: revisiting woodfuel sustainability in the developing world. Environ Res Lett 12:115002. https://doi.org/10.1088/1748-9326/aa83ed
57. Sedano F, Mizu-Siampale A, Duncanson L, Liang M (2022) Influence of charcoal production on forest degradation in Zambia: a remote sensing perspective. Remote Sens 14:3352. https://doi.org/10.3390/rs14143352
58. Schoene D, Killmann W, von Lüpke H, LoycheWilkie M (2007) Definitional issues related to reducing emissions from deforestation in developing countries. Food and Agriculture Organization of the United Nations, Rome
59. Pande RR, Kalamkar VR, Kshirsagar M (2019) Making the popular clean: improving the traditional multipot biomass cookstove in Maharashtra, India. Environ Dev Sustain 21:1391–1410. https://doi.org/10.1007/s10668-018-0092-4
60. Amrose S, Kisch GT, Hirubi C, et al (2008) Development and testing of the Berkeley Darfur stove. Lawrence Berkeley National Laboratory, Berkeley, CA
61. Clean Cooking Alliance (2023) Clean Cooking Catalog. http://catalog.cleancookstoves.org/. Accessed 1 Jun 2023
62. Kshirsagar MP, Kalamkar VR (2014) A comprehensive review on biomass cookstoves and a systematic approach for modern cookstove design. Renew Sustain Energy Rev 30:580–603. https://doi.org/10.1016/j.rser.2013.10.039
63. ISO (2012) IWA 11:2012. Guidelines for evaluating cookstove performance. https://www.iso.org/standard/61975.html. Accessed 3 Jun 2023
64. ISO (2018) ISO 19867-1:2018. Clean cookstoves and clean cooking solutions—Harmonized laboratory test protocols—Part 1: Standard test sequence for emissions and performance, safety and durability. https://www.iso.org/standard/66519.html. Accessed 3 Jun 2023
65. ISO (2019) ISO 19869:2019. Clean cookstoves and clean cooking solutions—Field testing methods for cookstoves. https://www.iso.org/standard/66521.html. Accessed 3 Jun 2023
66. Amegah AK, Jaakkola JJ (2016) Household air pollution and the sustainable development goals. Bull World Health Organ 94:215–221. https://doi.org/10.2471/BLT.15.155812
67. Bruce N, Rehfuess E, Mehta S et al (2006) Indoor air pollution. In: Jamison DT, Breman JG, Measham AR et al (eds) Disease control priorities in developing countries, 2nd edn. The

International Bank for Reconstruction and Development/The World Bank Group, Washington, DC, pp 793–815
68. Hanna R, Duflo E, Greenstone M (2016) Up in smoke: the influence of household behavior on the long-run impact of improved cooking stoves. Am Econ J Econ Policy 8:80–114. https://doi.org/10.1257/pol.20140008
69. Champion WM, Grieshop AP (2019) Pellet-fed gasifier stoves approach gas-stove like performance during in-home use in Rwanda. Environ Sci Technol 53:6570–6579. https://doi.org/10.1021/acs.est.9b00009
70. Wright C, Sathre R, Buluswar S (2020) The global challenge of clean cooking systems. Food Secur 12:1219–1240. https://doi.org/10.1007/s12571-020-01061-8
71. Ruiz-Mercado I, Masera O (2015) Patterns of stove use in the context of fuel–device stacking: rationale and implications. EcoHealth 12:42–56. https://doi.org/10.1007/s10393-015-1009-4
72. Carvalho RLT de (2016) Wood-burning stoves worldwide: technology, innovation and policy. PhD Thesis, Aalborg University
73. Illerup JB, Hansen BB, Lin W et al (2020) Performance of an automatically controlled wood stove: thermal efficiency and carbon monoxide emissions. Renew Energy 151:640–647. https://doi.org/10.1016/j.renene.2019.11.057
74. EEA (2019) Chapter 1.A.4 Small combustion 2019. In: EMEP/EEA air pollutant emission inventory guidebook 2019. European Environment Agency, Copenhagen, Denmark, pp 1–189
75. Anca-Couce A, Hochenauer C, Scharler R (2021) Bioenergy technologies, uses, market and future trends with Austria as a case study. Renew Sustain Energy Rev 135:110237. https://doi.org/10.1016/j.rser.2020.110237
76. Míguez JL, Morán JC, Granada E, Porteiro J (2012) Review of technology in small-scale biomass combustion systems in the European market. Renew Sustain Energy Rev 16:3867–3875. https://doi.org/10.1016/j.rser.2012.03.044
77. Kubica K, Paradiz B, Dilara P (2007) Small combustion installations: techniques, emissions and measures for emission reduction. Publications Office of the European Union, Luxembourg
78. Podschelne C, Ciolkosz D (2013) Residential wood heat in the Northeast. In: Jacobson M, Ciolkosz D (eds) Wood-based energy in the northern forests. Springer, New York, New York, NY, pp 111–124
79. Reichert G, Schmidl C (2023) SWOT analysis of non-technical and technical measures towards "(nearly) zero-emission stove technologies." Energies 16:1388. https://doi.org/10.3390/en16031388
80. van Loo S, Koppejan J (2008) The handbook of biomass combustion and co-firing. Earthscan, London; Sterling, VA
81. Ghorashi SA, Khandelwal B (2023) Toward the ultra-clean and highly efficient biomass-fired heaters. A review. Renew Energy 205:631–647. https://doi.org/10.1016/j.renene.2023.01.109
82. EPA (2023) EPA-certified wood stove database. https://cfpub.epa.gov/oarweb/woodstove/index.cfm?fuseaction=app.about
83. Schieder W, Storch A, Fischer D, et al (2013) Luftschadstoffausstoß von Festbrennstoff-Einzelöfen. Untersuchung des Einflusses von Festbrennstoff-Einzelöfen auf den Ausstoß von Luftschadstoffen. Umweltbundesamt, Vienna
84. Musil-Schäffer B, McCarry A, Schmidl C, Haslinger W (2010) European wood-heating technology survey: an overview of combustion principles and the energy and emissions performance characteristics of commercially available systems in Austria, Germany, Denmark, Norway and Sweden. New York State Energy Research and Development Authority (NYSERDA), Albany, NY
85. Priestley M, Kong X, Pei X et al (2023) Pros and cons of wood and pellet stoves for residential heating from an emissions perspective. Environ Sci Atmospheres 3:717–730. https://doi.org/10.1039/D2EA00022A
86. Feldmeier S, Schwarz M, Wopienka E, Pfeifer C (2021) Categorization of small-scale biomass combustion appliances by characteristic numbers. Renew Energy 163:2128–2136. https://doi.org/10.1016/j.renene.2020.10.111

87. Miedema JH, van der Windt HJ, Moll HC (2018) Opportunities and barriers for biomass gasification for green gas in the Dutch residential sector. Energies 11:2969. https://doi.org/10.3390/en11112969

Biomass for Industrial and District Heating

Isabel Malico

Abstract The industrial sector, the world's largest energy consuming end-user, is a major greenhouse gas emitter. It heavily relies on fossil fuels, with only a small contribution from renewables, and of these, only biomass (mainly primary solid biofuels) is not marginal at a global scale. Several factors contribute to the limited adoption of renewables within the industry. The sector's extraordinary diversity and complexity make a one-size-fits-all solution impossible. Industrial energy consumption varies significantly among different sub-sectors and even within each sub-sector, depending on production composition and industrial processes. Energy-intensive industries typically consume substantial amounts of process heat, while non-energy-intensive ones tend to rely more on electricity. Given the importance of energy-intensive industrial sub-sectors, finding solutions to decarbonise process heat is crucial. Process heat encompasses various applications, technologies, energy sources, temperatures and delivery methods. There is substantial demand for high-temperature process heat (>500 °C), with only a limited number of renewable energy options available, including bioenergy. Bioenergy holds the potential to contribute to the decarbonisation of industry but requires tailored solutions for each sub-sector and context. This chapter presents key commercially available biomass heat production systems, which vary in configuration, technologies and scale, with similarities to district heating systems, also discussed.

Keywords Process heat · High-temperature heat · Bioenergy · Conversion technologies · Combustion · Gasification · Co-combustion

I. Malico (✉)
IDMEC, Escola de Ciências e Tecnologia, Departamento de Engenharia Mecatrónica, Universidade de Évora, Rua Romão Ramalho 59, 7000-671 Évora, Portugal
e-mail: imbm@uevora.pt

A. C. Gonçalves and I. Malico (eds.), *Forest Bioenergy*, Green Energy and Technology,
https://doi.org/10.1007/978-3-031-48224-3_9

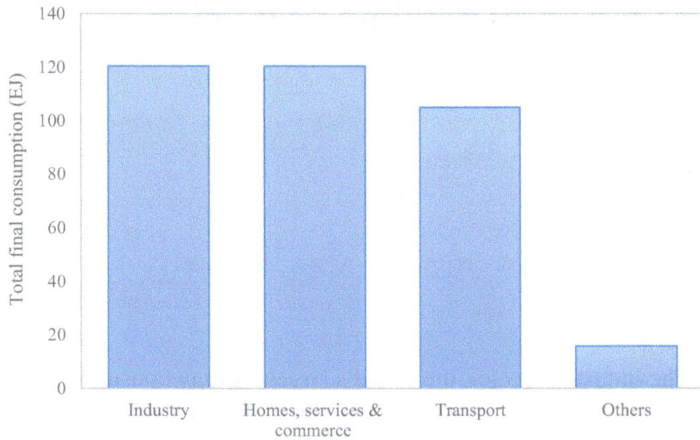

Fig. 1 Total final energy consumption in the world in 2020 by end-use. (*Data source* [1])

1 Introduction

In 2020, the industrial sector accounted for one-third of the world total final energy consumption, approximately 120 EJ (Fig. 1), slightly surpassing the combined consumption of the residential, commercial and public service sectors, which also totalled around 120 EJ [1].[1] With this consumption, industries were the world's largest energy consuming end-use sector, closely followed by the group formed by the residential, commercial and public service sectors (addressed in chapter "Biomass for Domestic Heat").

While the industrial and the residential, commercial and public services sectors have similar final energy consumptions, industries account for more than twice the share of total direct greenhouse gas emissions from end-use sectors compared to the residential, commercial and public buildings (37% versus 16%) (Fig. 2). The term "direct" excludes indirect emissions from the electricity and heat generation consumed in the end-use sectors. The discrepancy between these two sectors reflects the difficulties in the penetration of renewable energy sources in the industrial sector, a challenge addressed in this chapter.

Given the significant contribution of the industrial sector to global greenhouse gas emissions, decarbonising this sector becomes crucial in order to reduce greenhouse gas emissions and keep global warming well below the 2 °C threshold above pre-industrial levels, achieving climate goals [2].

The significance of the industrial sector in total final energy consumption varies across different world regions. China, a highly industrialised country, has the highest share (Fig. 3), with over half of its total final energy consumption (54%) attributed

[1] Note that these figures exclude the non-energy use of fossil fuels (for example, the fuels used as feedstocks to make products such as plastics and chemicals or bitumen used as road surface).

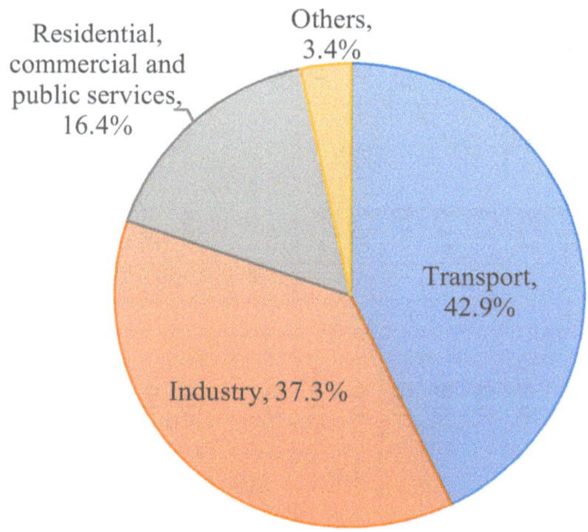

Fig. 2 Share of the different end-use sectors in the greenhouse gas emissions from end-use sectors in the world in 2020. (*Data source* [1])

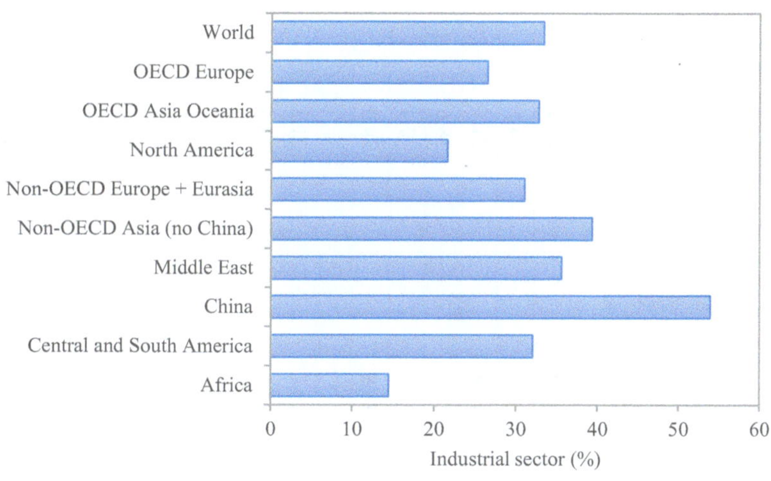

Fig. 3 Share of the industrial sector in the total final energy consumption for different regions in 2020. (*Data source* [1])

to industry in 2020 [1]. In contrast, Africa has the lowest share, with less than 15% of its total final energy consumption dedicated to industrial activities [1].

In 2020, biofuels and waste accounted for just 8.4% of the energy consumed in the global industrial sector [1] (Table 1). Despite this relatively modest contribution from bioenergy, it stands out as the sole renewable energy source (RES) with substantial direct use by the industry worldwide. The industrial sector primarily relies on fossil fuels to meet its energy needs, with nearly 60% of the total energy used in global industry in 2020 derived directly from fossil fuels, mostly coal, followed by natural

Table 1 Share (in %) of the different energy sources in the world industrial energy consumption in 1990 and 2020. (*Data source* [1])

	Coal	Crude oil + oil products	Natural gas	Biofuels and waste	Other RES	Electricity	Heat
1990							
World	25.4	18.5	20.2	5.7	0.0	21.6	8.6
Africa	26.4	26.8	8.7	15.5	0.0	22.6	0.0
Central and South America	7.5	26.0	15.2	31.1	0.0	21.2	0.0
China	72.6	8.9	1.1	0.0	0.0	12.8	4.6
Middle East	0.4	41.9	49.4	0.0	0.0	8.3	0.0
Non-OECD Asia excluding China	38.5	20.8	4.5	21.5	0.0	14.5	0.2
Non-OECD Europe and Eurasia	14.3	12.8	21.7	0.0	0.0	19.0	32.2
North America	14.0	16.4	38.6	4.7	0.0	26.2	0.2
OECD Asia and Oceania	22.1	35.8	7.4	2.8	0.1	31.8	0.0
OECD Europe	21.2	18.8	24.0	4.4	0.0	26.7	5.0
2020							
World	26.3	10.4	21.2	8.4	0.0	28.4	5.2
Africa	16.1	18.5	20.6	18.9	0.0	26.0	0.0
Central and South America	7.6	17.1	15.3	34.5	0.0	25.5	0.0
China	42.8	8.1	9.1	0.0	0.0	32.7	7.3
Middle East	2.0	14.8	70.3	0.0	0.0	12.8	0.0
Non-OECD Asia excluding China	35.0	11.7	9.8	21.1	0.0	22.0	0.5
Non-OECD Europe and Eurasia	20.0	7.8	25.7	2.2	0.0	22.3	22.1
North America	5.2	8.4	47.4	11.4	0.0	26.2	1.5
OECD Asia and Oceania	18.5	17.1	17.4	5.7	0.1	39.2	2.0
OECD Europe	8.4	10.1	30.9	10.7	0.1	34.1	5.7

gas [1]). The share of electricity used by the industries has been increasing in importance, and in 2020, this energy vector was the most used energy source, accounting for 28.4% of the industrial energy consumption [1]. Some of this electricity, and also of the derived heat, comes from renewable energy sources; therefore, the use of renewable energies in the industrial sector was higher than what can be directly observed in Table 1. However, global electricity generation continues to rely on fossil fuels (*cf.* chapter "Biomass for Power Production and Cogeneration"), as does the generation of derived heat (*cf.* Sect. 4).

In 2020, Central and South America (35%), non-OECD[2] Asia (excluding China) (21%) and Africa (19%) stood out as regions where biofuels and waste had the most significant share in the industrial energy consumption [1]. These regions also had important biomass consumptions in the residential sector (*cf.* chapter "Biomass for Domestic Heat"), and Central and South America was the world region with the highest share of biomass in the energy sector (*cf.* chapter "Biomass for Power Production and Cogeneration"). Combined, non-OECD Asia and Central and South America accounted for 60% of the biofuels and waste used by the industry in 2020, followed by North America (16%) and OECD Europe (13%) [1].

The relative importance of fossil fuels in the industrial sector has been declining, albeit at a slow rate (averaging 0.4% per year over the last three decades [1]). However, between 1990 and 2020, the consumption of fossil fuels by the industry increased 45%, from 48 to 70 EJ, following the growth in energy consumption by the industry [1]. During this period, crude oil and oil products were the only fossil energy sources to experience a decrease in consumption, while the industrial demand for coal and natural gas increased, both in quantity and share (even though, in the last decade, global consumption of coal by the industry has been decreasing (Fig. 4)).

The consumption of biofuels and waste by the global industry in 2020 (10 EJ) was approximately 2.5 times higher than in 1990. Moreover, there was and even more significant relative growth in other renewable energy sources, such as geothermal and solar, which had a more than six-fold increase over the last 30 years [1]. However, despite the importance of promoting the adoption of these RES by the industry, they still have no expression in the global industrial energy consumption.

The large majority of the biofuels and waste consumed by the industrial sector in 2020 were primary solid biofuels, accounting for an average of 93% worldwide. However, in certain regions, industrial waste also constituted a significant share [1]. When only the renewable fraction of waste is considered, worldwide, solid biofuels represented nearly the entire biomass consumption by the industry worldwide.

The remainder of this chapter is structured as follows. It begins with a description of how energy is consumed in the industrial sector. Given that the industrial sector predominantly consumes energy in the form of heat on a global scale, and forest biomass is particularly well-suited for heat generation, Sect. 3, focus on presenting the most relevant biomass heat production systems used in this sector. While these systems vary in terms of configuration, technologies and scale, some share significant similarities with those used in district heating. Consequently, district heating systems

[2] OECD stands for Organisation for Economic Cooperation and Development.

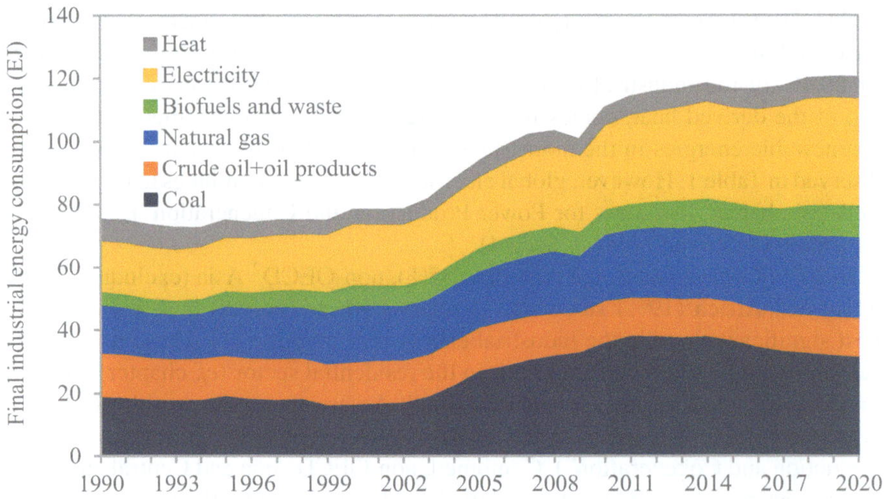

Fig. 4 World final energy consumption in the industrial sector by energy source from 1990 to 2020. (*Data source* [1])

are discussed at the end of this chapter. On the other hand, the topic of combined heat and power (CHP) generation, which is very important for both applications, will not be addressed in this chapter; instead, it will be covered in chapter "Biomass for Power Production and Cogeneration", when power production based on forest biomass is described. The chapter ends with some final conclusions.

2 Industrial Energy Consumption

Industries consume electricity for operating industrial equipment (e.g., motors, compressed air systems), as well as for lighting, space heating, cooling, ventilation and powering computers and other electric equipment. Additionally, they demand heat for process heating and water and space heating.

The structure of energy consumption in the industrial sector is highly complex, with significant variations among different industry sub-sectors and even within the same sub-sector, among industrial facilities [3, 4]. Typically, the majority of industrial facilities purchase electricity from electrical utilities or independent power producers. Some also generate electricity for self-use and/or for sale, often through CHP systems. While the purchase of derived heat also exists, it is not as common to sell heat off-site as it is for electricity or transport fuels [5], primarily due to the challenges associated with heat distribution.

The distribution of energy consumption across various end-uses (power, process heating, space heating and process and space cooling) in the industrial sector depends on the composition of industrial sector production and the specific characteristics of existent industrial processes. A comprehensive understanding of how energy is consumed in the world's industry is currently lacking, as there are no global official statistics that disaggregate the energy consumed in industry into its end-uses. In general, heat plays a significant role in the energy consumed by the industries. However, because heat generally does not require metering, and heat markets are often local and dispersed, there is a general gap in heat demand data [5]. This gap hinders demand-oriented energy policies for the sector [4].

Some countries are aware of the need for more information on the heat and cooling sectors and their importance for the energy transition towards a sustainable low-carbon economy. For example, the European Union (EU) defined its strategy for the heating and cooling sectors in 2016 [6], set indicative targets for the EU countries to increase the annual share of renewable energy sources in heating and cooling [7], and supported and published studies to increase the knowledge on the heating and cooling sectors (e.g., [8–10]). As a consequence, information on the characteristics of the consumption in the industrial sector exists for these regions. Presenting a description of this information does not characterise but helps to form a picture of the energy needs of the world's industry. In this regard, the next paragraphs briefly present the EU's industrial energy consumption.

In the European Union, the industrial sector was the third largest energy consumer, accounting for 26% of the total final energy consumption in 2021 [11]. Similarly to what happens in the world, the industry of the EU 27 Member States (EU27) rely on electricity and the direct use of fossil fuels (33% and 49% of the final industrial energy consumption, respectively) [11]. Natural gas (33%) was by far the most consumed fossil fuel [11]. The renewable contribution to the EU27 came primarily from primary solid biofuels (90.6%), with some contribution of renewable municipal waste (3.6%), biogas (2.2%), liquid biofuels (1.9%) and ambient heat (1.6%) [11].

Five industry sub-sectors contribute the most for the EU27's industrial energy consumption (Fig. 5): chemical and petrochemical (21.5%), non-metallic minerals (14.1%), paper, pulp and printing (13.5%), food, beverages and tobacco (11.6%), and iron and steel (10.2%). Understanding how energy is consumed in these industries and promoting energy efficiency and renewable energies is critical to be able to reduce the overall impact of the industrial sector on the environment.

A full end-use energy balance for the EU27 industry revealed that, in 2012, 57% of the energy was consumed for process heating, 10% for space heating, 3% for cooling and the remainder (30%) was mainly used for mechanical applications driven by electricity (Table 2) [12]. Another study, focusing on eight energy intensive sub-sectors that consumed 98% of the EU28 (EU27 + the United Kingdom) industrial final energy consumption in 2013, concluded that process heating consumed 66% of the total final energy consumption and electricity had a 26% share [13]. More recently, a study by TU Wien [10] focused on space heating concluded that 8.7% of the EU27 industrial energy consumption in 2017 was for space and water heating. These studies show that most of the energy used by the European Union's industry is in the form of heat, specially process heat.

Industrial processes vary significantly by industry sub-sector and so do their energy needs. The most energy intensive industry sub-sectors typically consume

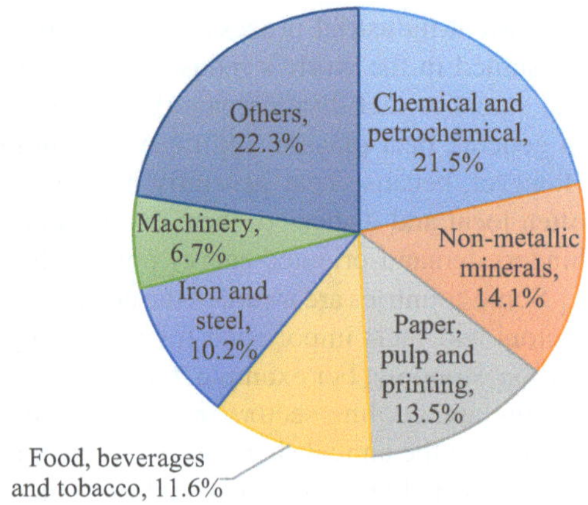

Fig. 5 Share of industry sub-sectors in the final industrial energy consumption in EU27 in 2021. (*Data source* [1])

Table 2 Share (in %) of various energy end-uses by industry sub-sector in EU27 in 2012. (*Data source* [12])

	Process heating	Space heating	Non-heating and cooling	Cooling
Chemical and petrochemical	61.7	4.1	28.8	5.4
Non-metallic minerals	76.3	4.7	18.4	0.6
Paper, pulp and printing	60.9	5.4	33.2	0.5
Food, beverages and tobacco	43.6	21.9	20.2	14.4
Iron and steel	85.2	1.5	13.3	0.1
Machinery and transport	14.5	27.3	56.7	1.5
Non-ferrous metals	40.6	3.7	55.5	0.2
Total industry	57.1	10.1	29.9	3.0

a large share of heat (Table 2). For example, 87% and 66% of the energy consumed by the iron and steel and chemical and petrochemical sectors in the EU27 in 2012 was heat. On the other hand, non-energy intensive industries, like the manufacturing of machinery, generally consume more electricity than heat.

The direct use of fossil fuels, especially natural gas, dominates the heat consumption in the European industry (Fig. 6). Natural gas accounted for 36% of the final energy consumption for process heating in the EU27 countries in 2012, followed by coal and other fossil fuels. In general, approximately three quarters of the energy demand for process heating was met with fossil fuels. Biomass provided 12% of the energy used for process heating and was the only renewable energy source used directly by the industry with some expression.

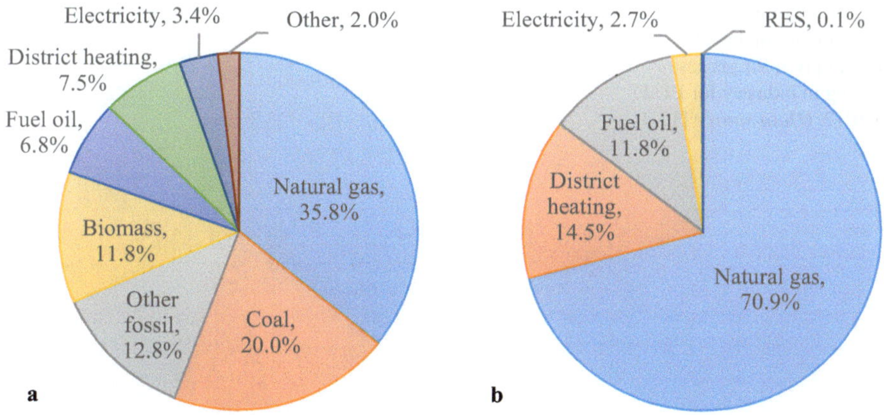

Fig. 6 Share of energy sources in the final energy consumption **a** for process heating and **b** in space and water heating in industry for EU27 in 2012. (*Data source* [10, 12])

The importance of the direct use of fossil fuels is even larger when it comes to industrial space and water heating (Fig. 6b). In 2012, natural gas and fuel oil accounted, respectively, for 71 and 12% of the energy needs for this end-use in EU27 countries. Moreover, the direct use of renewable energy sources for space and water heating in the EU27 industry was negligible and the only way that renewables penetrated was indirectly through electricity and district heating, which are partially produced from renewable energy sources. Of the two, the largest share in industrial heating belongs to district heating, which accounted for, respectively 15% and 8% of the energy consumed for space and process heating in the EU27 in 2012.

The energy sources used for process heating are much more diversified than for space and water heating, reflecting the very large diversity of technologies used for process heating. Indeed, the term process heat refers to a huge variety of applications, using different technologies and energy carriers (e.g., steam, liquid water, air) at different temperature levels. The latter is of particular importance when addressing the decarbonisation of the industrial sector.

More than half of the process heat consumed by the EU27 industry in 2012 was above 500 °C (Fig. 7); the same occurring in EU28 in 2015 [14]. High-temperature heat represented the large majority of the process heat consumed by the iron and steel (94%), non-metallic minerals (72%), and chemical and petrochemical (66%) industries in EU27 in 2012 [12]. In contrast, other industry sub-sectors that are also large energy consumers mostly consumed heat below 200 °C (in the paper, pulp and printing, and the food, beverages and tobacco, the share of process heat below 200 °C in the total process heat consumed was, respectively, 94% and 83% [12]).

The need for high-temperature heat was essentially met by the direct use of fossil fuels (91%) [12]. Moreover, supplying process heat at temperatures above 500 °C represented 84% of the coal, 76% of the "other fossil fuels" and 55% of the natural gas consumed by the EU27 industry in 2012 [12].

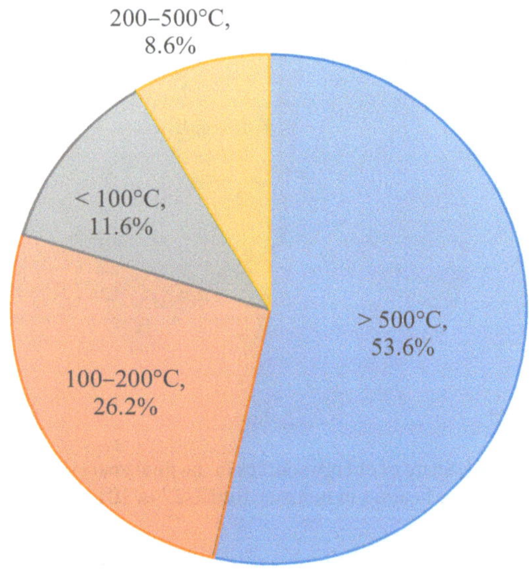

Fig. 7 Share of temperature levels in the final energy consumption for process heating in industry for EU27 in 2012. (*Data source* [12])

Not all the renewable energy sources and technologies are able to provide high-temperature process heat [15]. In principle, various solid biomass products are able to provide high-temperature heat [4, 15] but in practice, in the EU27 in 2012, 83% of the biomass used by the industry met process heat needs with temperatures below 200 °C and only 17% was used for process heat at temperatures above this level [12] Furthermore, solid biomass was mainly used by industry sub-sectors that generate residual biomass, such as the pulp, paper and printing, and the wood and wood product industries (see chapter "Sources and Distribution of Forest Biomass for Energy" for a description of the secondary wood residues generated by the wood-based industries). Together these sub-sectors consumed 85% of the final energy consumption of solid biomass in the EU27 industry in 2021 (Fig. 8). On the other hand, with a 6% share, the non-metallic mineral sector does not generate residual solid biomass, but still consumes a noteworthy proportion of solid biomass.

Process heat was virtually the only final energy use of biomass in the EU27 industry in 2012 [12]. This does not mean that industry does not generate electricity from biomass. However, the electricity generated by the industry is accounted for in transformation and not final energy consumption (*cf*. chapter "Biomass for Power Production and Cogeneration"). For example, the pulp and the wood-based panels industries commonly consume solid biomass in CHP systems for the production of heat and power [16–19].

Other than temperature, the way heat is delivered to the load is also important and diversified. Some industrial processes are continuous and require large amounts of energy to heat large volumes of materials, while others operate in batch mode, heat small quantities of materials and require precise temperature control [20]. In certain industries heat is provided directly to the material, in others indirectly. For example, in blast furnaces used in the steel industry, the flue gases are in direct contact with

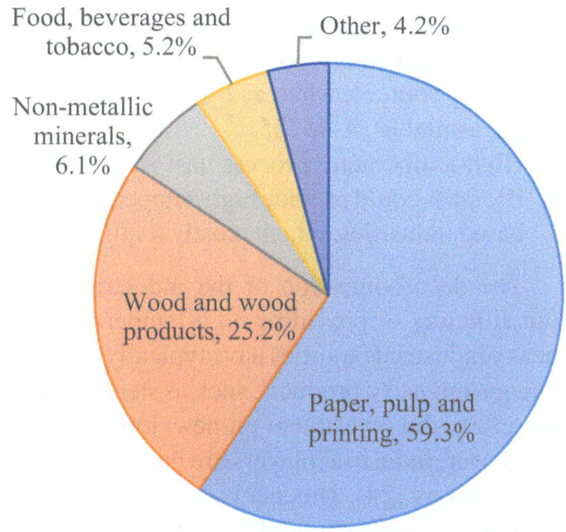

Fig. 8 Share of industry sub-sectors in the final energy consumption of solid biomass in industry for EU27 in 2021. (*Data source* [11])

the iron ore and are used for its reduction [21], whereas in the production of food and beverages, direct heating with solid fuels such as coal is generally not suitable since the flue gases contain pollutants that contaminate the products [22].

Other analyses characterise the industrial energy consumption in different regions and help form a picture of the needs of the world industry and the pathways available to decarbonise this sector. For example, the United States Energy Information Administration (EIA) regularly publishes the results of its Manufacturing Energy Consuming Survey [23] and the Australian Renewable Energy Agency recently published a report on the renewable options for industrial process heat where the consumption of the Australian industrial sector is analysed [24]. Complementing these analyses, several studies focus on specific industry sub-sectors, such as the iron and steel [25, 26], chemical [27, 28], cement [29, 30] or pulp and paper [31, 32] industries.

Even though a detailed characterisation of the world industrial energy needs is not available, the following can be stated:

- Industry is very diverse and the energy needs of the different industry sub-sectors and facilities are varied and complex.
- A few energy-intensive industry sub-sectors account for a large share of the world energy consumption and greenhouse gas emissions (of particular relevance are the chemical, iron and steel and cement industry[3]) but, a non-negligible part of the industrial energy consumption is dispersed by very different industries.
- Energy-intensive industry sub-sectors typically consume a large share of heat, mainly process heat.

[3] In 2020, these three sectors accounted for almost 60% of the world industrial energy consumption and more than 70% of the industrial CO_2 emissions [33]. The emissions from industrial processes are included in this value, which for some industrial processes (e.g., cement and lime production) are important [34].

- Demand for high-temperature heat in industry is substantial.
- High-temperature process heat can only be supplied by certain technologies and energy sources, whereas for the lower temperatures (200 °C) much more options are available [9, 15, 35].
- High-temperature process heat is today largely supplied by fossil fuels.
- Biomass could provide high-temperature process heat, but is mainly used in forest-based industries, which mostly require low and medium temperature process heat.

The decarbonisation of the industrial sector is a challenging task. Adding to the difficulty of providing high-temperature heat through low-carbon technologies, heavy industrial facilities have typically long lifetimes, are capital intensive and many energy-intensive products, such as steel, compete on global competitive markets and, therefore, the investment on new, low-carbon technologies poses real risks [9, 33]. Moreover, industrial players are generally averse to risk and have short payback time expectations [24]. Bioenergy may be part of the solution to decarbonise industrial heat, with several studies focused on the topic [4, 15, 21, 30, 35–37], but the solutions are dependent on the industry sub-sector and specific context with no one-size-fits-all solution.

3 Biomass Systems for Industrial Heating

As seen previously, globally, in the industrial sector most of the energy is consumed for process heating, but space heating is also worthy of reference and is relatively more important in the less energy-intensive sub-sectors, such as machinery and transport equipment or the food, beverages and tobacco (e.g., [4, 10]). Several technologies are available for the production of space heating within industrial facilities (e.g., combustion-based equipment fuelled by renewable and non-renewable fuels, ambient pumps, solar thermal systems). Alternatively, heat can be supplied via district heating (see Sect. 4 for a description of district heating systems based on biomass).

Because of the diversity of existing industrial processes and heating principles, a high diversity of technologies (e.g., boilers, kilns, blast furnaces, ovens, dryers) are available to generate process heat, varying in size from small-scale systems of a few kilowatts to large-scale systems of the order of megawatts. Among the factors that are important for the choice of the technology used are: the characteristics of the industrial process and heat demand, properties of the available fuels, costs and performance of technologies and legislation [38–40].

Combustion-based process heating systems are responsible for the generation of a large share of the energy used by the industry and are employed in almost every industry segment [20]. In this type of process heating system, heat is generated by the combustion of a fuel (usually with air, but other oxidants are also used) and distributed to the process. Biomass is not the most used fuel but common in certain industrial sub-sectors (*cf.* Sect. 2). The systems can be categorised into two groups: direct heating systems, where flue gases are in direct contact with the material being processed, and

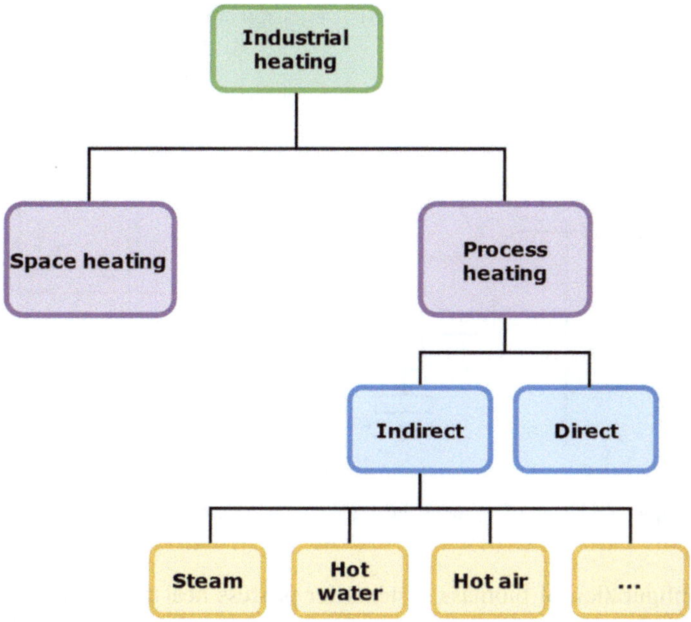

Fig. 9 Industrial heating

indirect heating systems, where flue gases transfer the heat to a heat transfer fluid (e.g., steam, hot water, hot air), which then delivers heat to the production process (Fig. 9).

Typically, high-temperature process heat is generated in direct systems [36]. There are many types of equipment used, depending on the specificities of the industrial process. Examples include furnaces used in the chemical and petrochemical [41–45], food, beverage and tobacco [46], iron and steel [47, 48], non-ferrous metals [49], non-metallic minerals [50–53] and pulp and paper [18] sub-sectors. Currently, most of these applications rely on fossil fuels, but some involve biomass co-firing with other fuels (e.g., in cement kilns [53]) or 100% biomass firing (e.g., in small blast furnaces [54]).

Although direct heating also provides heat at low and medium temperatures, such as in the case of the cork industry [55], most technologies used for the lower temperatures involve indirect heating, usually with steam as the heat transfer fluid [36]. For the generation of low- and medium-temperature process heating, diverse biomass conversion technologies are available and commonly used.

In contrast to many small-scale biomass energy conversion systems used for residential heating (*cf.* chapter "Biomass for Domestic Heat"), typically, industrial systems are automatically fed, involve advanced process control systems and pollution control equipment. Figure 10 presents an example of a possible layout of a solid biomass system used to indirectly provide process heat to an industrial process.

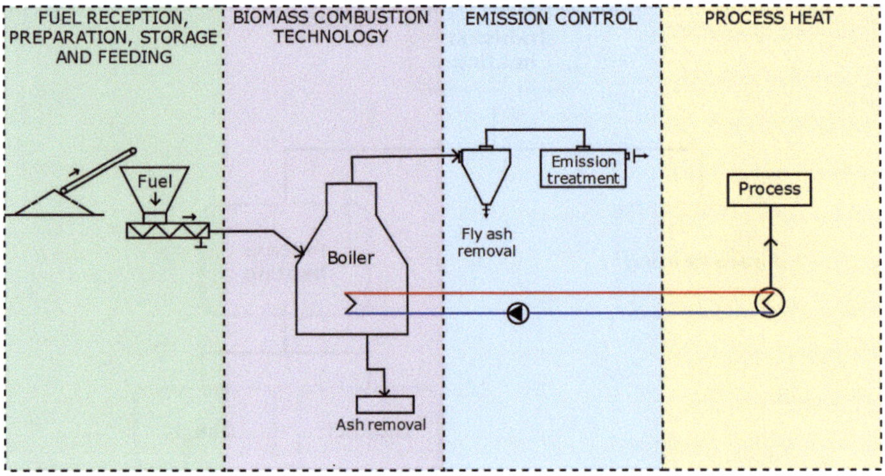

Fig. 10 Example of a layout of a process heat generation system fired with solid biomass

The configuration of biomass systems for process heat generation varies based on factors such as the scale of the system, operational requirements and the characteristics of the fuel feedstock. Typically, these systems integrate different facilities, components and equipment.

Usually, facilities that receive, store, preprocess and/or automatically feed the fuel into the energy conversion system are required. The logistics of biomass delivery and reception are intricately linked to the layout of the facility and the chosen storage methods. Unloading and/or transporting fuel to storage or processing areas can adopt fully automated, fully manual or hybrid approaches involving a combination of automation and manual intervention.

To ensure compliance with environmental and health standards, potentially optimise costs and align with the requirements of the energy conversion process, biomass often undergoes pre-processing before storage and energy conversion (see chapter "Forest Biomass as an Energy Resource" for a description of biomass pre-processing methods).

The storage of solid biomass can encompass both indoor and outdoor facilities, with some applications requiring both long- and short-term storage solutions, while others only short-term storage before directly feeding biomass into the conversion equipment.

The biomass feeding systems should be automatic and equipped with metering capabilities, enabling precise control over the amount of biomass supplied to the conversion equipment. This control ensures efficient conversion and consistent generation of heat, important in industrial contexts.

Moreover, environmental protection policies generally impose limits on pollutant emissions arising from combustion. Consequently, it becomes imperative to implement mitigation measures and control systems that align with legal mandates. This measures and systems can involve the use of advanced combustion technologies, air pollution control equipment (e.g., scrubbers, fabric filters, electrostatic precipitators, cyclones) and optimisation of operational practices to minimise emissions [38, 56, 57]. Choosing the right technology depends on factors such as the specific biomass material, combustion equipment and desired energy output.

Another integral part of the heat generation system fired with solid biomass is the collection of ash formed during combustion. A portion of the inorganic matter content of the biomass fuel is removed from the system in the form of solid ash particles and agglomerates that are collected at the bottom of the combustion chamber. Additionally, small ash particles and inorganic vapours are caught up by the combustion gases and are transported through the flue gas duct, being collected in specific equipment (e.g., electrostatic precipitators, fabric filters, cyclone separators).

At the centre of solid biomass systems designed for process heat generation is the equipment that converts the fuel into thermal energy. Most of the systems employed by the industry are based on the direct combustion of biomass, but some industries use systems based on gasification [15]. Both combustion and gasification technologies will be described below.

Combustion

Medium- and large-scale systems for biomass combustion rely on several technologies that are commercially available and mature. Most of these technologies can be categorised into three groups: fixed bed, fluidised bed and pulverised fuel combustion, depending on the flow conditions inside the combustion chamber [38, 40].

In fixed bed combustion, biomass fuel is fed onto a grate, where it burns with the primary air that flows through the bed, supplied through the grate from below (Fig. 11). Secondary air enters the combustion chamber above the bed to support further combustion of the gases and particulate matter that are formed during the initial burning process. The name "fixed-bed" originates from the fuel forming a bed that remains on the grate due to gravity, in contrast to what happens in fluidised bed combustors, where the particles of fuel are suspended.

Fix bed combustion systems can have different configurations and are further categorised into grate furnaces (overfeed stokers) and underfeed stokers, depending on the way fuel is fed into the combustion chamber. As the name indicates, in overfeed stokers, the biomass is supplied from above the grate, while in underfeed stokers from below. Different technologies of both underfeed and overfeed stokers exist, each with its own design and operational characteristics. Examples of mature technologies are fixed, moving, travelling, vibrating or rotating grate firing furnaces, cigar burners or horizontal-feed, side-ash discharge underfeed stoker [40, 57–59].

Similar to fixed bed combustion, fluidised bed combustion also involves the upward stream of primary combustion air supplied to the combustion chamber from the bottom. However, in fluidised bed combustion, this air is introduced at a sufficiently high velocity to maintain the bed in suspension, creating a "fluidised" state

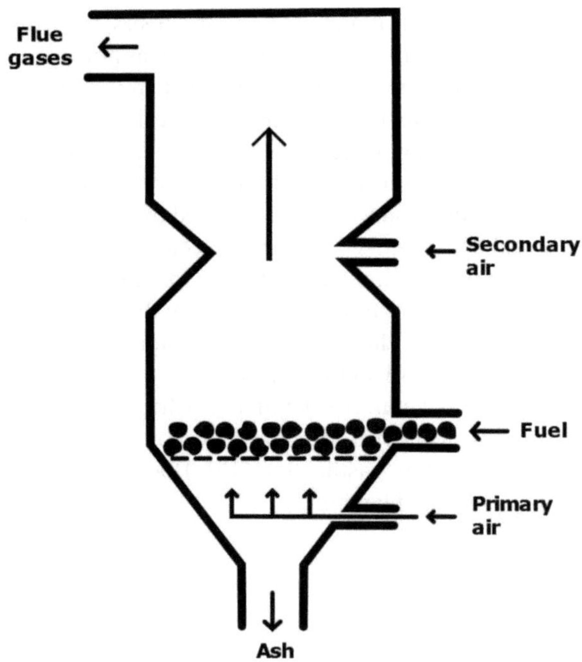

Fig. 11 Sketch of a grate furnace, a type of fixed bed combustor

(Fig. 12). In this technology, fuel is fed into a bed of suspended heated granular material (e.g., silica sand, dolomite), which constitutes the majority of the bed, usually accounting for 90–98% of the mixture of fuel and bed material [40]. The combination of intense heat transfer and mixing enhances combustion, promoting complete and efficient combustion while allowing low excess air. This reduces the volume of flue gases, allowing a more compact design, which is particularly beneficial for large-scale applications.

Fluidised bed combustion can be categorised into bubbling fluidised bed and circulating fluidised bed [60, 61]. The primary distinction between these two lies is the fluidisation velocity, which is notably higher for circulating fluidised combustion. As a consequence, in this technology, the bed material, which is smaller than in bubbling fluidised bed, is carried with the flue gases. Larger particles tend to either remain fluidised near the furnace bottom or get transported after undergoing size reduction due to the chemical reactions, thermal stresses and mechanical stresses [62]. The operation of circulating fluidised bed furnaces involves a cyclone directly linked to the combustion chamber, which separates and captures particles contained in the flue gases and recycle them to the fluidised bed for complete combustion.

Fluidised bed systems operate at atmospheric pressures, although variations that operate at elevated pressures have been developed. They are characterised by higher efficiencies, but are more complex and have higher associated costs [20]. The current research emphasis is primarily directed towards conventional atmospheric fluidised

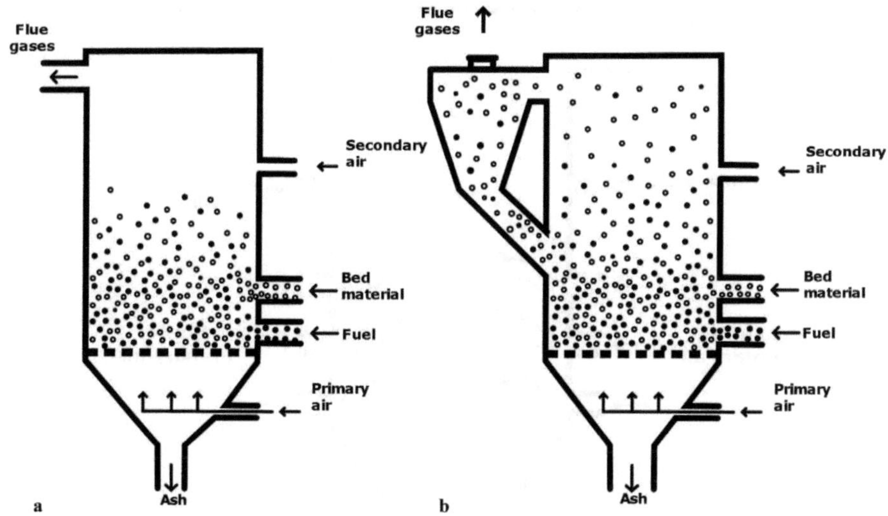

Fig. 12 Sketch of **a** bubbling fluidised bed furnace and **b** circulating fluidised bed furnace

bed combustion with limited attention given to the investigation of pressurised fluidised bed systems [63].

Pulverised fuel combustion, also called entrained flow or dust combustion, is most widely employed in coal-fired power stations [64], but not so common for biomass combustion. In pulverised fuel combustion furnaces, finely pulverised fuel alongside air are introduced in the combustion chamber (Fig. 13). Within the furnace, these particles heat up, releasing combustible gases and quickly reacting with oxygen, because of the small particle sizes. This technology is characterised by high efficiencies and allows very good load control and fast load changes [40, 65].

Disadvantages of pulverised fuel combustion are the requirement to burn biomass with low moisture content (<20 wt % wb) and small particle sizes (<5 mm) [38] and sensitivity to changes in fuel quality [57]. When pulverised fuel combustors are fired with solid biomass, if the particles are not already small because they originate from a specific industrial process, energy must be spent in grinding and drying the feedstock.

Table 3 presents the typical capacities, fuel requirements and performance for fixed bed and fluidised bed biomass combustion, the two types of technologies mostly used for the generation of process heat from biomass. The values presented are for reference and should be read with care since many of the parameters may be dependent on a specific technology within these combustion typologies. For example, the fuel delivery system also influences the size of the particles and grate furnaces with pneumatic conveyers for fuel delivery require particle sizes up to 5 mm, while sliding bar conveyers allow fuel 100 times larger [38].

Biomass fixed bed and fluidised bed combustion are the most commonly used technologies for the generation of heat. Biomass fixed bed combustion is typically

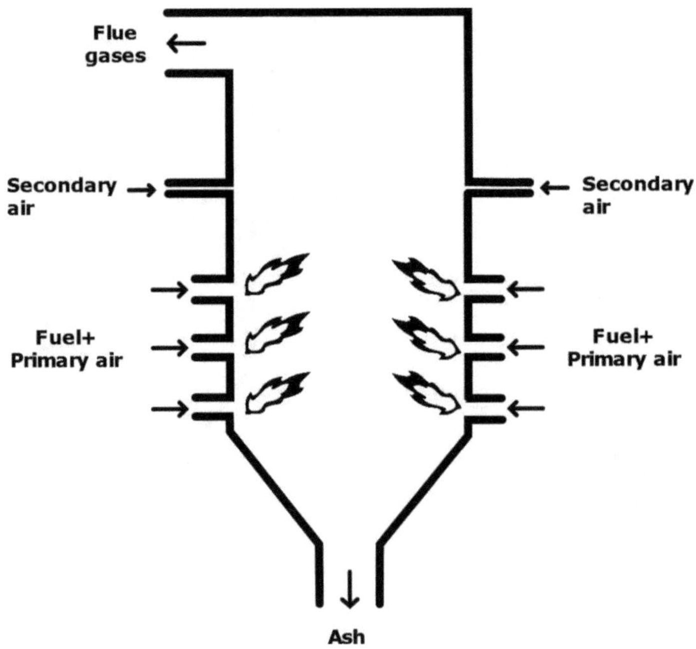

Fig. 13 Sketch of a pulverised fuel combustor

used for capacities below 20 MW_{th}, while fluidised bed combustion is preferred for capacities above 20–30 MW_{th} [36, 64]. Generally, both technologies offer considerable flexibility regarding moisture and ash content [65]. However, while fixed bed combustion systems are flexible in terms of fuel particle size, fluidised bed systems require smaller particle sizes [38, 65]. Additionally, fixed bed combustion allows for the mixture of various types of wood fuels [65] and is commonly used for low-grade fuels (e.g., demolition wood) [64]. Generally, these systems have lower capital and operational costs than fluidised bed systems, but are characterised by lower efficiencies [40]. Biomass pulverised combustion, although occasionally used in industry sub-sectors like chipboard manufacturing [40], is not as widely adopted for process heat. Instead its primary and more common application is in thermal power stations [68] (*cf.*, chapter "Biomass for Power Production and Cogeneration").

Gasification

An alternative approach to the conventional method of direct combustion involves gasification. During gasification, biofuels are converted into a low molecular weight combustible gaseous fuel mixture commonly referred to as syngas (for further details

Table 3 Comparison between fixed bed, bubbling fluidised and circulating fluidised bed combustion (based on [38, 40, 57, 60, 65–67])

	Fixed bed combustion	Bubbling fluidised combustion	Circulating fluidised combustion
Typical thermal capacity	100 kW_{th}–20 MW_{th}	5–15 MW_{th}	15–100 MW_{th}
Fuel flexibility	Good; all wood fuels and most types of biomass	Good; various types of biomass	Good; various types of biomass
Flexibility to fuel particle size	High; allow varying particle sizes with a minimum size of 5 mm	High but for smaller sizes (<25 mm)	High but for smaller sizes (<50 mm)
Maximum moisture content	60%	60%	60%
Maximum ash content	50%; low for underfeed stokers	50%	50%
Partial load operation	Good	Requires special technology	Requires special technology
Combustion efficiency	94–97%	~99%	Up to >99%
Capital costs	Medium to low	High (but lower than CFB)	High
Operation and maintenance costs	Medium to low, depending on the technology	High	High

CFB Circulating fluidised combustion

on the composition and denomination of the gaseous fuels that results from gasification, please consult chapter "Forest Biomass as an Energy Resource"). One of the advantages of syngas over the original solid biofuels is its flexibility and the wide array of potential applications. These applications range from the generation of heat to the production of advanced biofuels and chemicals, passing through the generation of combined heat and power.

Within the scope of industrial heating only (cogeneration will be explored in chapter "Biomass for Power Production and Cogeneration"), once generated in the gasifier, syngas can be burned to provide process heat. This forms a closed-coupled biomass gasification-combustion system, a technology commercially available [69]. When a high-quality clean gas is essential for the process, syngas will undergo treatment to eliminate tars and particulate matter before combustion. This results in a much cleaner fuel than the original solid biomass.

Gasifiers can be categorised as either directly heated or indirectly heated, depending on the method they employ to supply heat for the endothermic gasification process. In directly heated gasifiers, also known as autothermal gasifiers, heat is generated through the partial oxidation of biomass. Conversely, in indirectly heated gasifiers, also referred to as allothermal gasifiers, heat is provided indirectly either by the gasifying agent or through heat exchangers. The way heat is provided to the gasification reactions is determinant for the quality of the syngas, with indirectly heated gasifiers typically yielding syngas with a higher heating value [70].

Another common way of classifying gasification conversion technologies is based on their fundamental operation principle, including fixed bed, fluidised bed or entrained flow designs [71]. The fluid dynamics within the gasifier has a strong influence on the mixing between solid and gas and on the performance of gasifiers [72]. In this context, both fixed bed and fluidised bed systems, the main categories of gasifiers, employ similar equipment to direct combustion systems [69].

In fixed bed gasification systems, the fuel is fed from the top and is piled on a grate, moving downwards as it suffers chemical reactions (Figs. 14 and 15). On the other hand, the gasifying gas that passes through the biomass feedstock (almost always air [73]) may be introduced at diverse positions within the gasifier, resulting in different gas flow directions.

Fixed bed gasifiers are the classical and still the most commonly used technologies for gasification [71] and are well-suited for small-scale heat and/or power generation [74]. They represent a straightforward, cost-effective and well-established technology; however, they typically yield syngas with lower heating value than other configurations [69].

In the updraft gasifiers, also called counter-current gasifiers, the gasifying agent is introduced at the bottom and, as a consequence, the gasification process proceeds downwards (Fig. 14a). The syngas leaves the gasifier at the top and ash falls from the grate to the bottom of the gasifying chamber. This type of gasifier is efficient [74] and allows using biomass with high moisture content [71], but it has a drawback in that the syngas generated typically contains 10–20% tar, necessitating significant cleaning

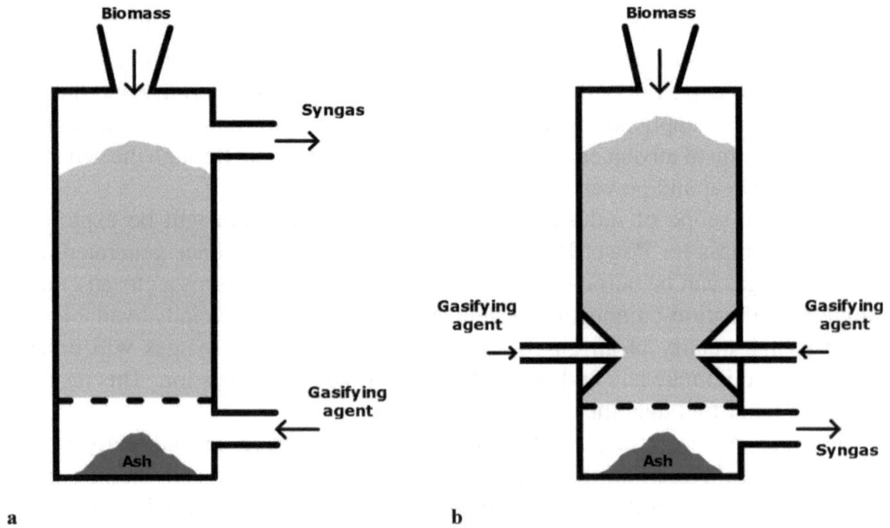

Fig. 14 Sketch of fixed bed **a** updraft gasifier and **b** downdraft gasifier

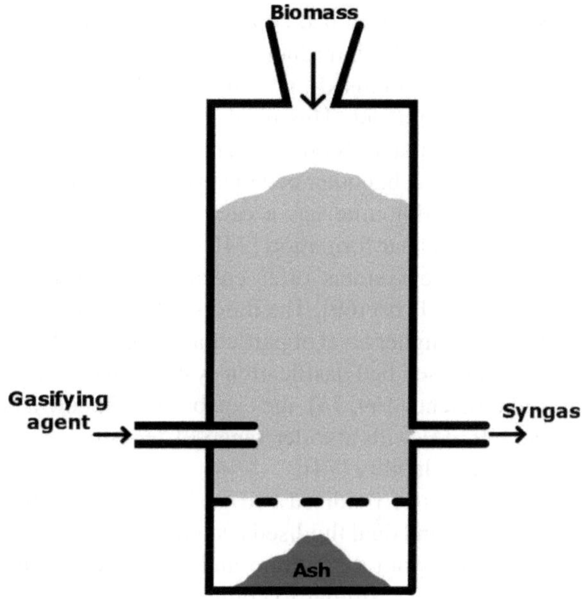

Fig. 15 Sketch of a fixed bed crossdraft gasifier

and processing before it can be used for purposes other than direct combustion [71]. Because of their inherent disadvantages, updraft gasifiers are not so widely used today [74].

In contrast to updraft gasifiers, downdraft gasifiers, also called co-current gasifiers, introduce the gasifying gas more or less at the middle section of the gasifying chamber (Fig. 14b). In this configuration both the biomass and the gasifying agent move downwards and ash falls from the grate to the bottom of the gasifier. As a consequence of the direction of the gas flow, syngas flows towards the bottom of the reactor, leaving the gasifier at a high temperature. Two types of downdraft gasifiers exist: with throat, as represented in Fig. 14b, or without throat; the former producing syngas with lower tar content [74]. While downdraft gasifiers tend to produce syngas with much lower tar content (less than 0.1%) than updraft gasifiers, they can be more complex and costlier to operate [71]. Another disadvantage is the fact that they require feedstock with low moisture content [71] and low ash content [73].

Downdraft gasifiers are widely used and are the most common technology for small-scale power generation [75]. Other main applications of the syngas produced with this technology are in boilers, dryers or direct fired rotary kilns [74].

Another type of fixed-bed gasifier is the crossdraft gasifier, also called cross-flow gasifier. In this configuration, the gasifying agent enters the reactor on one side and syngas leaves on the other side (Fig. 15). One of the main advantages of this type of configuration over the other fixed-bed types of gasifiers is the fast response time to load changes; however, crossdraft gasifiers are not widely applied and research is scarce [74].

Similar to fluidised bed combustion, fluidised bed gasification systems generate combustible gas by introducing biomass into a heated bed of suspended granular material that is fluidised by an upward flow of gas, which in the case of gasifiers is the gasifying agent. This results in an effective mixing between the gas and the different solid materials present in the gasifying chamber. The most common inert bed material is silica, but other bed materials might be an option for specific applications. For example, dolomite has a catalytic effect on the gasification process, helping reduce tar and char formation [74].

While these systems offer enhanced performance, they come with increased complexity and cost [69]. The fluidised bed design results in gas with relative low tar content but a higher level of particulates compared to fixed-bed systems [69]. Advantages of fluidised bed gasification systems over fixed bed systems include improved overall efficiency [69, 74], the capability to handle a broader range of biomass feedstocks [69, 74] with a wider range of feedstock particle size [74]. Moreover, they offer good scalability [74].

Three types of fluidised bed gasifiers exist: bubbling fluidised bed, circulating fluidised bed and dual fluidised bed gasifiers. In all, the gasifying agent is introduced in the reactor from the bottom and is evenly distributed in the gasifying chamber (Figs. 16 and 17). Similar to combustion, the primary difference between bubbling and circulating fluidised bed gasification lies in the velocity of the gasifying gas, which is higher for circulating fluidised bed gasification. Both are equipped with cyclones to separate solid particles from the syngas.

Circulating fluidised bed gasifiers are characterised by higher conversion efficiencies than bubbling fluidised bed gasifiers [74]. They are mainly used in the industrial sector (e.g., pulp and paper, cement sub-sectors) and for electricity generation [74].

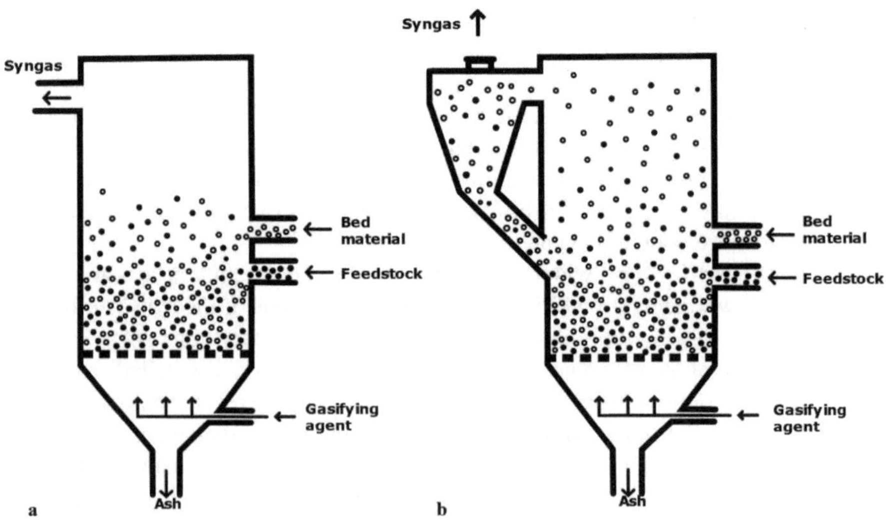

Fig. 16 Sketch of **a** bubbling fluidised bed gasifier and **b** circulating fluidised bed gasifier

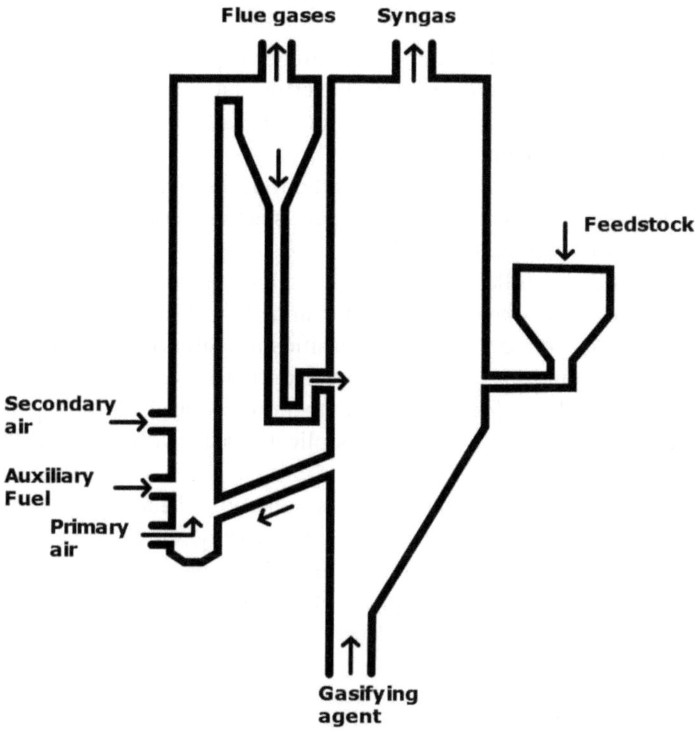

Fig. 17 Sketch of a dual fluidised bed gasifier

A dual fluidised bed gasifier is a type of allothermal gasifier with two interconnected fluidised beds: a fluidised bed gasifier, which converts biomass into raw syngas, and a fluidised bed combustor, which generates the necessary heat for the gasification reactions (Fig. 17). The two fluidised beds can be independently controlled, but are typically linked though a non-mechanical valve, which ensures the continuous circulation of the bed material that acts as a heat-carrier between the two fluidised bed reactors [72]. The reactors can be of different types, but a widely used configuration is that the gasifier operates as a bubbling fluidised bed gasifier and the combustor as a circulating fluidised bed combustor [72].

A mixture of residual char, tar and bed material coming from the gasifier enters the fluidised bed combustor, where the residual char and tar are oxidised in the presence of an oxidiser, generating heat and rising the temperature of the bed material. If needed, additional fuel may be incorporated into the fluidised bed combustor to control and maintain the temperature of the reactor [76]. Downstream of the combustor, a cyclone is utilised to separate the heat-carrying material from the flue gases. The heat-carrying material is then returned to the gasifier, while the flue gases are directed towards a heat recovery system. Biomass feedstock is introduced in the gasifier and is heated in contact with the hot bed material and with the gasifying agent, most often preheated steam. The syngas produced by a dual fluidised bed gasifier is characterised by low nitrogen and tar contents, higher hydrogen content and higher heating value [72]. However, dual fluidised bed gasifiers have higher capital and operation costs. The technology is especially interesting for converting biomass into second-generation fuels like Fischer–Tropsch Diesel or substitute natural gas due to the higher heating values and hydrogen contents [77].

Table 4 presents a comparison between the most commonly used biomass gasifiers. Other types of gasifiers were developed, such as entrained-flow reactors, but are not widely used with biomass [78].

Biomass gasification followed by syngas combustion has the potential to generate high-temperature process heat and is demonstrated in several industrial sub-sectors [79]. In comparison to direct combustion, gasification offers benefits such as a shorter response time to variable loads and more precise control over the combustion process [80]. This precision allows for improved temperature control and heat quality, which can be advantageous in specific industrial applications. However, it typically involves higher capital costs [80].

Using gasification for heat generation is generally cheaper than for producing electricity, primarily because the requirements for syngas quality are less stringent. However, utilising biomass gasification exclusively for process heat generation is one of the less economically valuable applications of syngas. As a result, gasification is often employed in combined heat and power applications [73].

Table 4 Comparison between downdraft, updraft and fluidised bed gasification with air as a gasifying agent (based on [71])

	Updraft gasification	Downdraft gasification	Bubbling fluidised bed gasification	Circulating fluidised bed gasification
Typical capacity (MW_e)[a]	<20	<10	10–100	10–100
Flexibility to fuel particle size (mm)	2–50	10–300	<5	<15
Moisture content (%)	<60	<20	<55	<55
Tar levels ($g{\cdot}Nm^{-3}$)	10–150	0.015–0.3	3–40	4–20
LHV ($MJ{\cdot}m^{-3}$)	5–6	4–6	4–7	4–6
Carbon conversion efficiency (%)	40–85	<85	70–90	80–90
Cold gas efficiency (%)	20–60	65–90	70–90	50–70
Capital costs	Low	Low	High	High

LHV Lower heating value

[a] Today most syngas is used for combined heat and power generation

Co-combustion

Co-combustion, also known as co-firing, refers to the simultaneous combustion of two or more different types of fuels in the same plant [81]. In the context of bioenergy, it involves combining biomass with another fuel, such as coal or natural gas. Co-combustion of solid biomass with coal is a process mainly used for the production of electricity, which will be addressed in chapter "Biomass for Power Production and Cogeneration". However, it is also used in industrial heating applications (for example, in the cement industry).

Co-combustion with coal has the potential to be implemented in existing coal-fired plants with minimal adjustments, leading to improved environmental outcomes [82, 83]. It is a reliable solution that, compared to using single coal firing, leads to the reduction in net CO_2, SO_x and NO_x emissions [38, 84, 85] and might result in a reduction of costs [84]. In comparison to dedicated biomass plants, co-combustion with coal offers advantages such as reduced costs [83, 85] and improved conversion efficiency [38, 84, 85] without depending on a continuous supply of biomass [84], which might be a limited resource. The technologies used for co-firing biomass with coal in power and CHP plants will be described in chapter "Biomass for Power Production and Cogeneration".

In the specific context of heat only generation, biomass can be favourably co-fired with coal in some high-temperature process heat applications such as in cement kilns, allowing, for example, for the combustion of contaminated waste wood [84, 86]. Major cement manufacturers are already actively incorporating solid biomass and other alternative fuels for co-firing to achieve cost-effective solutions [86]. The temperature requirements in the key energy-intensive processes of this sub-sector often exceed 1000 °C [13], a level that cannot be reached through conventional raw biomass combustion [15]. As a result, adoption of, for example, co-combustion and/or oxygen-enrichment is needed [15]. While recommendations suggest replacing up to 20% of fossil fuels with biomass, higher substitution rates were already successfully achieved [53]. The cement industry does not face significant technical obstacles to integrate higher levels of solid biomass [87]. However, constrains arise from the need for biomass pre-treatment, economic considerations and the local availability of biomass resources [53].

Similar to the challenges faced by the cement industry, the iron and steel sector represents another hard-to-abate industrial sub-sector with the potential to reduce carbon emissions through biomass co-firing. The utilisation of biomass as a renewable energy source in iron and steel making is among the few technically and economically viable options for curbing CO_2 emissions in the short and medium term [88]. For example, in the iron-making process, which typically relies on carbon-containing fuels, biomass can be co-fired with coke and coal in blast furnaces [21]. Wood-based feedstocks are the most suitable biomass types [35, 89], but the use of raw wood is inefficient and it is better to use charcoals, semi-charcoals or torrefied biomass [88]. The injection of biomass in blast furnaces presents, according to Suopajärvi et al. [21], the most substantial potential for biomass to replace fossil fuels within the iron and steel industry. Furthermore, biomass can be integrated into various processes to reduce the reliance on fossil-based reducing agents, such as incorporating biomass into coal blends for cokemaking [21, 88]. However, it is important to note that further research is needed and that presently biomass cannot generally compete with fossil fuels in economic terms [48].

4 Biomass Systems for District Heating

District heating is an efficient energy system, characterised by centralising thermal energy conversion within a facility and then distributing the heat produced to a group of users through a network of underground pipes. Various energy sources, such as natural gas, biomass or waste heat, are utilised to generate the heat in district heating systems. The medium for conveying thermal energy typically consists of hot water, which can be readily transported over considerable distances [88].

By centralising energy conversion, these systems can employ advanced technologies and optimise the combustion process to minimise emissions and enhance energy efficiency [90]. Furthermore, the network design enables waste heat recovery from, for example, CHP plants or industrial processes, making efficient use of heat that might be otherwise lost [91]. This heat recycling can be combined with renewable energies, substituting for fossil fuels and, therefore, minimising the environmental impact of heating.

District heating systems are versatile and can serve a diverse range of users, including residential, commercial and industrial facilities. According to the IEA energy balances, in 2020, industry was the main user for the 15.7 EJ of derived heat supplied worldwide, followed by the residential sector (Fig. 18). China (38%), Russia (33%) and Europe (20%) were responsible for more than 90% of the production of derived heat in the world [1].

Market penetration of district heating systems varies from one country to another. In nations where district heating, regardless of the energy source, is prevalent, it supplies heat to approximately half of the building stocks, driven by strong driving forces [91]. In contrast, in countries with low awareness or competitiveness, the presence of such systems is scarce [91]. Europe is the region where more district heating systems are implemented.

District heating is particularly well suited for the dense urban environment, where a concentrated user base can benefit from the shared energy infrastructure [92]. This centralised approach not only enhances energy efficiency, but also simplifies maintenance and infrastructure management. On the other hand, the economic competitiveness of district heating systems depends on the international fuel prices, concentration of heat demands and energy and environmental policies [91].

The strongest argument for the implementation of district heating systems has been the recuperation of the unavoidable heat losses from thermal power plants through the use of CHP systems, being district heating often associated to CHP plants [91].

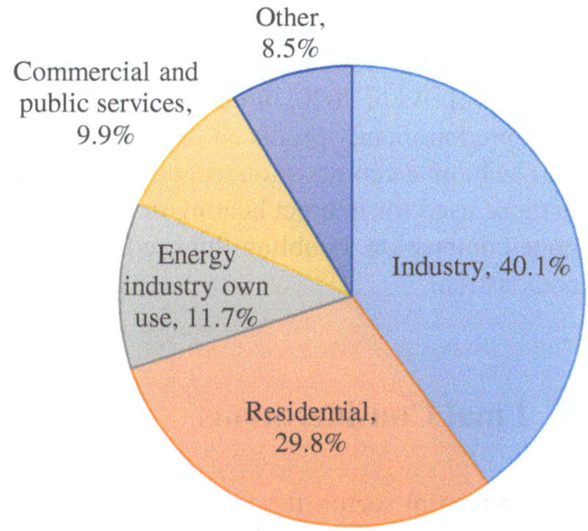

Fig. 18 Share of different users in the derived heat supplied in the world in 2020. (*Data source* [1])

In 2020, 73% of the derived heat supplied worldwide was generated in CHP plants [1].

The potential for integrating renewable energy sources in district heating is large, but, currently, most of the global derived heat generation relies on fossil fuels, especially coal and natural gas (Fig. 19), which are the two dominating energy sources used in CHP plants (*cf.* chapter "Biomass for Power Production and Cogeneration").

Despite the low share of biofuels in district heating worldwide, the supply of bioenergy by district heating is common in some countries [91]. This supply relies predominantly on CHP systems and is mostly located in the European Union [90, 91]. Additionally, biomass heat-only plants also exist, but they are primarily used in small-scale district heating systems [92]. Sweden serves as an exemplary model for a

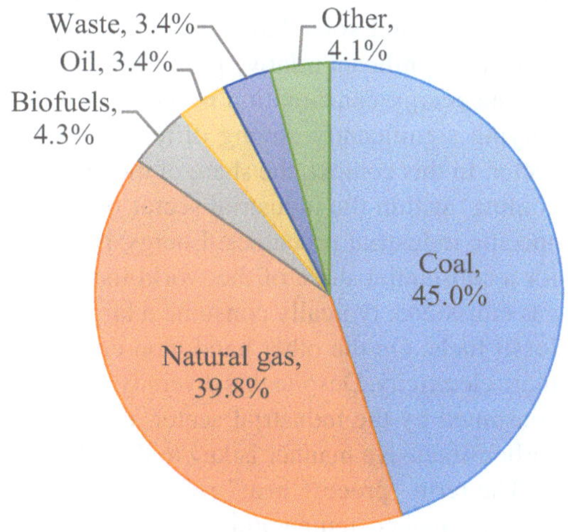

Fig. 19 Share of different energy sources in the derived heat supplied in the world in 2020. (*Data source* [1])

nation with a significant district heating system based on forest biomass. In Sweden, district heating plays a crucial role, providing over half of the heat in the residential sector [92]. As of 2020, biomass accounted for almost half of the supply of derived heat, predominantly produced in CHP plants [1].

The biomass combustion technologies used in district heating systems are similar to those used for indirect heating in industrial applications described above: mostly grate combustion, bubbling fluidised bed combustion and circulating fluidised bed combustion.

5 Final Considerations

The industrial sector, the largest energy consuming end-use sector, accounted for approximately one-third of the world total final energy consumption (around 120 EJ). Moreover, industries also significantly contributed to the global direct greenhouse gas emissions, reflecting the limited adoption of low carbon technologies in the sector. Indeed, industrial energy consumption is dominated by fossil fuels, mainly coal and natural gas. Together, these two fossil fuels represented almost half of the energy sources used by the industry in 2020.

Electricity is another of the significant energy sources in industrial facilities (28.4% of the energy consumption in industry in 2020) and its role has been increasing over time. The use of electricity is a potential indirect way of incorporating renewable energy sources in the industrial sector, but, currently, the global electricity generation is also still heavily reliant on fossil fuels.

In 2020, biofuels and waste represented only 8.4% of the energy consumed by the industrial sector. Despite the low share, the only RES with an expressive direct consumption in the industry is biomass, mostly primary solid biofuels, which represent 93% of industrial biofuel and waste consumption. The other RES like geothermal and solar have almost no expression in the industrial sector, despite showing significant relative growths.

The energy consumption in the industrial sector is highly diverse and complex, varying significantly among different sub-sectors and even within the same sub-sector. In this context, the share of the different energy end-uses (power, heating and cooling) within the industrial sector depends on its composition of production and specific industrial processes. Energy-intensive industry sub-sectors, which account for a substantial share of the world industrial energy consumption and greenhouse gas emissions, typically consume a large share of heat, especially process heat, and fossil fuels. On the other hand, non-energy intensive industries generally consume more electricity. Despite this diversity, globally, heat plays a crucial role in the energy consumed by the industrial sector and the ability to provide process heat in a less carbon-intensive manner is key to decarbonise the world's industry.

The term "process heat" refers to a wide variety of applications, technologies, energy carriers, temperatures and modes of delivering the heat to the materials being processed. Particularly important for the decarbonisation of the industry is the temperature level at which heat is delivered to an industrial process. The demand for process heat at high temperatures (> 500°C) is substantial and not all technologies and energy carriers are able to provide it. Within the RES, biomass is one of the few renewable energy options for the direct supply of high-temperature heat. On the other hand, for the lower temperature process heat applications, much more technological options are available.

The diversity of industrial processes, energy needs and technology requirements make decarbonising the industrial sector challenging. High-temperature heat, which is essential for many industrial processes, is currently largely supplied by fossil fuels. Additional challenges for the transition to low-carbon technologies within the industrial sector are the long lifetimes of the industrial facilities, high capital costs, global market competition and industry aversion to risk. Bioenergy is one of the potential solutions, but its applicability and adequacy vary depending on the industrial sub-sector and specific context.

Combustion-based process heating systems, used in diverse industrial facilities, generate a large share of the energy consumed in the industrial sector. Most of them are fired by fossil fuels, but biomass is also used. There are two broad categories of systems: direct heating systems, where flue gases are in direct contact with the material being processed, and indirect heating systems, where flue gases transfer heat to a heat transfer fluid, which then supplies heat to the production process. Typically, high-temperature process heat is generated through direct systems. Various types of equipment are in operation, tailored to specific industrial processes, such as furnaces in chemical, non-metallic minerals, iron and steel and other industrial sub-sectors. Most of the direct systems use fossil fuels, although biomass co-firing or 100% biomass firing systems are also deployed in specific industries. On the other hand, for low and medium process temperatures, indirect heating systems are mostly used, often involving steam as the heat transfer medium. Diverse biomass conversion technologies for generating heat at these temperature levels are available and used, with a strong deployment in the forest-based industries due to the availability of secondary woody residues.

Modern industrial energy systems are characterised by automation, advanced process control systems and pollution control mechanisms. They are required to comply with environmental and health standards, and integrate advanced combustion technologies, air pollution control equipment and operational optimisations.

The industrial sector relies on various medium- and large-scale biomass combustion technologies, which vary in size, fuel requirements and performance characteristics. These technologies can be categorised into fixed bed, fluidised bed and pulverised fuel combustion based on flow conditions in the combustion chamber. Fixed bed and fluidised bed combustion are the most used in the industrial sector. Fixed bed combustion, typically used for the lower scales, offers flexibility in terms of fuel type and particle size, while fluidised bed combustion, mostly used for the larger capacities, requires smaller particles. Fixed bed systems generally have lower capital and operational costs but lower efficiencies compared to fluidised bed systems. Biomass pulverised combustion is primarily used in thermal power plants rather than for industrial process heat.

A commercially available alternative to conventional direct combustion involves gasification followed by combustion. In the gasifier, the solid biofuels are converted into a gaseous fuel mixture known as syngas, which is then burned. Syngas offers flexibility and has diverse uses, including the generation of heat or combined heat and power. When high-quality clean gas is required, syngas undergoes treatment to remove tars and particulate matter before combustion. Gasifiers can be categorised as directly heated or indirectly heated. In directly heated gasifiers, heat for the gasification reactions is generated through the partial oxidation of biomass, while indirectly heated gasifiers use the gasifying agent or heat exchangers for heating. The method of heat supply significantly affects syngas quality, with indirectly heated gasifiers typically yielding syngas with a higher heating value. Gasification technologies can also be classified based on their operation principles into fixed bed and fluidised bed gasifiers. Fixed bed gasifiers are the classical and still mostly used technology, being well-suited for small scale heat and/or power generation. For the larger scales, fluidised bed gasifiers are used. While they offer improved efficiency and scalability, they are more complex and costlier compared to fixed-bed systems.

An alternative to 100% biomass-firing is co-combustion, which involves simultaneously burning multiple types of fuels in the same plant. Even though co-combustion of biomass and coal is most commonly used for electricity generation, it also finds applications in industrial heating. Co-combustion can be integrated in existing coal-fired plants with minimal modifications. In heat-only applications, co-firing biomass with coal can be advantageous in high-temperature industrial processes, such as in cement kilns, where temperature requirements exceed what raw biomass combustion can achieve.

Since the combustion technologies employed in district heating systems are similar to those used in industrial applications, this sector is also addressed in this chapter. District heating is an efficient energy system that centralises thermal energy conversion and distributes heat through an underground network to a wide range of users, including residential, commercial and industrial facilities. Centralisation allows for advanced technologies and optimised combustion processes, reducing emissions and improving energy efficiency among other advantages. Despite their potential for renewable energy integration, many district heating systems worldwide still rely on fossil fuels, mainly coal and natural gas, although some countries have successfully implemented biomass-based district heating systems.

Acknowledgements The work was supported by *Fundação para a Ciência e a Tecnologia*, through IDMEC, under LAETA [project UIDB/50022/2020]. Isabel Malico expresses her gratitude to Lucas Saffian for the valuable help drawing the sketches included in this chapter.

References

1. IEA (2022) Energy statistics data browser. https://www.iea.org/data-and-statistics/data-tools/energy-statistics-data-browser. Accessed 8 May 2023
2. Paltsev S, Morris J, Kheshgi H, Herzog H (2021) Hard-to-abate sectors: the role of industrial carbon capture and storage (CCS) in emission mitigation. Appl Energy 300:117322. https://doi.org/10.1016/j.apenergy.2021.117322
3. Kesicki F, Yanagisawa A (2015) Modelling the potential for industrial energy efficiency in IEA's world energy outlook. Energ Effi 8:155–169. https://doi.org/10.1007/s12053-014-9273-7
4. Malico I, Nepomuceno Pereira R, Gonçalves AC, Sousa AMO (2019) Current status and future perspectives for energy production from solid biomass in the European industry. Renew Sustain Energy Rev 112:960–977. https://doi.org/10.1016/j.rser.2019.06.022
5. Seyboth K, Beurskens L, Langniss O, Sims REH (2008) Recognising the potential for renewable energy heating and cooling. Energy Policy 36:2460–2463. https://doi.org/10.1016/j.enpol.2008.02.046
6. European Commission (2016) Communication from the Commission to the European Parliament, the Council, the European Economic and Social Committee and the Committee of the Regions. An EU strategy on heating and cooling. European Commission, Brussels, Belgium
7. Union E (2018) Directive (EU) 2018/2001 of the European Parliament and of the Council of 11 December 2018 on the promotion of the use of energy from renewable sources. Off J Eur Union L328:82–209
8. Tilia, TU Wien, IREES, et al (2022) District heating and cooling in the European Union. Overview of markets and regulatory frameworks under the revised Renewable Energy Directive. European Commission, Brussels, Belgium
9. Trinomics, Öko-Institut, DTU (2022) Policy support for heating and cooling decarbonisation: roadmap. European Commission, Brussels, Belgium
10. TU Wien, e-think, Fraunhofer ISI, et al (2022) Renewable space heating under the revised Renewable Energy Directive: ENER/C1/2018 494: final report. European Commission, Brussels, Belgium
11. Eurostat (2023) Eurostat database. https://ec.europa.eu/eurostat/web/main/data/database. Accessed 18 Jul 2023
12. Fraunhofer ISI, Fraunhofer ISE, IRESS, et al (2016) Mapping and analyses of the current and future (2020–2030) heating/cooling fuel deployment (fossil/renewables). Work package 1: final energy consumption for the year 2012. European Commission
13. Chan Y, Kantamaneni R, Allington M (2015) Study on energy efficiency and energy saving potential in industry and on possible policy mechanisms. ICF Consulting Limited, London, UK
14. Fleiter T, Elsland R, Rehfeldt M, et al (2017) Profile of heating and cooling demand in 2015. Heat Roadmap Europe
15. Taibi E, Gielen D, Bazilian M (2012) The potential for renewable energy in industrial applications. Renew Sustain Energy Rev 16:735–744. https://doi.org/10.1016/j.rser.2011.08.039
16. Berglin N, Berntsson T (1998) CHP in the pulp industry using black liquor gasification: thermodynamic analysis. Appl Therm Eng 18:947–961. https://doi.org/10.1016/S1359-4311(98)00038-6
17. Jönsson J, Algehed J (2010) Pathways to a sustainable European kraft pulp industry: trade-offs between economy and CO_2 emissions for different technologies and system solutions. Appl Therm Eng 30:2315–2325. https://doi.org/10.1016/j.applthermaleng.2010.01.025
18. Suhr M, Klein G, Kourti I et al (2015) Best available techniques (BAT) reference document for the production of pulp, paper and board. Publications Office of the European Union, Luxembourg
19. Stubdrup KR, Karlis P, Roudier S, Sancho LD (2016) Best available techniques (BAT) reference document for the production of wood-based panels. Publications Office of the European Union, Luxembourg

20. U.S. Department of Energy, Industrial Heating Equipment Association (2004) Improving process heating system performance. A sourcebook for industry. U.S. Department of Energy, Washington, D.C.
21. Suopajärvi H, Umeki K, Mousa E et al (2018) Use of biomass in integrated steelmaking—status quo, future needs and comparison to other low-CO_2 steel production technologies. Appl Energy 213:384–407. https://doi.org/10.1016/j.apenergy.2018.01.060
22. Danish Energy Agency (2022) Technology data for industrial process heat. Danish Energy Agency
23. US EIA (2018) Manufacturing energy consumption survey (MECS). https://www.eia.gov/consumption/manufacturing/. Accessed 20 Aug 2023
24. Lovegrove K, Alexander D, Bader R, et al (2019) Renewable energy options for industrial process heat. Australian Renewable Energy Agency, O'Connor
25. Pardo N, Moya JA (2013) Prospective scenarios on energy efficiency and CO_2 emissions in the European iron and steel industry. Energy 54:113–128. https://doi.org/10.1016/j.energy.2013.03.015
26. He K, Wang L (2017) A review of energy use and energy-efficient technologies for the iron and steel industry. Renew Sustain Energy Rev 70:1022–1039. https://doi.org/10.1016/j.rser.2016.12.007
27. Worrell E, Phylipsen D, Einstein D, Martin N (2000) Energy use and energy intensity of the U.S. chemical industry. University of California, Berkeley, CA
28. Saygin D, Patel MK, Worrell E et al (2011) Potential of best practice technology to improve energy efficiency in the global chemical and petrochemical sector. Energy 36:5779–5790. https://doi.org/10.1016/j.energy.2011.05.019
29. Madlool NA, Saidur R, Hossain MS, Rahim NA (2011) A critical review on energy use and savings in the cement industries. Renew Sustain Energy Rev 15:2042–2060. https://doi.org/10.1016/j.rser.2011.01.005
30. Usón AA, López-Sabirón AM, Ferreira G, Sastresa EL (2013) Uses of alternative fuels and raw materials in the cement industry as sustainable waste management options. Renew Sustain Energy Rev 23:242–260. https://doi.org/10.1016/j.rser.2013.02.024
31. Fleiter T, Fehrenbach D, Worrell E, Eichhammer W (2012) Energy efficiency in the German pulp and paper industry—a model-based assessment of saving potentials. Energy 40:84–99. https://doi.org/10.1016/j.energy.2012.02.025
32. Lipiäinen S, Kuparinen K, Sermyagina E, Vakkilainen E (2022) Pulp and paper industry in energy transition: towards energy-efficient and low carbon operation in Finland and Sweden. Sustain Prod Consumption 29:421–431. https://doi.org/10.1016/j.spc.2021.10.029
33. IEA (2022) World energy outlook 2022. International Energy Agency, Paris
34. Cui D, Deng Z, Liu Z (2019) China's non-fossil fuel CO_2 emissions from industrial processes. Appl Energy 254:113537. https://doi.org/10.1016/j.apenergy.2019.113537
35. Lenz V, Szarka N, Jordan M, Thrän D (2020) Status and perspectives of biomass use for industrial process heat for industrialized countries. Chem Eng Technol 43:1469–1484. https://doi.org/10.1002/ceat.202000077
36. Saygin D, Gielen DJ, Draeck M et al (2014) Assessment of the technical and economic potentials of biomass use for the production of steam, chemicals and polymers. Renew Sustain Energy Rev 40:1153–1167. https://doi.org/10.1016/j.rser.2014.07.114
37. Olsson O, Schipfer F (2021) Decarbonizing industrial process heat: the role of biomass. IEA
38. Nussbaumer T (2003) Combustion and co-combustion of biomass: fundamentals, technologies, and primary measures for emission reduction. Energy Fuels 17:1510–1521. https://doi.org/10.1021/ef030031q
39. Demirbas A (2005) Potential applications of renewable energy sources, biomass combustion problems in boiler power systems and combustion related environmental issues. Prog Energy Combust Sci 31:171–192. https://doi.org/10.1016/j.pecs.2005.02.002
40. van Loo S, Koppejan J (2008) The handbook of biomass combustion and co-firing. Earthscan, London; Sterling, VA

41. EIPPCB (2007) Reference document on best available techniques in the production of polymers. EIPPCB, Seville
42. EIPPCB (2007) Reference document on best available techniques for the manufacture of large volume inorganic chemicals - ammonia, acids and fertilisers. EIPPCB, Seville
43. Brinkmann T, Giner Santonja G, Schorcht F et al (2014) Best available techniques (BAT) reference document for the production of chlor-alkali. Publications Office of the European Union, Luxembourg
44. Barthe P, Chaugny M, Roudier S, Delgado Sancho L (2015) Best available techniques (BAT) reference document for the refining of mineral oil and gas industrial emissions. Publications Office of the European Union, Luxembourg
45. Falcke H, Holbrook S, Clenahan I, et al (2017) Best available techniques (BAT) reference document for the production of large volume organic chemicals. Publications Office of the European Union, Luxembourg
46. Giner Santonja G, Karlis P, Stubdrup KR et al (2019) Best available techniques (BAT) reference document for the food, drink and milk industries. Publications Office of the European Union, Luxembourg
47. Remus R, Aguado-Monsonet MA, Roudier S, Sancho L (2013) Best available techniques (BAT) reference document for iron and steel production. Publications Office of the European Union, Luxembourg
48. Mousa E, Wang C, Riesbeck J, Larsson M (2016) Biomass applications in iron and steel industry: an overview of challenges and opportunities. Renew Sustain Energy Rev 65:1247–1266. https://doi.org/10.1016/j.rser.2016.07.061
49. Cusano G, Rodrigo Gonzalo M, Farrell F et al (2017) Best available techniques (BAT) reference document for the non-ferrous metals industries. Publications Office of the European Union, Luxembourg
50. EIPPCB (2007) Reference document on best available techniques in the ceramic manufacturing industry. EIPPCB, Seville
51. Scalet BM, Garcia Muñoz M, Sissa AQ et al (2013) Best available techniques (BAT) reference document for the manufacture of glass. Publications Office of the European Union, Luxembourg
52. Schorcht F, Kourti I, Scalet BM et al (2013) Best available techniques (BAT) reference document for the production of cement, lime and magnesium oxide. Publications Office of the European Union, Luxembourg
53. Mikulčić H, Klemeš JJ, Vujanović M et al (2016) Reducing greenhouse gasses emissions by fostering the deployment of alternative raw materials and energy sources in the cleaner cement manufacturing process. J Clean Prod 136:119–132. https://doi.org/10.1016/j.jclepro.2016.04.145
54. Pinto RGD, Szklo AS, Rathmann R (2018) CO_2 emissions mitigation strategy in the Brazilian iron and steel sector–from structural to intensity effects. Energy Policy 114:380–393. https://doi.org/10.1016/j.enpol.2017.11.040
55. Nepomuceno Pereira R, Malico I, Mesquita P, et al (2017) Energy use of cork residues in the Portuguese cork industry. In: Proceedings of the 12th conference on sustainable development of energy, water and environment systems—SDEWES 2017. SDEWES Center, Dubrovnik; Croatia
56. Beauchemin PA, Tampier M (2008) Emissions from wood-fired combustion equipment. Envirochem Services Inc., North Vancouver, B. C.
57. Yin C, Rosendahl LA, Kær SK (2008) Grate-firing of biomass for heat and power production. Prog Energy Combust Sci 34:725–754. https://doi.org/10.1016/j.pecs.2008.05.002
58. Energy and Environmental Analysis, Eastern Research Group (2007) Biomass combined heat and power catalog of technologies. Environmental Protection Agency, U. S
59. Mladenović R, Dakić D, Erić A et al (2009) The boiler concept for combustion of large soya straw bales. Energy 34:715–723. https://doi.org/10.1016/j.energy.2009.02.003
60. Koornneef J, Junginger M, Faaij A (2007) Development of fluidized bed combustion—an overview of trends, performance and cost. Prog Energy Combust Sci 33:19–55. https://doi.org/10.1016/j.pecs.2006.07.001

61. Scala F (2013) Fluidized bed technologies for near-zero emission combustion and gasification. Woodhead Publishing, cambridge, UK
62. Bayham SC, Breault R, Monazam E (2016) Particulate solid attrition in CFB systems—an assessment for emerging technologies. Powder Technol 302:42–62. https://doi.org/10.1016/j.powtec.2016.08.016
63. Leckner B (2016) Developments in fluidized bed conversion of solid fuels. Therm Sci 20:1–18. https://doi.org/10.2298/TSCI150703135L
64. Kleinhans U, Wieland C, Frandsen FJ, Spliethoff H (2018) Ash formation and deposition in coal and biomass fired combustion systems: progress and challenges in the field of ash particle sticking and rebound behavior. Prog Energy Combust Sci 68:65–168. https://doi.org/10.1016/j.pecs.2018.02.001
65. Obernberger I (2009) Reached developments of biomass combustion technologies and future outlook. In: Proceedings of the 17th European biomass conference. Hamburg, Germany, pp 20–37
66. Saidur R, Abdelaziz EA, Demirbas A et al (2011) A review on biomass as a fuel for boilers. Renew Sustain Energy Rev 15:2262–2289. https://doi.org/10.1016/j.rser.2011.02.015
67. da Costa TP, Quinteiro P, Tarelho LA da C, et al (2018) Environmental impacts of forest biomass-to-energy conversion technologies: grate furnace versus fluidised bed furnace. J Clean Prod 171:153–162.https://doi.org/10.1016/j.jclepro.2017.09.287
68. de Souza-Santos ML (2004) Solid fuels combustion and gasification. Modeling, simulation, and equipment operations. Marcel Dekker, New York, NY
69. Peterson D, Haase S (2009) Market assessment of biomass gasification and combustion technology for small- and medium-scale applications. National Renewable Energy Laboratory, Golden, CO
70. Göransson K, Söderlind U, He J, Zhang W (2011) Review of syngas production via biomass DFBGs. Renew Sustain Energy Rev 15:482–492. https://doi.org/10.1016/j.rser.2010.09.032
71. Thomson R, Kwong P, Ahmad E, Nigam KDP (2020) Clean syngas from small commercial biomass gasifiers; a review of gasifier development, recent advances and performance evaluation. Int J Hydrogen Energy 45:21087–21111. https://doi.org/10.1016/j.ijhydene.2020.05.160
72. Hanchate N, Ramani S, Mathpati CS, Dalvi VH (2021) Biomass gasification using dual fluidized bed gasification systems: a review. J Clean Prod 280:123148. https://doi.org/10.1016/j.jclepro.2020.123148
73. Bain RL, Broer K (2011) Gasification. In: Brown RC (ed) Thermochemical processing of biomass. Conversion into fuels, chemicals and power. Wiley, Chichester, pp 47–77
74. Sansaniwal SK, Pal K, Rosen MA, Tyagi SK (2017) Recent advances in the development of biomass gasification technology: a comprehensive review. Renew Sustain Energy Rev 72:363–384. https://doi.org/10.1016/j.rser.2017.01.038
75. Dimpl E (2011) Small-scale electricity generation from biomass. Experience with small-scale technologies for basic energy supply. Part I: biomass gasification, 2nd ed. GIZ Hera, Eschborn
76. Schmid JC, Benedikt F, Fuchs J et al (2021) Syngas for biorefineries from thermochemical gasification of lignocellulosic fuels and residues—5 years' experience with an advanced dual fluidized bed gasifier design. Biomass Conv Bioref 11:2405–2442. https://doi.org/10.1007/s13399-019-00486-2
77. Karl J, Pröll T (2018) Steam gasification of biomass in dual fluidized bed gasifiers: a review. Renew Sustain Energy Rev 98:64–78. https://doi.org/10.1016/j.rser.2018.09.010
78. Claude V, Courson C, Köhler M, Lambert SD (2016) Overview and essentials of biomass gasification technologies and their catalytic cleaning methods. Energy Fuels 30:8791–8814. https://doi.org/10.1021/acs.energyfuels.6b01642
79. IRENA (2015) A background paper to "renewable energy in manufacturing". IRENA, Abu Dhabi
80. Podschelne C, Ciolkosz D (2013) Residential wood heat in the Northeast. In: Jacobson M, Ciolkosz D (eds) Wood-based energy in the northern forests. Springer, New York, New York, NY, pp 111–124

81. Hansson J, Berndes G, Johnsson F, Kjärstad J (2009) Co-firing biomass with coal for electricity generation—an assessment of the potential in EU27. Energy Policy 37:1444–1455. https://doi.org/10.1016/j.enpol.2008.12.007
82. Khorshidi Z, Ho MT, Wiley DE (2014) The impact of biomass quality and quantity on the performance and economics of co-firing plants with and without CO_2 capture. Int J Greenhouse Gas Control 21:191–202. https://doi.org/10.1016/j.ijggc.2013.12.011
83. Roni MS, Chowdhury S, Mamun S et al (2017) Biomass co-firing technology with policies, challenges, and opportunities: a global review. Renew Sustain Energy Rev 78:1089–1101. https://doi.org/10.1016/j.rser.2017.05.023
84. Demirbaş A (2003) Sustainable cofiring of biomass with coal. Energy Convers Manage 44:1465–1479
85. Sahu SG, Chakraborty N, Sarkar P (2014) Coal–biomass co-combustion: an overview. Renew Sustain Energy Rev 39:575–586. https://doi.org/10.1016/j.rser.2014.07.106
86. Rahman A, Rasul MG, Khan MMK, Sharma S (2015) Recent development on the uses of alternative fuels in cement manufacturing process. Fuel 145:84–99. https://doi.org/10.1016/j.fuel.2014.12.029
87. Pardo N, Vatopoulos K, Krook-Riekkola A, Moya JA (2012) Heat and cooling demand and market perspective. Publications Office of the European Union, Luxembourg
88. Babich A, Senk D (2013) Biomass use in the steel industry: back to the future? Stahl Eisen 133:57–67
89. Suopajärvi H, Pongrácz E, Fabritius T (2013) The potential of using biomass-based reducing agents in the blast furnace: a review of thermochemical conversion technologies and assessments related to sustainability. Renew Sustain Energy Rev 25:511–528. https://doi.org/10.1016/j.rser.2013.05.005
90. Lake A, Rezaie B, Beyerlein S (2017) Review of district heating and cooling systems for a sustainable future. Renew Sustain Energy Rev 67:417–425. https://doi.org/10.1016/j.rser.2016.09.061
91. Werner S (2017) International review of district heating and cooling. Energy 137:617–631. https://doi.org/10.1016/j.energy.2017.04.045
92. Ericsson K, Werner S (2016) The introduction and expansion of biomass use in Swedish district heating systems. Biomass Bioenerg 94:57–65. https://doi.org/10.1016/j.biombioe.2016.08.011

Biomass for Power Production and Cogeneration

Isabel Malico

Abstract Despite intensive efforts to decarbonise the power sector and the growing contribution of renewables to global electricity generation, fossil fuels, especially coal, continue to dominate as the most commonly used energy sources in this sector. The power industry accounts for a substantial portion of the world total energy supply and remains the largest contributor to CO_2 emissions. In 2020, renewable energy sources accounted for 28% of the electricity generation, with only 2% of the electricity produced derived from biofuels. Despite this relatively small share, the role of bioenergy in the power sector holds the potential to contribute to grid stability, a critical factor as the share of intermittent renewables in the energy mix increases. Additionally, co-combustion of biomass in coal power plants offers a cost-effective means of reducing carbon emissions, particularly in regions heavily reliant on coal. Several commercial technologies are available for converting biomass into electricity. While the efficiency of dedicated biomass-to-electricity plants is relatively low, combined heat and power systems that utilise waste heat achieve significantly higher overall efficiencies. The choice of technology depends on factors like capacity, efficiency and economic viability. This chapter provides an overview of commonly used conversion technologies for power generation from solid biomass.

Keywords Electricity · Combined heat and power · Bioenergy · Conversion technologies · Co-combustion

1 Introduction

Electricity plays a fundamental role in modern societies, and its share in global final energy consumption is excepted to increase in the future [1]. Moreover, the energy used in electricity generation represents an important proportion of the global total energy supply. In 2020, 16% of the total energy supplied worldwide, which

I. Malico (✉)
IDMEC, Escola de Ciências e Tecnologia, Departamento de Engenharia Mecatrónica, Universidade de Évora, Rua Romão Ramalho 59, 7000-671 Évora, Portugal
e-mail: imbm@uevora.pt

A. C. Gonçalves and I. Malico (eds.), *Forest Bioenergy*, Green Energy and Technology,
https://doi.org/10.1007/978-3-031-48224-3_10

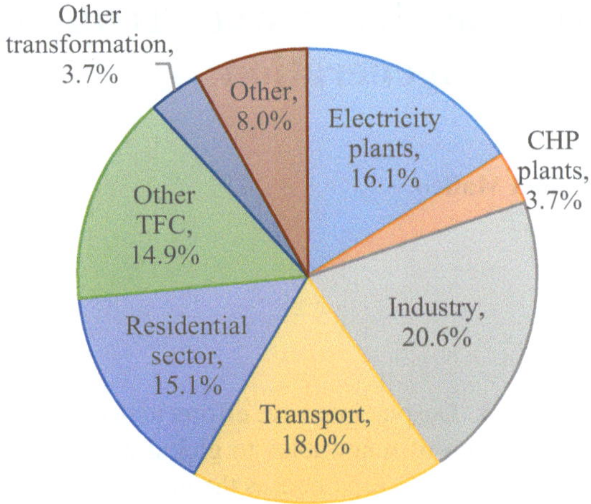

Fig. 1 Share of the different energy uses in the total energy supply in the world in 2020. (*Data source* [2]). TFC—Total final consumption

corresponds to about 94 EJ, was used in the conversion of primary forms of energy into electricity in plants that are designed to produce electricity by both main activity producers that generate electricity for sale to third parties as their primary activity or autoproducers that generate electricity wholly or partially for their own use as an activity which supports their primary activity (Fig. 1, [2]).[1] Additionally, 4% of the total energy supplied worldwide, which corresponds to around 22 EJ, was used in the conversion of primary forms of energy into electricity and heat in CHP (combined heat and power) plants, which are designed to produce both heat and electricity, by main activity producers or autoproducers [2].[2]

In 2020, 69% of the world total energy supply was made available for final consumption, namely for non-energy use and the energy end-use sectors (21% for industry, addressed in chapter "Biomass for Industrial and District Heating", 18% for transport, 15% for the residential sector, addressed in chapter "Biomass for Domestic Heat", 8% for other end-use sectors and 7% for non-energy use) [2]. The rest of the total energy supplied was mainly used to support operations of the energy sector (6%) and in other conversions of primary forms of energy into secondary and further transformation (e.g., in heat plants, oil refineries or charcoal production plants).

The relative importance of the shares of the energy used in the conversion of primary forms of energy in electricity and CHP plants, in the total energy supply, varies among the different world regions (Fig. 2). The groups constituted by the Organisation for Economic Cooperation and Development (OECD) countries in Asia and Oceania, by non-OECD countries in Asia (excluding China) and by countries in the Middle East use more than 20% of their energy supply in the conversion of

[1] Note that these figures exclude the energy own use by the electricity plants. They correspond to energy lost during conversion of primary energy products into electricity.

[2] Note that these figures exclude the energy own use by the CHP plants. They correspond to energy lost during conversion of primary energy products into electricity and heat (for the latter, fuel inputs for the production of heat consumed within the autoproducer's establishment are excluded and accounted in the final consumption of fuels).

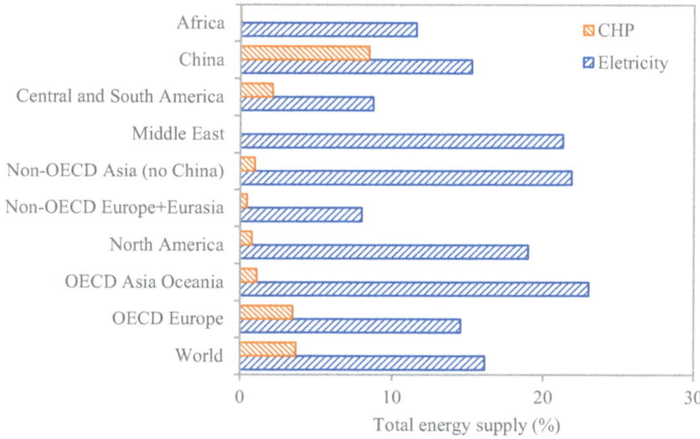

Fig. 2 Share of the energy used in the conversion of primary forms of energy to electricity in electricity plants and electricity and heat sold by CHP plants in the total energy supply in the world in 2020. (*Data source* [2])

primary forms of energy into electricity in electricity plants. On the other extreme, in the groups formed by non-OECD countries in Europe and Eurasia and by the countries in Central and South America, the energy used in the conversion of primary forms of energy into electricity in electricity plants is lower than 10% of the total energy supply.

In China, the share of the energy used in the conversion of primary energy forms into electricity and heat sold by CHP plants in the total energy supplied is 8.5%, the highest percentage among the country groups represented in Fig. 2, while in the groups of countries of the Middle East, Africa, non-OECD Europe and Eurasia and North America, the percentage of the total energy supply which was used in the conversion of primary forms of energy into electricity and heat sold in CHP plants was less to 1%.

As for the share of the different energy sources transformed in electricity and CHP plants in 2020, coal was the most consumed fuel (39% in electricity and 63% in CHP plants), followed by natural gas (22% in electricity and 29% in CHP plants) (Fig. 3). Biofuels and waste played a minor role in the overall electricity production in electricity plants (3%), but are relatively more important in the production of electricity and heat sold by CHP plants (7%).

Globally, fossil fuels continue to dominate the electricity sector, which was the largest source of global energy-related CO_2 emissions in 2020, totalling 36% or 12.3 Gt CO_2 emissions [3]. (It is important to note that this dominance varies by region, as some countries have made substantial progress in adapting renewable energy sources, resulting in higher renewable energy shares). Coal stands out as the major contributor to these emissions, being responsible for around three-quarters of the total, despite generating just over one-third of the world's electricity in 2020 [3].

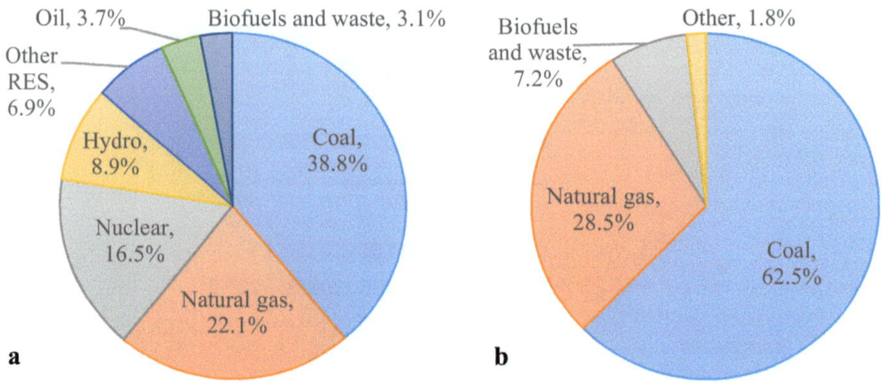

Fig. 3 Share of the various energy sources transformed in **a** electricity plants and **b** CHP plants in the world in 2020. (*Data source* [2])

Natural gas was the second largest contributor to both global emissions and electricity generation in the same year [3].

Notably, the power sector is a substantial consumer of both coal and natural gas on a global scale. In 2020, approximately 63% of the world's annual coal consumption and 38% of its natural gas were utilised in electricity and CHP plants [2]. This underscores the critical role of the power sector in the overall demand for these fossil fuels, highlighting the importance of transitioning to cleaner and more sustainable energy sources in this sector to reduce emissions and mitigate environmental impacts.

In 2020, 26,833 TWh of electricity were produced worldwide, more than the double of that produced in 1990 [2]. Electricity generated from coal accounted for the largest share in the world total electricity generation (35%), followed by electricity generated from natural gas (24%), hydro power (17%) and nuclear power (10%) (Table 1). The relative importance of coal in the total electricity generation has been slowly decreasing in the last decades, but that of natural gas increasing (in 1990 the shares of coal and natural gas power generations were, respectively, 37% and 15%). The fossil fuel that has significantly lost importance in the electricity generation worldwide was oil, with a contribution of 11% to the electricity generated worldwide in 1990 and of only 2.5% in 2020 [2].

Coal is a very important energy source for electricity generation in Asia, which is the region that produces more electricity from this fossil fuel (more than half of the worldwide electricity produced from coal in 2020 was in China) [2]. In 2022, several regions (mostly European Union and India, and to a lesser extent, China) increased coal power generation due to a higher electricity demand in the aftermath of the Covid-19 pandemic, concerns about the natural gas prices and energy security and weather-related reasons [1, 4].

The share of electricity generated by renewable energy sources worldwide has increased in the last thirty years. In 1990, it was 20% and in 2020 28% [2]. Wind and solar photovoltaic (PV) had virtually no expression in 1990 but in 2020 contributed, respectively, 6.0% and 3.1% to the global generation of electricity. However, hydro

Table 1 Share (in %) of the electricity generated from various energy sources in the world total electricity generation in 1990 and 2020 (*Data source* [2])

	Coal	Natural gas	Biofuels	Nuclear	Hydro	Wind	Solar PV	Other sources
1990								
World	37.2	14.7	0.9	16.9	18.4	0.0	0.0	11.8
Africa	51.8	14.2	0.1	2.7	18.2	0.0	0.0	13.0
Central and South America	3.2	8.6	1.5	1.9	71.5	0.0	0.0	13.4
China	72.2	0.4	0.0	0.0	19.5	0.0	0.0	7.8
Middle East	0.0	50.9	0.0	0.0	5.3	0.0	0.0	43.7
Non-OECD Asia excluding China	41.1	9.1	0.1	6.3	24.0	0.0	0.0	19.4
Non-OECD Europe and Eurasia	23.1	38.1	0.0	11.3	14.0	0.0	0.0	13.6
North America	46.9	10.6	2.0	18.0	16.0	0.1	0.0	6.5
OECD Asia and Oceania	23.2	16.7	0.8	21.5	11.9	0.0	0.0	25.8
OECD Europe	37.9	6.5	0.4	29.6	17.4	0.0	0.0	8.2
2020								
World	35.2	23.6	2.1	10.0	16.6	6.0	3.1	3.4
Africa	29.4	41.1	0.2	1.2	17.9	2.1	1.4	6.7
Central and South America	4.8	18.0	6.2	1.9	54.2	6.7	2.0	6.3
China	63.3	3.3	1.7	4.7	17.4	6.0	3.3	0.3
Middle East	0.1	72.5	0.0	0.5	1.9	0.2	0.8	24.0
Non-OECD Asia excluding China	54.9	19.4	2.1	2.7	12.9	2.5	2.7	2.8
Non-OECD Europe and Eurasia	21.0	38.6	0.3	18.4	18.7	1.0	0.8	1.2
North America	17.4	37.0	1.2	17.8	13.8	7.6	2.5	2.7
OECD Asia and Oceania	34.6	33.8	1.9	10.1	6.8	1.8	6.2	4.9
OECD Europe	13.3	20.6	5.1	20.7	17.9	14.1	4.6	3.7

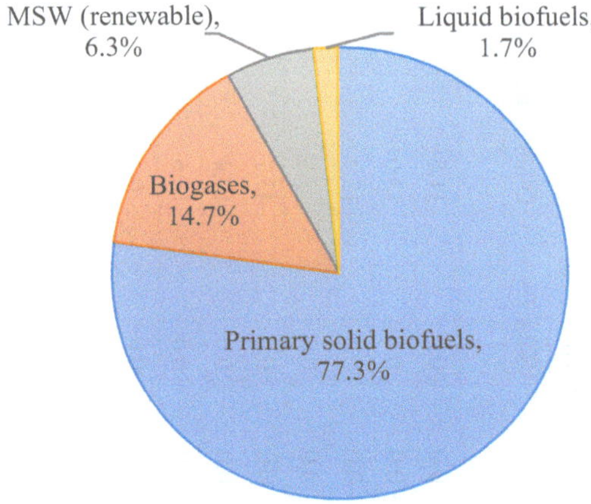

Fig. 4 Share of the various biomass sources in the bioelectricity generated in the world in 2020. (*Data source* [2]). MSW (renewable)—Renewable fraction of the municipal solid wastes

power is still the dominant renewable energy source globally (share of 16.6% in 2020). As far as biofuels are concerned, 2.1% of the electricity generated in the world in 2020 was based in this renewable energy source, with the electricity produced from biofuels having increased around 5.5 times since 1990 [2].

It was in Central and South America and in OECD Europe that the electricity generated from biofuels had the most important share in the total electricity generated in 2020 (6.2% and 5.1%, respectively). These are the two world regions where the share of electricity generated from fossil fuels was the lowest (in Central and South America more than half of the electricity generated is hydroelectricity, while in OECD Europe the sources are more diversified, with nuclear power having the highest share). Moreover, almost one-third of the electricity generated from biofuels worldwide was in OECD Europe; China being the second largest producer of electricity from biofuels (23%) [2]. Moreover, all the regions saw a substantial increase in the electricity generation from biofuels from 1990 to 2020, except North America, where, in 2020, the electricity generated by biofuels was 87% of that produced in 1990 [2].

Most of the bioelectricity generated worldwide in 2020 came from solid biofuels (Fig. 4). However, the relative importance of solid biofuels in this energy-use sector is lower than in others with relevance in terms of bioenergy consumption. (*cf.* chapters "Biomass for Domestic Heat" and "Biomass for Industrial and District Heating", in the residential and industrial sectors, almost all the consumption of biomass refers to solid biofuels). Particularly important for biopower production is biogas, with OECD Europe being by far the largest producer (73%) of the global biogas electricity in 2020 [2]).

Industry was the end-use sector that consumed the largest share of electricity (42%), followed by households (28%) and the commercial and public service sector (20%) in 2020 (Fig. 5).

Despite the small share of bioenergy in the global electricity generation, it can contribute as a flexible resource in the renewable power supply system. Presently, it

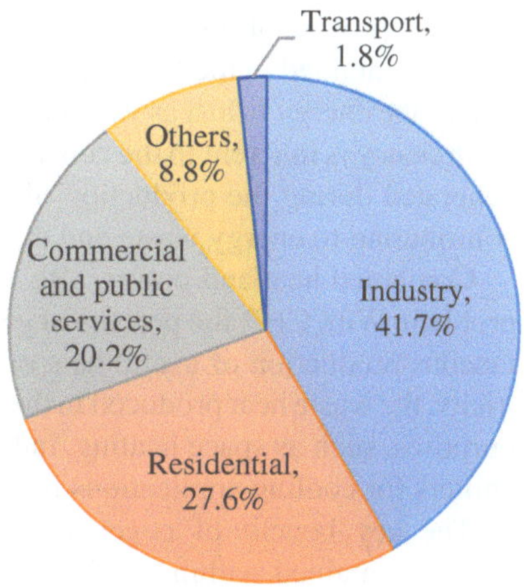

Fig. 5 Share of the different end-use sectors in the final consumption of electricity in the world in 2020. (*Data source* [2])

is already used for grid balancing and holds the potential to have a more significant contribution to balancing future grids [5]. As the share of renewable energy sources in electricity generation grows, the challenge of ensuring a secure energy supply increases. Conventional dispatchable energy sources will face increased operational costs, pushing them out of the market and potentially leading to a capacity-based market [5]. Bioenergy can play a role in grid stability, including as a form of energy storage [5].

The remainder of this chapter is structured as follows. First the most relevant technologies used in dedicated biomass-fired power generation, including both electricity-only and combined heat and power production, are described. Next, co-combustion of biomass is covered. Finally, the chapter will conclude with some key insights and considerations.

2 Technologies for Electricity and Combined Heat and Power Production

Chapter "Biomass for Industrial and District Heating" introduced commercial primary conversion technologies that effectively convert solid biomass into either heat or fuels. In the context of heat generation through combustion, heat can either be used directly or, alternatively, directed towards the production of electricity (Fig. 6). To achieve the latter, a variety of secondary energy conversion technologies come into play, including conventional steam turbines, steam engines, organic Rankine cycles (ORC), Stirling engines, reciprocating internal combustion engines, gas turbines and micro-turbines, among others. The choice of which technology to employ depends on a multitude of factors intricately linked to the primary conversion technologies in use. This section will present the main secondary conversion technologies used for biopower production.

In the process of producing electricity from biomass or any other fuel, not all the energy contained in the fuel is converted into electricity; typically a significant fraction of the energy is lost as waste heat during the electricity generation process. This inefficiency is inherent to the conventional power generation systems, where the heat generated during the production of electricity is often rejected to the environment, contributing to energy waste and environmental concerns.

Combined heat and power, also known as cogeneration, offers a solution to this problem. With CHP, the process of generating electricity is combined with the simultaneous production of useful heat or steam. This means that while generating electricity, the waste heat produced in the process is captured and used for various heating purposes, such as space heating, industrial processes, or even for driving absorption chillers for cooling applications [6].

The key benefit of cogeneration is the improved efficiency in utilising the energy contained within the fuel [6]. By using the otherwise wasted heat, cogeneration systems can achieve overall energy efficiency levels that far surpass those of conventional power generation systems. This not only reduces energy costs but also minimises greenhouse gas emissions, making cogeneration a more sustainable approach to energy production [6].

The primary conversion technologies commercially available for biomass conversion into power or combined heat and power are combustion and gasification (Fig. 6).

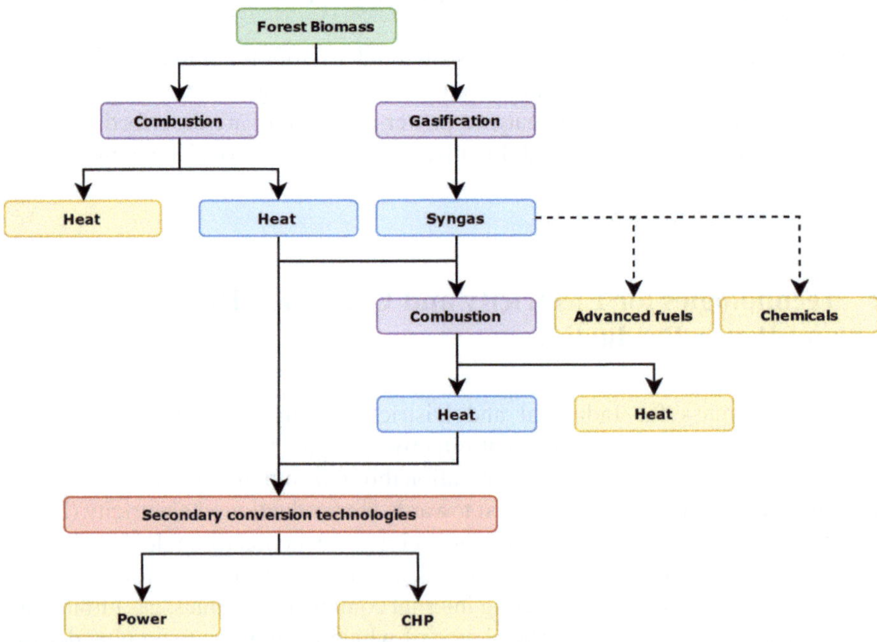

Fig. 6 Biomass conversion routes based on combustion and gasification. The dashed lines refer to processes that are not commercial and are not presented in detail in this book

Combustion is commonly used to generate heat, power or CHP, whereas gasification is primarily employed for small-scale CHP production [7]. Nevertheless, direct-fired systems that rely on combustion stand as the predominant conversion technology [8]. One of the key advantages of using gasification for power production, in comparison to direct biomass combustion, lies in its higher electrical efficiency, particularly for smaller-scale systems [9]. However, syngas produced through gasification often contains high levels of tar and char, requiring gas cleaning before its utilisation in reciprocating internal combustion engines or gas turbines [10].

Most secondary conversion technologies used in biopower production rely on heat engines, which are systems design to convert thermal energy into mechanical work. Heat engines can be categorised into two main types based on the working fluid they utilise: external combustion engines and internal combustion engines (or closed thermal cycles and open cycles, respectively). In internal combustion engines, the reactants and combustion products themselves serve as the working fluid [11]. Examples of such engines include directly fired gas turbines, compression-ignition engines and spark-ignition engines. On the other hand, external combustion engines use a different fluid as the working medium. Steam turbines and Stirling engines fall into this category.

External combustion engines are particularly suited for applications involving direct biomass combustion because they protect the engine from potential damage caused by fly-ash particles and metals present in the flue gases [12]. Conversely, internal combustion engines require gas cleaning when solid biomass combustion is used as the heat source [12]. An alternative to directly burning solid biomass in internal combustion engines is the combustion of syngas produced by gasification. The next sub-sections present commonly used secondary conversion technologies used in the conversion of solid biomass into power or CHP.

Steam Turbines

Conventional steam turbines represent the most widely adopted technology in both biomass-fired power plants [13] and combined heat and power plants [8]. These steam turbines are well-established and are commonly employed in facilities with a capacity of less than 50 MW_e [8], offering economic viability even at capacities as low as 1 MW_e [14]. The relatively smaller size of biomass-fired power plants, in comparison to their coal-fired counterparts, for example, is typically attributed to the availability of local biomass resources [15]. Nevertheless, numerous dedicated biomass power plants and CHP plants with capacities above 100 MW_e are currently operational (e.g., [16]). In this case, biomass has to be sourced from a wider region and/or imported.

The electrical efficiency of biomass-fired steam turbines varies depending on the installed capacity, with larger capacities generally yielding higher efficiencies, whereas smaller capacities tend to exhibit lower efficiencies [16, 17]. Typically, the electrical efficiencies of existent biomass plants fall within the range of 20–40% [5].

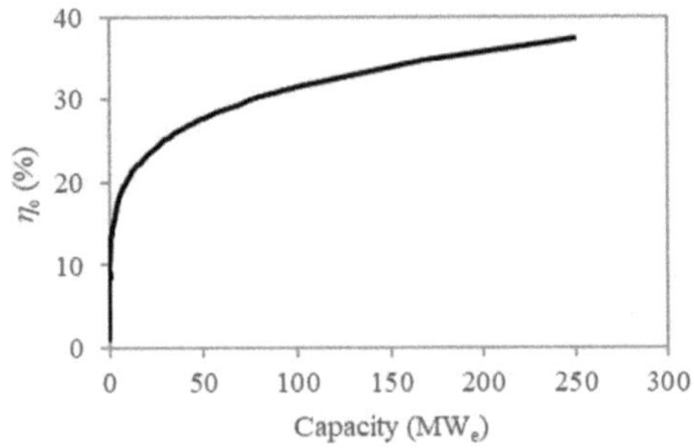

Fig. 7 Electrical efficiency of biomass-fired power plants with steam turbines as prime mover versus plant capacity (adapted from [16, 17, 20–23])

The lower limit typically corresponds to smaller capacity power plants and CHP power plants, while the largest power plants rarely exceed 35% efficiency [5]. On the other hand, industrial CHP plants with steam extraction may have efficiencies below 20% [5]. The electrical efficiency decreases significantly with capacity for small, decentralised power plants with a steam turbine (Fig. 7). For example, for microturbines with 30 kW_e, typical efficiencies drop down to 6–8% [18]. However, if these turbines are used in CHP applications, the overall efficiency, which is the sum of the electrical and heating efficiencies, can reach 90% [19].

In direct-fired biomass power plants with a steam turbine prime mover (heat engine), biomass is burned within a boiler, generating high-pressure, high-temperature steam that drives a turbine, usually coupled with a generator to produce electricity (Figs. 8 and 9). In CHP applications, steam may be extracted from the turbine at intermediate pressures and temperatures to be directly used or to provide heat, for example, for an industrial process, district heating, space heating or cooling [8]. The major solid biomass conversion technologies associated with steam turbines are fixed bed and fluidised bed boilers and co-combustion [8].

The Rankine cycle is the thermodynamic cycle that models a simple vapour power plant, which is based on the following working principle (Fig. 8a): Liquid water is converted to high-pressure, high temperature steam in a boiler and passes through the turbine, where it expands to a lower pressure, transferring work to the shaft of the turbine that is connected to an electricity generator. Afterwards, the vapor leaving

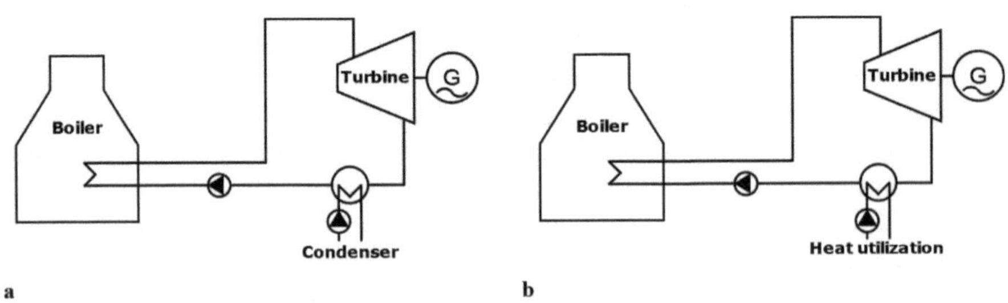

Fig. 8 Basic layout for **a** condensing steam turbine and **b** back-pressure steam turbine

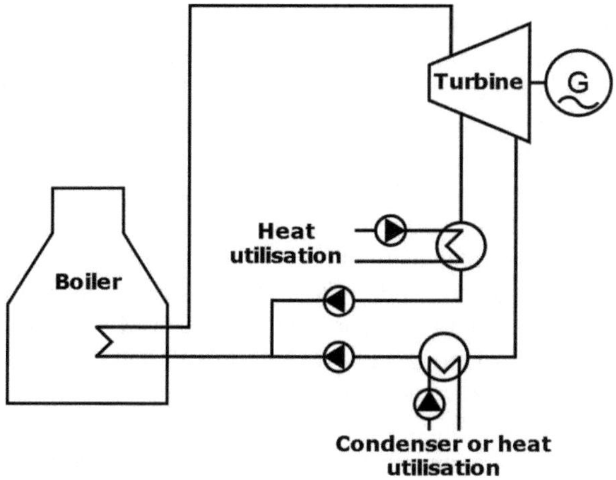

Fig. 9 Basic layout for an extraction steam turbine

the turbine passes through a condenser, where in condenses. To complete the cycle, a pump feeds the liquid water at elevated pressure back into the turbine.

Steam turbines are characterised by a variety of designs and levels of complexity, allowing them to be tailored to specific applications and performance requirements [8]. There are three basic types of steam turbines: condensing steam turbines, extraction steam turbines and back-pressure turbines. The condensing steam turbine is used for electricity-only applications and expands the steam to low pressure (vacuum conditions), exhausting directly to condensers (Fig. 8a). On the other hand, the other two steam turbine types are used for CHP. Extraction turbines, extract part of the steam at intermediate pressure for heat utilisation, while the rest of the steam is either expanded to low pressure to a condenser or delivered for heat utilisation at low pressure (Fig. 9). Alternatively, in back-pressure turbines, also called non-condensing turbines, the steam is exhausted at a pressure high enough to be used by the process, leaving the turbine at conditions close to those required by the specific CHP application (Fig. 8b).

Compared to a condensing power plant, the utilisation of heat in steam turbines used for CHP typically results in a reduction in electrical efficiency of around 10%, since only a portion of the enthalpy difference in the turbine is utilised for power generation [12]. Nevertheless, co-generation enhances the overall efficiency, allowing it to reach levels up to 90% [19].

The heat to power ratio, a crucial factor in CHP system selection, represents the ratio of thermal energy to electricity needed by an energy-consuming facility. A wide range of heat-to-power ratio are possible [24]. Small CHP plants, characterised by low electrical efficiencies, should be operated in a heat controlled mode, whereas large CHP plants are typically operated in an electricity-controlled mode [12]. It is essential for the heat-to-power ratio of the facility to align with the characteristics of the intended CHP system.

Steam turbines can be used in an organic Rankine cycle, which closely resembles the conventional steam cycles in its operating principles. However, in contrast to the use of water of the conventional cycle, the ORC employs organic fluids (e.g., siloxanes, synthetic oils or hydrocarbons, such as cyclohexane, decane and toluene) with a lower boiling point as its working fluid [25, 26]. This choice enables ORC systems to function more efficiently at relatively low temperatures, typically ranging from 70 to 300 °C [12]. As a consequence they offer an attractive solution for the recovery of waste heat [27], being also capable of utilising a dedicated heat source, such as forest biomass [28]. Commercially available biomass ORC combined heat and power plants are utilised for capacities up to 8 MW_e (CHP applications dominate the market) [29]. These systems represent the most widely adopted biomass technology based on combustion under 1 MW_e [30]. ORC offers advantages such as superior performance at partial loads [25] and improved electrical efficiencies when compared to conventional steam turbines of equivalent capacities [12], albeit still within the relatively modest efficiency range of 10–20% [25].

Biomass-fuelled power plants utilising the organic Rankine cycle typically incorporate a secondary circuit for heating the organic fluid, as opposed to direct heating by the flue gases (Fig. 10). In this configuration, hot flue gases are used to heat thermal oil within an atmospheric liquid tube boiler. This thermal oil, in turn, provides the necessary heat for evaporating the organic fluid in the evaporator. Following expansion in the turbine, where it generates work to drive the generator shaft and produce electricity, the organic fluid transfers energy to the CHP application. In order to enhance efficiency, a regenerator can be introduced after the turbine to preheat the organic fluid before it enters the evaporator. Additionally, an economiser can be employed to recover heat from the flue gases, which still have a relatively high enthalpy (please note that the economiser is not shown in the figure).

Because the steam boiler used in the conventional Rankine cycle is replaced by an atmospheric liquid tube boiler in an ORC, the specific investment costs [31] and maintenance costs [12] are lower for the ORC. However, generally, conventional steam turbines gain economic advantage for larger capacities [31].

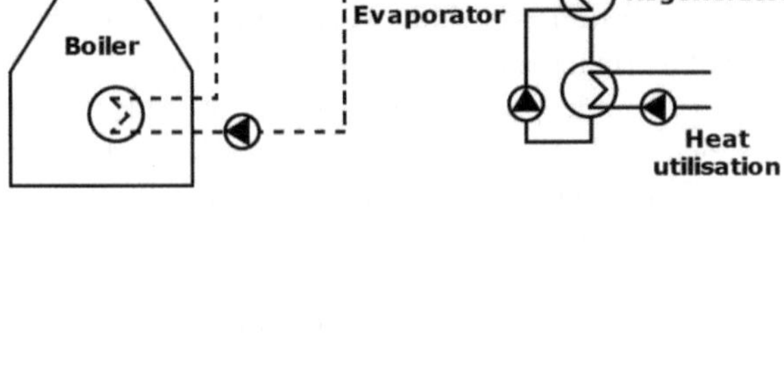

Fig. 10 Layout for an organic Rankine cycle biomass steam turbine

Reciprocating Internal Combustion Engines

Reciprocating internal combustion engines (ICE)[3] are a mature and widely used technological solution in power generation and land, sea and air transportation [6]. What fundamentally distinguishes the design and operation of these engines from most other engine types is the fact that the combustion chamber is the primary location where the engine produces work. As a consequence, combustion occurs intermittently. ICEs can be categorised into two main types: spark-ignition engines, also known as Otto engines or gasoline engines (although they can run on various other fuels), and compression-ignition engines, commonly referred to as Diesel engines. Both types of engines operate with gaseous and liquid fuels and not with solid fuels [8]. Consequently, solid biomass needs to undergo a conversion process to transform it into gaseous or liquid biofuels before it can be utilised in an ICE.

Historically, diesel engines were the preferred choice for power generation. However, due to environmental concerns related to particulate matter and NO_x emissions, their usage has been progressively limited, especially in industrialised countries, despite their higher efficiency [6]. As a result, spark-ignition engines have now become the preferred choice for stationary power applications with higher duty cycles [6].

The economic viability of internal combustion engines for on-site power generation frequently depends on the efficient utilisation of the thermal energy, which typically accounts for a large part of the fuel energy input [6]. When reciprocating combustion engines are used in CHP applications, heat can be recovered from the hot flue gases, the jacket cooling water and engine lubrification oil (at different temperatures). This offers flexibility compared to, for example, gas turbines [8].

Gasifiers coupled with reciprocating internal combustion engines and steam turbines used in an ORC are the dominant technologies in use for biomass-based CHP installations smaller than 1 MW_e [30]. Moreover, when considering a wide range of system sizes, ICEs powered by syngas find primarily application in small-scale systems [8]. The major solid biomass conversion technologies associated with reciprocating internal combustion engines are fixed bed and fluidised bed gasifiers [8].

Reciprocating internal combustion engines are simple, robust, low-cost, start quickly, follow load well and present good efficiencies at partial loads [8, 11]. Additionally, they are able to handle moderately well syngas impurities, which leads to most straightforward and cost-effective cleaning systems when compared to other secondary technologies coupled with gasification [10]. However, ICEs are characterised by high operational and maintenance costs and high levels of NO_x emissions

---

[3] Reciprocating internal combustion engines are commonly referred to as internal combustion engines. Technically, an internal combustion engine is one in which combustion takes place inside the engine itself, with the working fluids being the reactants and combustion products [11]. According to this definition, directly fired gas turbines and others engines fall into the category of internal combustion engines. However, the term "internal combustion engine" is widely used when referring to spark-ignition and compression-ignition engines. In this book, the acronym "ICE" is used when referring to reciprocating internal combustion engines, because of its widespread usage.

[10]. The complexity of gasification system operation has limited the uptake of this technology and justifies the current low cumulative installed capacity [32]. Despite these disadvantages, gasification combined with ICEs is commercially available [12] and represents the most prevalent option for the gasification route [33].

Electrical efficiencies of systems integrating gasifiers and internal combustion engines fall within the typically range of 15–40% [34] and are higher than those of conventional steam turbines, especially at lower capacities [9, 20, 21].

Other Technologies

Stirling engines present another commercially available option as a prime mover for biomass-based combined heat and power applications. When fuelled by biomass, they are reciprocating external combustion engines, which operate on a closed cycle. The heat generated by the combustion of biofuels (e.g., solid forest biomass or syngas) is transferred to the working fluid, typically air or helium, that produces the mechanical work [12].

Unlike reciprocating internal combustion engines, Stirling engines operate with continuous combustion, which enables more complete and cleaner combustion, resulting in lower emissions [8]. Another advantage inherent in external combustion, as opposed to the internal combustion characteristic of ICEs, is the capability to use the hot flue gases generated by the combustion of solid biofuels as a heat source. However, practical implementation requires that these flue gases are as clean as possible to prevent corrosion or fouling of the heat exchanger [12]. Because of these problems, the use of low-quality fuels can lead to considerable operational difficulties. Additionally, Stirling engines are characterised by lower maintenance requirements [35]. Still, their high specific investment costs pose challenges for economic viability, favouring CHP applications with a continuous demand profile [36].

Stirling engines are well-suited for small-scale capacities, ranging from around 1 kW_e to slightly over 100 kW_e and typically offering electrical efficiencies in the range of 15–30% [12]. They are particularly well suited for residential and commercial applications, thanks to their power outputs and heat-to-power ratio [35]. However, despite being a technological promise, Stirling engines using biomass as a heat source still have limited wide spread adoption due to unresolved technical problems and the application of Stirling engines on the market is more common with other heat sources, such as solar or gas [36].

Another commercial technology available for small plants is steam engines, which are used in a Rankine cycle and share the same system layout with the steam turbine (Fig. 8, but with the steam turbine replaced by the steam engine). They represent a mature technology for small power plants, ranging from 20 kW_e to 5 MW_e [37], and offer efficiencies that are on par with or slightly higher than conventional steam turbines (typical efficiencies are 6–20%) [12]. Notably, they exhibit superior part load efficiency compared to steam turbines, although, in certain countries, they have been supplanted by more cost-effective alternatives [37] and the market availability is low [36, 38].

The utilisation of solid biomass in gas turbines is presently in the development phase and has not yet become a commercial reality [37]. Gas turbines operate according to the Brayton cycle and can function as either an internal combustion engine (directly fired gas turbines), where flue gases flow through the turbine, or as an external combustion engine (externally fired gas turbines), where combustion is external to the power cycle and the thermal energy from biomass combustion is transferred to the working fluid through a heat exchanger. Furthermore, these turbines can operate in simple or combined-cycle modes.

In directly fired gas turbines, syngas must undergo a purification process before being used in the turbine to avoid damaging the blades of the turbine, which increases the operational costs [8]. Moreover, because syngas has a low heating value, conventional natural gas turbines must be retrofitted to operate on low-heating value fuels, which is not an easy task and increases the capital cost of the turbine [8]. Because of these challenges, even though being a proven technology in power generation when using high-grade fuels, low heating-value fuels have not been demonstrated in gas turbines [8].

Externally fired gas turbines, also called indirect gas turbines, offer advantages over directly fired gas turbines, including lower syngas cleaning costs, the ability to use low-pressure combustion or the possibility to use lower quality fuels [39]. As a result, the focus of research is on biomass-fuelled externally fired gas turbines, with a particular emphasis on their potential for CHP applications [39]. The complexity of the high-temperature heat exchange between the flue gas and the working fluid is a major barrier to the commercialisation of this technology [12, 37].

Other potential method for power generation from solid biomass that has not reached commercialisation is the integration of a gasification systems with fuel cells [40, 41]. Unlike the technologies mentioned earlier, fuel cells do not rely on combustion. They offer the potential of achieving high efficiencies; however, significant challenges must be addressed before this technology can be commercialised. These challenges encompass issues such as cost, durability and the necessity for thorough syngas clean up upstream the fuel cell [41].

3 Co-combustion

In the energy sector, similarly to process heat applications (*cf.* chapter "Biomass for Industrial and District Heating"), co-combustion (also called co-firing) with coal might be an interesting option. This approach gains particular significance when considering that coal remains a central player in global electricity production (Fig. 3), contributing substantially to worldwide energy-related CO_2 emissions.

Co-combustion of biomass with coal involves simultaneous combustion in power or CHP plants. Many existing coal power plants can partially or entirely be fired with biomass [42] and, as of 2013, approximately 230 power and CHP plants, from 50 to

700 MW_e, used co-combustion, mostly in northern Europe and the United States of America [43]. Moreover, biomass co-combustion initiatives around the world have predominantly focused on retrofitting existing coal power plants [42]. This approach leverages the advantages of co-combustion, which allows a cost-effective transition from power plants originally designed for coal to a renewable energy source without the need for entirely new infrastructure [42, 44, 45]. While not entirely carbon neutral, co-combustion can significantly reduce carbon emissions compared to exclusively using fossil fuels [46–48], thereby, allowing for a reduction in the carbon footprint associated with coal-based energy production. Furthermore, because of the typically composition of biomass, co-combustion most often leads to a decrease in SOx and NOx emissions when compared to coal-firing [46–48].

Co-combustion is gaining popularity for CHP applications and can be applied to a large variety of combustor types [24]. The type of combustion technology used is essentially dependent on the original technology installed in the coal power plant. Most biomass co-combustion facilities are large pulverised coal power boilers [12], but bubbling and circulating fluidised bed boilers, stoker boilers and cyclone boilers are also used [45].

Compared to the option of 100% biomass firing in dedicated power plants, co-combustion of coal with biomass reduces specific costs [45, 48], increases the overall conversion efficiency [46–48] and security of fuel supply [42]. Typical electrical efficiencies of biomass co-firing power plants are in the range of 36–44%, depending on the efficiency of the coal-fired power plant [16]. When raw biomass is co-fired in a coal power plant, there is a decrease in efficiency compared to 100% coal-firing [49]. However, if torrefied biomass is used instead, there is no decrease in efficiency, due to its similar properties to coal [49]. In recent years, there has been considerable attention given to the thermal treatment of biomass materials, with the aim of improving their properties for power production [42].

However, it is crucial to critically assess co-combustion from the perspectives of resource and energy efficiencies. This is especially important when it involves large scale power generation, long-distance biomass transport and potentially inefficient conversion processes with limited or no utilisation of heat [50].

Three options are available for co-combustion: (i) direct co-firing, where solid biomass and coal are fed into a furnace (Fig. 11a); (ii) parallel co-firing, where an additional, separate biomass boiler produces steam which is used within the coal plant steam and power generation systems (Fig. 12); and (iii) indirect co-firing, where biomass is initially gasified and both the syngas produced and the coal are burned in the furnace (Fig. 11b) [12, 45, 47, 51].

Direct co-firing can be achieved in several ways. The simplest method involves pre-mixing solid biomass with coal in the coal conveying system and processing it through existing coal milling and firing systems. This approach is cost-effective but only suitable for low co-firing ratios (10–12%) [42]. Another approach involves the separate handling and injection of biomass into the coal firing system, with options including injection into the pulverised coal pipework, modification of existing pulverised coal burners or utilisation of new, purpose-designed biomass burners. This allows for higher co-firing levels, but has higher capital costs [42]. Pulverised

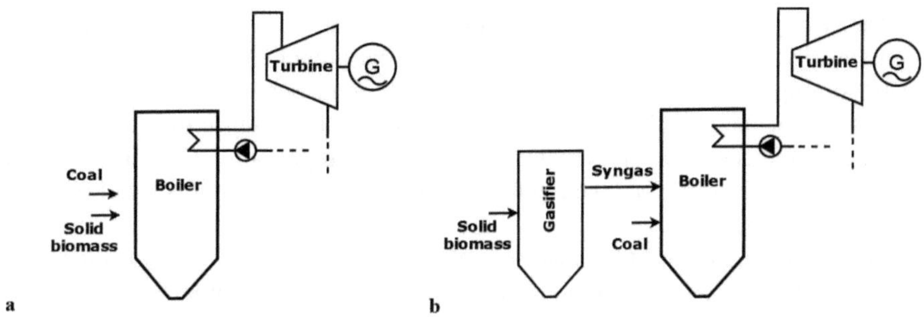

Fig. 11 Partial layout of **a** direct biomass co-firing system and **b** indirect biomass co-firing system

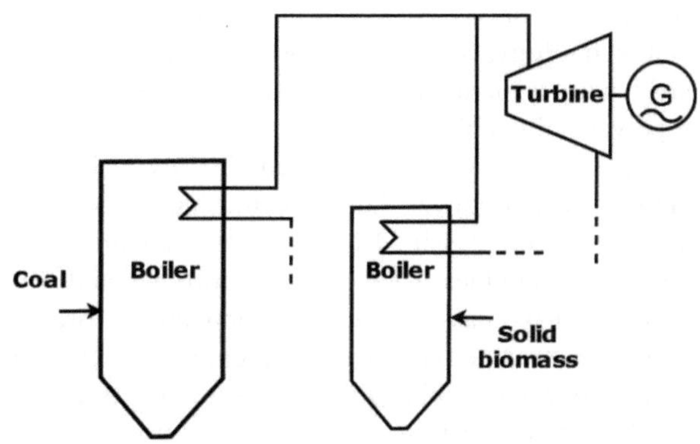

Fig. 12 Partial layout of parallel biomass co-firing system

combustion allows for co-firing ratios in the range of 1–40%, while fluidised bed combustion is the most suitable technology allowing for co-firing ratios up to 95% [45].

4 Final Considerations

Electricity plays a critical role in modern societies and its significance is expected to increase as its uses diversify. The generation of electricity accounts for a substantial fraction of the global total energy supply and represents the largest source of energy-related CO_2 emissions worldwide. In 2020, approximately 20% of the world total energy supply was used in power plants for electricity production or in CHP plants, which produce both electricity and heat simultaneously. Moreover, during the same year, electricity generation contributed to 36% of all energy-related CO_2 emissions worldwide.

Despite all the efforts to decarbonise the power sector, fossil fuels, especially coal, continue to be the primary energy sources for global electricity generation. In 2020, coal and natural gas combined accounted for nearly 60% of the world electricity production. However, renewable energy sources, including biomass, have been steadily growing and constituted 28% of the global electricity production in 2020. Among these RES, biofuels made a small contribution, comprising approximately 2% of the global electricity production. The importance of biofuels in the generation of electricity varies by region. It is more important in Central and South America and in Europe, but even in these regions the contribution is small (6% and 5%, respectively).

The conversion of solid biomass into electricity requires the use of primary and secondary conversion technologies. Primary conversion technologies convert solid biomass into heat or fuels, while secondary conversion technologies typically convert this heat or fuels into power.

The primary conversion technologies commercially available for biomass conversion into power or CHP are combustion and gasification. Combustion, the predominant technology, is commonly used for heat, power or CHP generation. In contrast, gasification is mainly used for small-scale CHP, even though it is also used in heating or power-only applications.

Most secondary conversion technologies in biopower production are based on heat engines, which convert heat into mechanical work and then into electricity. Heat engines can be classified into external and internal combustion engines. External combustion engines, such as steam turbines and Stirling engines, are suitable for direct solid biomass combustion applications, protecting the engine from flue gas contaminants. Internal combustion engines, such as directly fired turbines or reciprocating internal combustion engines, require flue gas or syngas cleaning, depending on whether combustion or gasification is used.

Conventional power generation systems, which only generate electricity, waste a significant portion of the energy contained in the fuel as heat. Combined heat and power, also known as cogeneration, address this inefficiency by capturing waste heat for useful purposes, such as space heating or industrial processes. CHP systems enhance energy efficiency and reduce both costs and greenhouse gas emissions, making them a sustainable energy production approach.

The most widely used secondary conversion technologies for the conversion of biomass into power or CHP are conventional steam turbines, used in plants with capacities typically ranging from 1 to 50 MW_e (larger plants with hundreds of MW_e are operational, though). Their electrical efficiency depends on the installed capacity, with larger capacities generally having higher efficiencies, and typical efficiency values lying between 20 and 40%. In CHP applications, overall efficiency can reach up to 90%. Organic Rankine cycle systems are a commercial alternative for smaller capacities (up to 8 MW_e), providing improved performance at partial loads and electrical efficiencies within a range of 10–20% (typical larger than that of conventional steam turbines of equivalent capacity). ORC systems have lower investment and maintenance costs, making them suitable for small to medium-sized biomass CHP installations, while conventional steam turbines are economically favourable for larger capacities.

Reciprocating internal combustion engines are another commercially available alternative, mainly used with gasifiers for smaller biomass-based CHP installations.

These systems offer electrical efficiencies ranging from 15 to 40%. Although ICEs have high operational and maintenance costs and NOx emissions, they offer advantages such as simplicity, robustness and lower investment cost, being a prevalent option for gasification-based CHP systems.

Other secondary conversion technologies for the conversion of solid biomass into electricity exist, some commercial, others still under-development. These include Stirling engines, steam engines and gas turbines or fuel cells.

Another possibility for the integration of biomass in the power sector is co-combustion, which involves the simultaneous combustion of biomass and other fuels. In regions heavily reliant on coal power, co-combustion of biomass with coal might offer a cost-effective solution for the transition from coal power plants to a renewable energy source without the need for an entirely new infrastructure. Among other advantages, this strategy significantly reduces carbon emissions compared to the exclusive use of fossil fuel, contributing to a reduction in the carbon footprint associated with coal-based energy production. Various combustion technologies can be adapted for co-combustion, with electrical efficiencies typically ranging from 36 to 44%, depending on the efficiency of the original coal-fired plant.

Acknowledgements The work was supported by *Fundação para a Ciência e a Tecnologia*, through IDMEC, under LAETA [project UIDB/50022/2020].

References

1. IEA (2022) World energy outlook 2022. International Energy Agency, Paris
2. IEA (2022) Energy statistics data browser. https://www.iea.org/data-and-statistics/data-tools/energy-statistics-data-browser. Accessed 8 May 2023
3. IEA (2021) World energy outlook 2021. International Energy Agency, Paris
4. IEA (2022) Coal 2022. Analysis and forecast to 2025. International Energy Agency, France
5. Arasto A, Chiaramonti D, Kiviluoma J, et al (2017) Bioenergy's role in balancing the electricity grid and providing storage options—an EU perspective. IEA Bioenergy
6. Darrow K, Tidball R, Wang J, Hampson A (2017) Catalog of CHP technologies. Environmental Protection Agency, U. S
7. Hrbek J (2020) Past, present and future of thermal gasification of biomass and waste. Acta Innovations 35:5–20. https://doi.org/10.32933/ActaInnovations.35.1
8. Energy and Environmental Analysis, Eastern Research Group (2007) Biomass combined heat and power catalog of technologies. Environmental Protection Agency, U. S
9. Itai Y, Santos R, Branquinho M et al (2014) Numerical and experimental assessment of a downdraft gasifier for electric power in Amazon using açaí seed (Euterpe oleracea Mart.) as a fuel. Renewable Energy 66:662–669. https://doi.org/10.1016/j.renene.2014.01.007
10. Bain RL, Broer K (2011) Gasification. In: Brown RC (ed) Thermochemical processing of biomass. Conversion into fuels, chemicals and power. Wiley, Chichester, pp 47–77
11. Heywood JB (2018) Internal combustion engine fundamentals, 2nd edn. McGraw-Hill
12. van Loo S, Koppejan J (2008) The handbook of biomass combustion and co-firing. Earthscan, London; Sterling, VA
13. Corradetti A, Desideri U (2007) Should biomass be used for power generation or hydrogen production? J Eng Gas Turbines Power 129:629–636. https://doi.org/10.1115/1.2718226

14. Pantaleo AM, Ciliberti P, Camporeale S, Shah N (2015) Thermo-economic assessment of small scale biomass CHP: steam turbines versus ORC in different energy demand segments. Energy Procedia 75:1609–1617. https://doi.org/10.1016/j.egypro.2015.07.381
15. Gonçalves AC, Malico I, Sousa AMO (2021) Energy production from forest biomass: an overview. In: Gonçalves AC, Sousa AMO, Malico I (eds) Forest biomass–from trees to energy. IntechOpen, London, UK, pp 1–23
16. IEA-ETSAP, IRENA (2015) Biomass for heat and power. Technology brief. IRENA
17. Brasil ACM, Brasil A, Malico I (2020) Evaluation of the electrical energy potential of woody biomass in the near region of the hydropower plant Tucuruí-Brazil. Waste Biomass Valor 11:2297–2307. https://doi.org/10.1007/s12649-018-0407-6
18. Pritchard D (2002) Biomass combustion gas turbine CHP. Talbott's heating
19. Eisentraut A, Brown A (2012) Technology roadmap: bioenergy for heat and power. International Energy Agency, Paris
20. Bridgwater AV (1995) The technical and economic feasibility of biomass gasification for power generation
21. Dornburg V, Faaij APC (2001) Efficiency and economy of wood-fired biomass energy systems in relation to scale regarding heat and power generation using combustion and gasification technologies. Biomass Bioenerg 21:91–108
22. Bridgwater AV, Toft AJ, Brammer JG (2002) A techno-economic comparison of power production by biomass fast pyrolysis with gasification and combustion. Renew Sustain Energy Rev 6:181–246. https://doi.org/10.1016/S1364-0321(01)00010-7
23. Caputo AC, Palumbo M, Pelagagge PM, Scacchia F (2005) Economics of biomass energy utilization in combustion and gasification plants: effects of logistic variables. Biomass Bioenerg 28:35–51. https://doi.org/10.1016/j.biombioe.2004.04.009
24. Vatopoulos K, Andrews D, Carlsson J et al (2012) Study on the state of play of energy efficiency of heat and electricity production technologies. Publications Office of the European Union, Luxembourg
25. Tańczuk M, Ulbrich R (2013) Implementation of a biomass-fired co-generation plant supplied with an ORC (Organic Rankine Cycle) as a heat source for small scale heat distribution system—a comparative analysis under Polish and German conditions. Energy 62:132–141. https://doi.org/10.1016/j.energy.2013.09.044
26. Algieri A, Morrone P (2014) Energetic analysis of biomass-fired ORC systems for micro-scale combined heat and power (CHP) generation. A possible application to the Italian residential sector. Appl Therm Eng 71:751–759. https://doi.org/10.1016/j.applthermaleng.2013.11.024
27. Lecompte S, Huisseune H, van den Broek M et al (2015) Review of organic Rankine cycle (ORC) architectures for waste heat recovery. Renew Sustain Energy Rev 47:448–461. https://doi.org/10.1016/j.rser.2015.03.089
28. Tchanche BF, Gr L, Frangoudakis A, Papadakis G (2011) Low-grade heat conversion into power using organic rankine cycles—a review of various applications. Renew Sustain Energy Rev 15:3963–3979. https://doi.org/10.1016/j.rser.2011.07.024
29. Tartière T, Astolfi M (2017) A world overview of the organic rankine cycle market. Energy Procedia 129:2–9. https://doi.org/10.1016/j.egypro.2017.09.159
30. Prando D, Renzi M, Gasparella A, Baratieri M (2015) Monitoring of the energy performance of a district heating CHP plant based on biomass boiler and ORC generator. Appl Therm Eng 79:98–107. https://doi.org/10.1016/j.applthermaleng.2014.12.063
31. Malico I, Nepomuceno Pereira R, Gonçalves AC, Sousa AMO (2019) Current status and future perspectives for energy production from solid biomass in the European industry. Renew Sustain Energy Rev 112:960–977. https://doi.org/10.1016/j.rser.2019.06.022
32. Strzalka R, Schneider D, Eicker U (2017) Current status of bioenergy technologies in Germany. Renew Sustain Energy Rev 72:801–820. https://doi.org/10.1016/j.rser.2017.01.091
33. Buragohain B, Mahanta P, Moholkar VS (2010) Biomass gasification for decentralized power generation: the Indian perspective. Renew Sustain Energy Rev 14:73–92. https://doi.org/10.1016/j.rser.2009.07.034

34. Chowdhury N-R (2014) Advances and trends in woody biomass gasification. PhD Thesis, Instituto Superior Técnico
35. Kuhn V, Klemeš J, Bulatov I (2008) MicroCHP: overview of selected technologies, products and field test results. Appl Therm Eng 28:2039–2048. https://doi.org/10.1016/j.applthermaleng.2008.02.003
36. Schneider T, Ruf F, Müller D, Karl J (2021) Performance of a fluidized bed-fired Stirling engine as micro-scale combined heat and power system on wood pellets. Appl Therm Eng 189:116712. https://doi.org/10.1016/j.applthermaleng.2021.116712
37. Salomón M, Savola T, Martin A et al (2011) Small-scale biomass CHP plants in Sweden and Finland. Renew Sustain Energy Rev 15:4451–4465. https://doi.org/10.1016/j.rser.2011.07.106
38. Anca-Couce A, Hochenauer C, Scharler R (2021) Bioenergy technologies, uses, market and future trends with Austria as a case study. Renew Sustain Energy Rev 135:110237. https://doi.org/10.1016/j.rser.2020.110237
39. Zare V (2020) Role of modeling approach on the results of thermodynamic analysis: concept presentation via thermoeconomic comparison of biomass gasification-fueled open and closed cycle gas turbines. Energy Convers Manage 225:113479. https://doi.org/10.1016/j.enconman.2020.113479
40. Gadsbøll RØ, Thomsen J, Bang-Møller C et al (2017) Solid oxide fuel cells powered by biomass gasification for high efficiency power generation. Energy 131:198–206. https://doi.org/10.1016/j.energy.2017.05.044
41. Pongratz G, Subotić V, von Berg L et al (2022) Real coupling of solid oxide fuel cells with a biomass steam gasifier: operating boundaries considering performance, tar and carbon deposition analyses. Fuel 316:123310. https://doi.org/10.1016/j.fuel.2022.123310
42. Livingston WR, Middelkamp J, Willeboer W, et al (2016) The status of large scale biomass firing. IEA Bioenergy
43. IEA-ETSAP, IRENA (2013) Biomass co-firing. Technology brief. IEA-ETSAP and IRENA
44. Khorshidi Z, Ho MT, Wiley DE (2014) The impact of biomass quality and quantity on the performance and economics of co-firing plants with and without CO2 capture. Int J Greenhouse Gas Control 21:191–202. https://doi.org/10.1016/j.ijggc.2013.12.011
45. Roni MS, Chowdhury S, Mamun S et al (2017) Biomass co-firing technology with policies, challenges, and opportunities: a global review. Renew Sustain Energy Rev 78:1089–1101. https://doi.org/10.1016/j.rser.2017.05.023
46. Demirbaş A (2003) Sustainable cofiring of biomass with coal. Energy Convers Manage 44:1465–1479
47. Nussbaumer T (2003) Combustion and co-combustion of biomass: fundamentals, technologies, and primary measures for emission reduction. Energy Fuels 17:1510–1521. https://doi.org/10.1021/ef030031q
48. Sahu SG, Chakraborty N, Sarkar P (2014) Coal–biomass co-combustion: an overview. Renew Sustain Energy Rev 39:575–586. https://doi.org/10.1016/j.rser.2014.07.106
49. Knapp S, Güldemund A, Weyand S, Schebek L (2019) Evaluation of co-firing as a cost-effective short-term sustainable CO_2 mitigation strategy in Germany. Energ Sustain Soc 9:32. https://doi.org/10.1186/s13705-019-0214-3
50. Mesarović M (2017) Challenging times for power plants. Termotehnika 43:1–22. https://doi.org/10.5937/termoteh1704001M
51. Saidur R, Abdelaziz EA, Demirbas A et al (2011) A review on biomass as a fuel for boilers. Renew Sustain Energy Rev 15:2262–2289. https://doi.org/10.1016/j.rser.2011.02.015

Conclusions and Future Research Needs

Ana Cristina Gonçalves and Isabel Malico

Abstract This is the last chapter of the book entitled "Forest Bioenergy: From Wood Production to Energy Use". The preceding chapters have covered the sources and distribution of forest biomass, its availability for different stand structures, and the methodologies and tools employed for its estimation. Furthermore, the diverse pathways for converting forest biomass into energy and biofuels have been explored, along with the associated technologies for biomass-to-energy conversion in the residential, industrial and power sectors. Forest biomass is a highly versatile source of energy, holding the potential to play a central role in a sustainable energy future. However, its use should guarantee the continued sustainability of forest stands, forests and the array of products they offer. It is equally essential to identify and prioritise the most effective pathways for the conversion of forest biomass into energy. A large body of work has been done; however, improvements can be achieved with whole system approaches that can accommodate the variability of forest biomass and conversion technologies and enable a comprehensive characterisation of the entire system. Accurate and precise statistics are of primordial importance for the whole system analysis.

Keywords Forest biomass · Residues · Forest plantations · Silviculture · Bioenergy · Conversion technologies · Biofuels · Bioheat · Biopower

Forests are crucial to the equilibrium of our planet and provide a wide range of products and services, including wood fuel. This biomass fuel has served mankind

A. C. Gonçalves (✉)
MED – Mediterranean Institute for Agriculture, Environment and Development &
CHANGE – Global Change and Sustainability Institute, Instituto de Investigação e Formação Avançada, Departamento de Engenharia Rural, Escola de Ciências e Tecnologia, Universidade de Évora, Apartado 94, 7002-544 Évora, Portugal
e-mail: acag@uevora.pt

I. Malico
IDMEC, Escola de Ciências e Tecnologia, Departamento de Engenharia Mecatrónica, Universidade de Évora, Rua Romão Ramalho 59, 7000-671 Évora, Portugal
e-mail: imbm@uevora.pt

A. C. Gonçalves and I. Malico (eds.), *Forest Bioenergy*, Green Energy and Technology,
https://doi.org/10.1007/978-3-031-48224-3_11

as the primary fuel for millennia and continues to be the main source of energy for heating and cooking for millions of people worldwide (*cf.* chapter "Introduction to Forest Bioenergy"). However, with the industrial revolution, biomass lost its leading role as the most widely used energy source. The twentieth century witnessed a diversification of end-use technologies and energy sources, with fossil fuels (coal, oil and natural gas) dominating the supply of energy today. Nevertheless, concerns over energy security, environmental impacts and deplection of fossil fuels, especially of oil, have given momentum to investment in renewable energy sources. In many high-income countries, where bioenergy had lost its importance, modern uses of biomass have gained more relevance and bioenergy is now seen as one of the key pathways to mitigate global climate change.

Today, bioenergy, the most widely used renewable energy source, accounts for 10% of the world non-food energy supply and its contribution is expected to grow in the future. Presently, it is primarily sourced from forest residues and waste, mainly providing heat in the residential sector, followed by process heat generation in the industrial sector. Additionally, the use of forest bioenergy to support the decarbonisation of the electricity supply and provide firm, low-carbon electricity has been gaining relevance in the last decades. On the other hand, in the future, forest biomass may become relevant in the production of second-generation liquid biofuels that do not conflict with food supply or affordability and that assist in decarbonising the transport sub-sectors where other renewable energy sources face difficulties in penetration (e.g., aviation, shipping).

Apart from its versatility as a fuel and the potential to easily substitute fossil fuels in numerous applications, forest biomass offers several other advantages. It is a renewable energy source produced in most countries around the world and primarily used locally, which helps increase energy security and meet the growing global energy demand. When sustainably produced and converted into energy using modern, efficient and clean equipment, forest biomass usually presents environmental benefits in comparison to fossil fuels, with many of its current applications being cost-effective.

Biomass for energy can be sourced from several land use systems, with forest systems presently being the primary suppliers of biomass for energy. It can be classified into three different categories: primary or direct, secondary or indirect and tertiary or recovered (*cf.* chapter "Sources and Distribution of Forest Biomass for Energy"). Primary forest biomass refers to the biomass directly exported from forest systems. It includes both products and residues (e.g., stemwood, tree tops, stumps, branches, leaves, trees without market value for timber and dead or damaged trees). Secondary biomass consists of residues from wood-based industrial facilities, while tertiary biomass encompasses the residues or waste from the economic or social activities outside the forest sector.

The availability of forest biomass for energy depends on factors such as the forest area, growing stock and the management system, which present significant variability worldwide. FAO statistics (*cf*. chapter "Sources and Distribution of Forest Biomass for Energy") provide insights into the distribution of forest area, growing stock, as well as production and consumption of wood fuel and industrial roundwood. While the production of the primary sources of bioenergy relies on the distribution of forest and their growing stock, the generation of secondary sources depends on the number, size and type of forest-based industries, as well as the quantity and quality of the wood received and processed, alongside with the industrial processes used. On the other hand, tertiary sources encompass a heterogeneous range of materials with no or variable amount of contaminants, with their availability dependent on the recovery rates of wood wastes from industrial production and urban collection.

Overall, there was an increase in forest area and growing stock in Europe, Asia, North America and Oceania, and a decrease in South America and Africa from 1961 to 2020 (*cf.* chapter "Sources and Distribution of Forest Biomass for Energy", Figs. 2.4 and 2.5), which may result in constraints on the availability of industrial and fuel wood in the latter two regions. Moreover, the share of roundwood used for wood fuel was larger in Africa and Asia, and this, along with the reduction of forest area and growing stock, may further decrease the availability of biomass for energy.

The recovery rates of primary sources vary from 0 to 80%, whereas the generation factors for secondary residues vary from 5 to 60%. On the other hand, concerning the energy valorisation of tertiary wood residues, some countries lack the collection and treatment of wood waste, while others sort and valorise almost all their municipal and industrial wood waste. Such variable rates emphasise the variability of available biomass for energy. Moreover, worldwide FAO statistics are based on information reported by each country. Yet, when no information is provided estimations are made. This can result in over or underestimations. The aforementioned highlights the need to promote accurate country level statistics for forest area, growing stock and production and consumption of roundwood and wood fuel.

Forest biomass (and carbon) sustainability has been object of forest research. The aim was to understand the dynamics of biomass per tree, per stand and per forest and which interactions influence biomass allocation and storage (*cf.* chapter "Stand Structure and Biomass"). The analysis with integrated approaches at the ecosystem level is suggested due to the interactions between the ecosystem components (e.g., trees, other flora, fauna soil, nutrient and water cycles), species, silvicultural systems, stand structure and disturbances (e.g., silvicultural practices, cuts, storms, fires) in the tree (and stand) growth and thus biomass allocation and storage. The analysis of biomass increases in complexity from tree to stand and to forest, due to the variability of growth and re-allocation of live to dead biomass and from dead biomass to soil organic matter. In general, the sustainability of a forest system is achieved when overall system biomass is maintained approximately constant (though variability between the biomass components in the forest system can exist) and the potential productivity of the system is maintained or enhanced. This can be achieved for most species and sites by choosing the best suited stand structure, silvicultural system, silvicultural practices and defining the maximum threshold for the export of woody products.

An ongoing discussion regarding the maintenance of the biomass in the forest systems or its export has been observed in recent years, which has resulted in policies and legal frameworks. However, there seem to be contradictory points of view. To promote the mitigation of the effects of climate change, the maintenance of biomass in the forest systems is suggested whether as live or dead biomass or soil organic matter. Inversely, the mitigation of the effects of greenhouse gas emissions and the use of woody products is promoted by the use of biomass for several uses with shorter or longer lifecycles. On one hand, the maintenance and the promotion of biomass stocks in the forest stands drives forest management towards stand structures and silvicultural systems where the export of biomass is minimised and the re-allocation of live to dead biomass and to soil organic matter is enhanced. On the other hand, the use of biomass in woody products of different lifecycles promotes the export of biomass from the forest stands. Traditionally, only timber was exported. As it corresponds to biomass low in nutrients, its effects on the forest system sustainability were low and short to medium term lasting. The gradual increase of biomass exports, in particular of components rich in nutrients, originates the reduction of the nutrients' stocks in the forest systems. Yet, even with the export of large amounts of biomass, it is possible to maintain the forest system productive potential and sustainability, as long as the guidelines for sustainable management are followed.

Though there is a large body of research done, there is still the need for further research on the dynamics of biomass: with aging at the tree level; per stand structure, silvicultural system and silvicultural practices; per disturbance, including their frequency, intensity and spatial distribution, whether natural or artificial; on the allocation and re-allocation mechanisms of biomass in the forest systems; and the refinement of methods and techniques that allow to collect data sets that minimise the errors in the acquisition of the variables. This would be a tool to develop biomass functions with high accuracy as it would enable to accommodate in the models both the variability and the interaction between the factors that influence the biomass storage and that reduce the model uncertainties.

Energy plantations (*cf.* chapter "Energy Plantations") are forest systems with the goal of producing biomass for energy that at the same time release pressure for biomass for energy in other forest systems. Thus, suiting the areas and productivities of energy plantations at local and global levels may provide biomass for energy and at the same time protect and conserve sensitive forest systems.

In a climatic change frame, the evaluation and monitoring of biomass (above ground live and dead, below ground and soil organic matter) has been gaining importance. This is related to the selection of the best suited species per site and the adaptation and development of models of silviculture that enable to reach stand structures, silvicultural systems and silvicultural practices that maintain the forest system sustainability while maintaining or promoting the biomass storage. Moreover, the goal is to develop forest systems that are resilient and/or resistant to the disturbances derived from the climate change.

One of the most important tools to evaluate, monitor and predict biomass dynamics in time and space are the biomass functions developed at tree or area level (*cf*. chapters "Modelling Biomass" and "Overview of the Biomass Models"). Though many models have been developed, the increasing knowledge of the forest systems, data bases available and mathematical methods and techniques enable the development of new biomass models that are able to accommodate the variability in time and space and reduce the uncertainties of the models.

Sustainably managing forest resources and evaluating the availability of forest biomass is key to promote bioenergy that is compatible with food, feed, fibre, timber and environmental conservation and protection. However, downstream of forest biomass production, promoting the choice of the most suitable bioenergy pathways to convert forest biomass into energy is also a fundamental aspect to achieving this goal.

As has been seen throughout this book, forest biomass is a very versatile energy source, and several useful pathways are available to convert it directly into energy or biofuels (*cf.* chapter "Forest Biomass as an Energy Resource"). Raw biomass with little pre-treatment can be used to directly generate heat and/or electricity, or it can be processed to upgraded biofuels (solid, liquid or gaseous), and only then be converted to energy.

Raw forest biomass presents some disadvantages when compared to fossil fuels. It typically has higher moisture content, lower heating value, a higher oxygen/carbon ratio, lower density and a hydrophilic nature. The conversion of forest biomass into upgraded biofuels results in solid, liquid or gaseous fuels that have improved characteristics and are more similar to fossil fuels, making them convenient to use. Among all the possible improved biofuels produced from forest biomass, upgraded solid biofuels, such as pellets and briquettes, are the ones gaining market share due to their convenience of use and the advantages they offer in terms of energy conversion.

A wide variety of biomass conversion technologies and energy system configurations are possible, resulting in a broad spectrum of products. Two main groups of technologies exist: biochemical and thermochemical. However, only some routes that involve the thermochemical conversion of forest biomass into energy or fuels are currently mature and commercially available.

There has been much effort to develop biochemical technologies to produce fuels from lignocellulosic biomass, including forest biomass. This is because, contrary to first generation biofuels, they would not directly compete with food and feed. However, the recalcitrant nature of lignocellulosic feedstocks poses a challenge to biological degradation. Consequently, the commercial biochemical production of biofuels, such as biogas and bioethanol from wood, still faces many challenges, and more research is essential to make biochemical conversion pathways attractive.

Compared to biochemical conversion technologies, thermochemical pathways are much more suitable for converting forest biomass into fuels, which explains their success. Nevertheless, the commercial initiatives to produce advanced liquid biofuels from lignocellulosic biomass through pyrolysis and gasification struggle to thrive and compete with cheaper fossil fuels. Among other efforts, further process development is required to make these second-generation biofuels cost-effective. It is expected that policies aimed at decarbonising the transport sector will drive the development of advanced liquid biofuels and create the market conditions for their commercialisation.

In contrast, the direct conversion of forest biomass into heat and/or power is a commercial reality in many applications, covering a broad spectrum of capacities and technologies. The most popular process is combustion, but gasification followed by combustion in the same facility is also a reality in many markets, especially in small-scale combined heat and power (CHP) units.

The choice of the most suitable pathway for the conversion of forest biomass into energy is a complex task that is influenced by (*i*) the quality, quantity and cost of the biomass feedstock; (*ii*) the availability, cost and performance of the bioenergy conversion technologies; (*iii*) the various possible end-uses competing for the limited forest biomass and their requirements and (*iv*) existent implementation barriers and incentives.

The residential sector has an important share of the world final energy supply, being the sector that consumes the most forest biomass worldwide (*cf.* chapter "Biomass for Domestic Heat"). It is characterised by two very distinct ways of consuming biomass: the traditional and the modern. Both rely on combustion as the primary conversion technology of biomass into heat, but, while in the traditional use of biomass, the direct combustion of biomass takes place in inefficient and polluting equipment that consumes biomass often produced in an unsustainable way and not commercialised, modern uses of biomass are identified by much more efficient and cleaner equipment burning sustainably sourced biomass.

Several efforts to phase out traditional biomass uses have been made in the last fifty years, but the transition from traditional biomass to modern energy systems has been difficult. Currently, almost one-third of the world population, mostly in rural regions of low- and middle-income countries, lacks access to modern energy services and consumes biomass in a traditional way. In these regions, open-fires and inefficient stoves, such as three-stone fires and mud stoves, often with no or inefficient chimneys, are the main technologies used for cooking. Additionally, in the colder regions, inefficient stoves are also used for space heating, further exacerbating the problem.

The traditional use of biomass has far-reaching and negative consequences, affecting both the environment and the well-being of the communities that depend on it. This practice gives rise to indoor pollution with consequent impacts on the health of those who are exposed to the products of incomplete combustion, characteristic of the basic equipment associated with the traditional use of biomass. Furthermore, pollutants resulting from this equipment also impact the climate, exacerbating global environmental challenges. The inherent inefficiency of the equipment used also results in excessive consumption of biomass, putting pressure on forest resources and potentially leading to forest degradation in many parts of the world. Notably, the burden of collecting and handling biomass often falls disproportionately on women and children within these communities. The inequitable distribution of labour further highlights the urgent need for sustainable alternatives that alleviate the environmental and social challenges posed by traditional biomass use.

Many programs aimed at providing access to modern energy services have primarily focused on promoting the transition from wood to gaseous fossil fuels, but they have not given as much attention to replacing traditional, inefficient biomass stoves with modern, efficient ones, or to designing strategies for ensuring a sustainable supply of wood. It is essential to recognise that a substantial number of people rely on traditional biomass use out of necessity; they simply cannot afford alternative fuels or do not have access to modern energy infrastructures. In light of this, it is imperative to adopt holistic approaches that support local forest tree stands and forest management while also addressing the specific energy requirements and practices of the communities in question. There exists significant potential to improve the traditional use of biomass and minimise its adverse impacts. This potential can be unlocked through the development and introduction of more efficient, safer and less polluting biomass stoves. These stoves should be tailored to meet the diverse needs of local populations, ensuring that they not only provide a sustainable source of energy but also prioritise the well-being of both people and the environment.

In recent years, there has been significant progress in the development and commercialisation of more efficient and cleaner biomass systems for household heating in numerous countries. However, technical improvements are still actively pursued, with a primary focus on enhancing efficiency and reducing the environmental impacts of the small-scale biomass conversion equipment used in the residential sector, which typically does not integrate air pollution control equipment. Another crucial aspect involves ensuring a reliable and sustainable supply of biomass feedstock to the residential market. To achieve this goal, ongoing research is dedicated to optimising the entire forest biomass supply chain and fostering forest biomass markets. Simultaneously, there is a critical need to assess the economic viability of biomass residential heating when compared to other energy sources. The integration of residential biomass heating systems with other renewable energy sources, such as solar and wind, is a growing area of interest. This integration can improve energy reliability while further minimising environmental impacts.

Alongside the ongoing evolution of the residential sector, the industrial sector, the world's largest energy consuming end-user in 2020, stands out as a key sector where forest biomass could make a more significant, or at least, eventually a distinct contribution beyond today's. Indeed, the environmental impacts of the industrial sector are substantial. Industry accounted for approximately one-third of the final energy consumption and contributed to two-fifths of the direct greenhouse gas emissions among end-use sectors in 2020. As a result, decarbonising the industry sector becomes crucial to reduce greenhouse gas emissions and keep global warming well below the 2 °C threshold above pre-industrial levels. Industry heavily relies on fossil fuels but reducing this dependency is very challenging (*cf.* chapter "Biomass for Industrial and District Heating").

A complete picture of how energy is utilised by the global, highly diverse industry is currently unavailable, due to the lack of official statistics providing a breakdown of energy consumption in the industry by its various end uses. This poses a challenge in developing demand-oriented energy policies for the industrial sector. In this context, conducting a precise and detailed characterisation of the energy demands within the industrial sector is important in order to exploit additional opportunities for solid biomass.

The integration of renewable energy sources in the industrial sector is not an easy task, which explains the low share of renewables in this sector. Solid biomass already meets a substantial portion of the energy needs of wood-based industries that generate residual biomass. However, its potential role in the industry could expand, especially in some sub-sectors that need large quantities of process heat and do not generate biomass residues, such as iron and steel or cement sub-sectors. Decarbonising processes that requires high temperatures is especially challenging, as not all renewable energy sources and technologies can provide the necessary high temperature heat. This is an area where solid biomass may have a distinctive role.

A wide variety of technologies are currently commercially available for converting solid biomass into heat for industrial processes (as well as district heating). Most of these technologies are currently used for low- and medium-temperature process heating, with a few exceptions for high-temperature applications. Combustion is the most widely utilised conversion process, typically providing heat to industrial processes through a heat transfer medium, often steam. Gasification followed by combustion is also commercially used in industrial applications.

The entire spectrum of temperatures required for industrial processes can be covered by different biomass products. While raw biomass is suitable for lower temperature processes, achieving the higher temperature levels may necessitate pre-processing the raw biomass. Some of the pre-processing technologies suitable to upgrade forest biomass feedstocks for specific industrial processes, such as torrefaction, have not yet reached full commercialisation and require further development to reduce costs and become more attractive to the industrial sector. In addition to heat generation, the development of pre-processing technologies also benefits applications like CHP or power production and the production of advanced biofuels, whether through thermochemical or biochemical conversion.

Combustion is the most mature of all the available energy conversion processes for heat and/or power generation. Nevertheless, there is active ongoing research focused on further developing solid biomass combustion technologies. Particularly relevant for medium- and large-scale systems is the further reduction of emissions and the enhancement of equipment versatility in handling various types of fuels, including wastes. Additionally, significant research efforts are dedicated to the further development of gasification technologies, not only for heat or CHP applications but also for the production of advanced biofuels.

While solid biomass is capable of providing high-temperature process heat, it is currently primarily used within the forest-based industries. These industries typically generate their biomass residues but generally do not require high-temperature heat. There are opportunities for these industries to incorporate other renewable energy sources to meet their energy needs and to sell their residues for other applications. However, determining whether this is the best option within a broader context requires a thorough assessment. Furthermore, for industrial sectors that do not generate their residual biomass, ensuring a reliable supply of biomass is critical. Given that biomass is a finite resource with competing uses, it is essential to have detailed knowledge of the forest biomass availability and maintain continuous resource monitoring. To optimise the allocation of biomass to various pathways for process heat, comprehensive studies are necessary to explore the most effective routes among the diverse options available for converting biomass.

Another sector that competes for the finite biomass resources is the power sector (*cf.* chapter "Biomass for Power Production and Cogeneration"). This is the sector that was considered for initial climate change mitigation and that has seen more developments in terms of modern renewable energy incorporation. Nowadays, solar and wind power are already the cheapest solutions in some regions of the world. However, the sector still heavy relies on fossil fuels, mainly coal, with most of the electricity in the world being obtained from fossil fuels. As a consequence, the sector accounts for a substantial part of the global greenhouse gas emissions (36% in 2020).

Among the various energy uses explored in this book, including industry, households and power generation, the latter is the sector where, globally, forest biomass plays a relatively minor role. Even though the production of biopower has been promoted in many regions, the share of biomass in the electricity generated worldwide is modest (around 2% in 2020). Nevertheless, substantial quantities of biomass are consumed by the power sector. Biomass technologies offer a distinct advantage by providing the high level of operational flexibility essential for effective power grid balancing. As the utilisation of wind and solar power sources grows, significantly improving the greenhouse gas balance of power-generation systems, a concurrent surge in the demand for grid balancing solutions emerges. This context underscores the role of bioenergy not only as an energy source but also as a viable means of energy storage, particularly in the context of ensuring grid stability.

Several technologies are available for converting solid biomass into electricity at various scales. Most of these technologies rely on heat engines that convert thermal energy, primarily generated through combustion systems but partially through coupled gasification-combustion systems, into electricity. During this process, a significant portion of the energy contained in biomass is not converted into electricity but is instead rejected as heat. In conventional power plants, which still dominate the industry, this thermal energy is lost to the environment, resulting in energy waste and reduced efficiency.

If, on the other hand, the heat rejected by the heat engine is used in CHP systems, the energy contained within the fuel is much more valorised and the overall efficiency of the process greatly surpasses that of conventional electricity generation. This results in benefits such as cost reduction and the minimisation of greenhouse gas emissions.

More than half of the biomass used for electricity production refers to CHP systems, which operate with high overall efficiencies. The industrial sector, with its high electricity and heat demands, currently plays a key role in CHP generation from forest biomass. Additionally, district heating systems based on biomass and CHP technologies provide an efficient means of converting biomass into energy.

Traditionally, biopower has been predominantly generated in medium- to large-scale systems, representing the conventional approach of the sector. These systems still offer potential for improvement, with ongoing research dedicated to increasing their performance. Simultaneously, research efforts are dedicated to the development of small- and micro-scale CHP technologies, with a specific emphasis on addressing the energy needs of residential and localised contexts. In addition to scalling down, ongoing research in biomass polygeneration explores innovative approaches aimed at efficiently converting biomass resources into multiple forms of energy, including electricity, heat, cooling and biofuels, all in one integrated process. Recent studies have primarily focused on optimising the utilisation of biomass resources and integrating diverse subsystems to enhance overall performance, taking into consideration both energy and exergy viewpoints. Furthermore, there is a growing emphasis on the integration of hybrid renewable energy systems, which combine biopower with other renewable sources. This integrated approach not only diversifies the energy mix but also contributes to enhanced reliability and sustainability.

Another way to use biomass for power generation is through co-combustion of biomass with other fuels, such as coal. This is a straightforward approach that allows power plants originally designed for coal to transition to a renewable energy source without major changes. Biomass co-combustion can significantly reduce carbon emissions, making coal-based energy cleaner. In parallel, there is active research into biomass-based carbon capture and storage or utilisation concepts, which presents a promising area for net-negative greenhouse gas emission.

Throughout this book, it becomes evident that forest biomass is highly diverse and versatile. It consists of different feedstocks associated with various production systems, which can be used for a wide range of energy purposes, by means of different technologies. This versatility is one of the primary advantages of biomass as an energy source, but it also presents challenges when determining the best pathways for bioenergy. Achieving this requires a comprehensive understanding of several aspects, including energy needs and demands, the advantages and limitations of available conversion technologies, biomass feedstock characteristics, biomass availability, environmental impacts and regulations, to name just a few.

The pathways to decarbonise various sectors are uncertain, as is the precise role that forest biomass will play in the future, despite different projections. Biomass energy markets are highly complex and can be influenced by numerous factors, including climate change, energy prices, government policies and consumer demands. Despite the uncertainties surrounding the future of bioenergy, forests biomass will continue to be integrated into the energy mix and play a significant role. Effectively harnessing the numerous benefits of forest biomass for energy while minimising the risks is a challenge that requires careful consideration.

Funding This work was funded by National Funds through FCT - Foundation for Science and Technology under the Project UIDB/05183/2020, through MED, and Project UIDB/50022/2020, through IDMEC, under LAETA.

MIX
Papier aus verantwortungsvollen Quellen
Paper from responsible sources
FSC® C105338

If you have any concerns about our products,
you can contact us on
ProductSafety@springernature.com

In case Publisher is established outside the EU,
the EU authorized representative is:
Springer Nature Customer Service Center GmbH
Europaplatz 3, 69115 Heidelberg, Germany

Printed by Libri Plureos GmbH
in Hamburg, Germany